71 Advances in Polymer Science

Fortschritte der Hochpolymeren-Forschung

Analysis/Reactions/Morphology

With Contributions by
P.-J. Madec, E. Maréchal, J. E. Mark,
T. Otsu, R. W. Richards, J. P. Queslel,
T. Sato, B. Tieke

With 98 Figures and 31 Tables

Springer-Verlag
Berlin Heidelberg GmbH

ISBN 978-3-662-15217-1 ISBN 978-3-540-39435-8 (eBook)
DOI 10.1007/978-3-540-39435-8

Library of Congress Catalog Card Number 61-642

Originally published by Springer-Verlag Berlin Heidelberg New York Tokyo in 1985
Softcover reprint of the hardcover 1st edition 1985

2154/3020-543210

Editors

Editorial

With the publication of Vol. 51 the editors and the publisher would like to take this opportunity to thank authors and readers for their collaboration and their efforts to meet the scientific requirements of this series. We appreciate the concern of our authors for the progress of "Advances in Polymer Science" and we also welcome the advice and critical comments of our readers.

With the publication of Vol. 51 we would also like to refer to a editorial policy: *this series publishes invited, critical review articles of new developments in all areas of polymer science in English (authors may naturally also include workes of their own).* The responsible editor, that means the editor who has invited the author, discusses the scope of the review with the author on the basis of a tentative outline which the author is asked to provide. The author and editor are responsible for the scientific quality of the contribution.

Manuscripts must be submitted in content, language, and form satisfactory to Springer-Verlag. Figures and formulas should be reproducible. To meet the convenience of our readers, the publisher will include "volume index" which characterizes the content of the volume.

The editors and the publisher will make all efforts to publish the manuscripts as rapidly as possible, i.e., at the maximum six months after the submission of an accepted paper. Contributions from diverse areas of polymer science must occasionally be united in one volume. In such cases a "volume index" cannot meet all expectations, but will nevertheless provide more information than a mere volume number.

Starting with Vol. 51, each volume will contain a subject index.

Editors Publisher

Table of Contents

Small Angle Neutron Scattering from Block Copolymers

R. W. Richards
Department of Pure and Applied Chemistry, University of Strathclyde, Glasgow, G1 1XL/U.K.

The theory and essential principles of small angle neutron scattering are presented together with their evolution to expressions useable in the interpretation of experimental data. The two components to the scattering law, single particle scattering and interference function scattering are identified and some classical expressions for the scattering of single particles are reviewed. Instrumental aspects are given as a general overview of the features of small angle diffractometers. The major part of the review is concerned with available experimental results on block copolymers in the solid state. The results are compared with theoretical predictions and consequently theories of microdomain formation are briefly considered. Apart from domain dimensions, and separations, results of the determination of block dimensions are reported as well as attempts at modelling the complete scattering envelope. The few experiments on block copolymers in solution and the application of contrast variation are reported and the review concludes with a report on continuing work on structural changes consequent on uniaxial extension of block copolymers.

1 Introduction

There are few techniques which find application over a broad field of scientific endeavour. Amongst them we can include X-ray scattering, nuclear magnetic resonance and latterly neutron scattering. An inspection of the catalogue of experiments carried out in any one year at a major European centre for neutron beam studies [1] reveals that neutron scattering has been used to investigate lattice dynamics and structure in a variety of inorganic materials, excitations in Fermi fluids, quantised motion of discrete groups in molecules, adsorption phenomena, Langmuir-Blodgett films, colloidal dispersions and the structure and dynamics of biological and synthetic macromolecules. The variety of research capable by neutron scattering is a testament to the intrinsic properties of the neutron. Neutrons are uncharged sub-atomic particles with a mass of 1 a.u. and a spin of 1/2, due to their electrical neutrality and large mass they are able to penetrate matter to a significant extent before being absorbed. Interaction with matter is via nuclear-neutron interaction and/or the magnetic moment of the neutron. Neutron wavelength used are generally in the range 0.2 to 2.0 nm. and in this region neutron energy is circa 0.1 $kJmol^{-1}$. The wavelengths are such that the range of scattering vectors (see below) available by neutron scattering enables the investigation of length scales up to circa 200 nm. to be explored, whilst the neutron energy approximates to that of many dynamic molecular processes and thus observation of the transfer of energy to and from the neutron allows such processes to be studied.

Although neutron scattering has only been applied to polymers since the early 1970's, the uniqueness of the technique has provoked much published work and many reviews at periodic intervals [2–8]. Much of the published work on synthetic polymers has been concerned with small angle neutron scattering (SANS), i.e. structural investigations of polymer melts, concentrated and dilute solutions, amorphous and semi-crystalline polymers in the solid state. These results have contributed greatly to discussions on the nature of semi-crystalline polymers [9], polymer blends and the development of new theories of polymer solutions and networks [10]. Some 150 papers have been published in the decade 1973–1983 which deal with SANS from polymers. Attention here is focused on the application of SANS to the morphological and conformational examination of block copolymers. Little work has been reported on these systems, however, as will be shown here the fundamental equations are such that much information, in principle, is obtainable. The majority of the published work is concerned with investigations in the bulk state of microphase separated block copolymers and consequently the pertinant points of domain formation theory are reviewed. Basic theory of SANS and scattering laws are given, since these are essential in illustrating the utilisation of the technique. Since detailed description of apparatus and data analysis are available in the literature, only a brief resumé is given. Experimental results are discussed for the solid state, for copolymers in solution, mixed with homopolymers, and in extension.

2 Theory of Neutron Scattering

Rigorous derivation of the equations used in small angle neutron scattering is not presented here. The interested reader can find such derivations in the texts by Marshall and Lovesey, Turchin [12] and Kostorz [13] as well as in some reviews. What is presented here is a precis of the derivation of the final equations which will bring out some of the factors to be considered, furthermore, only elastic scattering is considered, i.e. no transfer of energy to or from the incident neutrons. The intensity of neutrons (number of scattering events) scattered per second by any process will be given by:

$$I = I_0 N \sigma \tag{1}$$

Where I_0 is the incident neutron flux in neutrons per second per square centimetre, N is the number of nuclei in a volume V and σ the total scattering cross section per nucleus. A scattering experiment is usually designed to provide a detailed analysis of the scattering pattern as a function of incident beam characteristics (flux, wave vector, polarisation, energy) the scattering being observed at some angle, 2θ, to the incident wave vector and collected over a solid angle $\Delta\Omega$. Incident (k_0) and scattered (k) wave vectors are connected via the scattering vector, Q, by the relation:

$$Q = k_0 - k \tag{2}$$

Figure 1 shows the generalised scattering experiment and the scattering diagram.

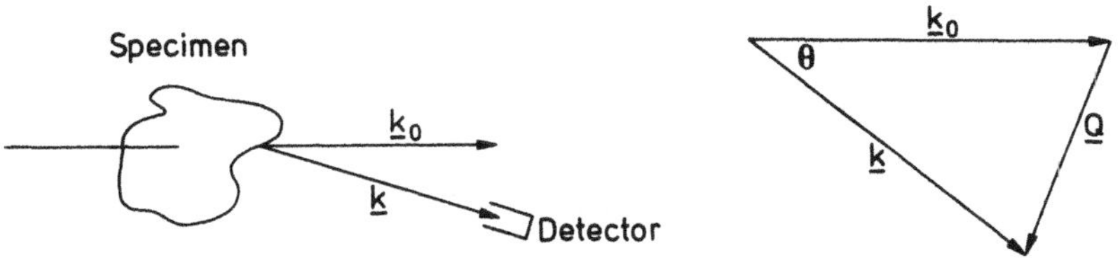

Fig. 1. Schematic diagram of a neutron scattering experiment and the associated scattering diagram

From the scattering diagram:

$$Q = (k^2 + k_0^2 - 2kk_0 \cos(2\theta))^{1/2} \tag{3}$$

For elastic scattering, $k = k_0$ whence:

$$Q = Q = (4\pi/\lambda) \sin(\theta) \tag{4}$$

since $k_0 = 2\pi/\lambda$ and λ is the incident neutron wavelength.

From this experimental arrangement then,

$$I = I_0 N \frac{d\sigma}{d\Omega} \Delta\Omega \tag{5}$$

where $d\sigma/d\Omega$ is the differential scattering cross section and is the parameter 'observed' in an elastic neutron scattering experiment and which must be related to features of the scattering material.

Fundamental neutron scattering theory shows that $d\sigma/d\Omega$ is given by:

$$d\sigma/d\Omega = 1/N(M_n 2\pi\hbar^2)^2 \,|\int V(r) \exp(iQ \cdot r\, d^3r)^2|$$

Where $V(r)$ is the Fermi pseudo potential describing neutron-nucleus interaction during scattering, M_n the neutron mass and r the vector separation of neutron and nucleus. Considering one nucleus then the scattering amplitude from that nucleus is:

$$-b = M_n/2\pi\hbar^2 \int V(r)\, d^3r \tag{6}$$

Where the integral is over the nuclear volume and thus $\exp(iQ \cdot r)$ tends to 1 since r is very small. The parameter b is the scattering length defined as the negative amplitude of the scattered wave. The scattering length is a property of the nucleus and varies from isotope to isotope, additionally for any one isotope with nuclear spin it may have one of two values dependent on the interaction between the nuclear and neutron spins. Consequently, the total differential scattering cross section for an assembly of many nuclei is the sum of the individual differential scattering cross sections,

$$d\sigma/d\Omega = 1/N \sum_r |b_r \exp(iQ \cdot r)|^2 \tag{7}$$

Temporarily we introduce another position vector r', then:

$$d\sigma/d\Omega = 1/N \sum_{r,r'} b_r b_{r'} \exp(iQ \cdot (r - r'))$$

$$= \sum_r b_r^2 + \sum_{r,r'} b_r b_{r'} \exp(iQ \cdot (r - r'))\; 1/N$$

$$\sum_r b_r^2 = N\langle b_r^2\rangle$$

and

$$\sum_{r,r'} b_r b_{r'} \exp(iQ \cdot (r - r')) = N\langle b_r\rangle^2 \sum \exp(iQ \cdot (r - r'))$$

$$= -N\langle b_r\rangle^2 + N\langle b_r\rangle^2 \sum \exp(iQ \cdot (r - r'))$$

Therefore:

$$d\sigma/d\Omega = \langle b^2\rangle - \langle b\rangle^2 + \langle b\rangle^2 \sum |\exp(iQ \cdot r)|^2 \tag{8}$$

Two contributions to Eq. (8) can be identified, the first of these is $(\langle b^2\rangle - \langle b\rangle^2)$ and represents the mean square deviation of individual scattering lengths from the average value. As such it is called the incoherent scattering cross section and constitutes a uniform, isotropic background scattering since it retains no information on the phase of the scattered neutrons. Incoherent scattering cross sections are tabulated as values of σ_{inc} where:

$$\sigma_{inc} = 4\pi(\langle b^2\rangle - \langle b\rangle^2) \tag{9}$$

The second contribution to Eq. (8) is the coherent scattering cross section and contains all the structural information pertaining to the scattering material since it retains the phase of the scattered neutrons in exp $(iQ \cdot r)$, and takes into account the interference effects for neutrons scattered from nuclei in different parts of the assembly. In summary, the elastic scattering of neutrons by matter has two components, an incoherent scattering cross section which is isotropic and retains no *structural* information on the sample, a coherent scattering cross section which provides structural information on the scattering specimen. As we shall see, the incoherent scattering background can be very troublesome and great efforts have to be made to minimise it or accurately account for it.

3 Small Angle Neutron Scattering

Small angle scattering is an ambiguous description since it does not specify the radiation or wavelength used. It is generally understood that small angle scattering relates to the scattering properties at small scattering vectors, Q, where Q should be small compared to the Q value for the first maximum in the structure factor of liquids [14] or the smallest reciprocal lattice vector for crystalline materials. As a rule of thumb small angle scattering will only resolve structural features whose length scales are larger than $2\pi/Q_{max}$, where Q_{max} is the maximum value of Q utilised.

Referring to Eq. (7) and considering only the coherent scattering cross section, the possibility of there being a finite volume over which the scattering event takes place is encompassed by replacing b_r with the scattering length density, $\varrho(r)$, at r. Thus on integrating over the whole sample volume then:

$$d\sigma/d = 1/N \left| \int_V \varrho(r) \exp(iQ \cdot r)\, d^3r \right|^2 \tag{10}$$

Consideration of Eq. (10) reveals that elastic coherent neutron scattering is the Fourier transform of the scattering length density variation from point to point in the sample. Since scattering length density is related to conventional mass density (vide infra), then the scattering length density variation is equivalent to the density autocorrelation function which may be described in terms of the pair correlation function. To proceed further we use a two phase model as defined by Kostorz [13], i.e. we have N_p particles embedded in a continuous matrix, where the respective scattering length densities are ϱ_p and ϱ_m, invoking the Babinet reciprocity principle [15] and from Eq. 10 then,

$$d\sigma/d\Omega = 1/N(\varrho_p - \varrho_m)^2 \left| \int_V \exp(iQ \cdot r)\, d^3r \right|^2 \tag{11}$$

The integral in Eq. (11) is the scattering law, S(Q), for the scattering system and contains all the information on the correlations within any one particle and correlations between particles dependent on the Q range explored. Restricting the treatment to correlations within one particle initially then the single particle form factor is:

$$F_p(Q) = \frac{1}{V_p} \int_V \exp(iQ \cdot r)\, d^3r$$

Where the division by the particle volume, V_p, ensures that at $Q = 0$, $F_p(Q) = 1$. Consequently for N_p particles:

$$d\sigma/d\Omega = V_p^2 N_p/N(\varrho_p - \varrho_m)^2 (C_p - C_m)^2 (F_p(Q))^2 \tag{12}$$

Where C_p is the concentration of observed species in the particles and C_m the concentration in the matrix. The inclusion of $(C_p - C_m)^2$ generalises the equation to circumstances where a labelled species is added to the system and is partitioned between the two phases. Equation 12 is the equation that will be used most frequently in the ensuing discussion, however, to be completely general the expression for the differential scattering should include an interference function, A(Q), descriptive the correlations between particles, thus:

$$d\sigma/d\Omega = V_p^2 N_p/N(\varrho_p - \varrho_m)^2 (C_p - C_m)^2 (F_p(Q))^2 A(Q) \tag{13}$$

Replacing in Eq. (5) and including the incoherent background scattering and dividing by the sample volume, V, then:

$$I(Q) = I_0 \frac{N}{V} \Delta\Omega \left[\frac{V_p^2/N_p}{N} (\varrho_p - \varrho_m)^2 (C_p - C_m)^2 (F_p(Q))^2 A(Q) + \frac{\sigma_{inc}}{4\pi}\right] \tag{14}$$

4 Cross Sections, Scattering Lengths and Scattering Length Density

Evidently the incoherent scattering cross section and scattering length are important parameters, in that they determine the scattering length density and consequently the magnitude of the coherent scattering, the level of background scattering and thus the signal to noise ratio. Neither scattering length nor incoherent cross section vary in a predictable manner across the periodic table, furthermore there is also an absorption cross section, σ_A, for each nucleus which is usually small but takes a large value

Table 1. Scattering lengths and cross sections for some elements

Element	10^{12}b (cm)	$10^{24}\sigma_{inc}$ (cm^2)	$10^{24}\sigma_A$ (cm^2)
^{1}H	−0.374	79.7	0.33
^{2}H	0.667	2.0	0
B	0.54 + 0.02i	0.7	755
^{12}C	0.665	0	0
^{14}N	0.94	0.40	1.88
^{16}O	0.58	0	0
F	0.57	0	0
Si	0.415	0	0.16
Cl	0.958	5.9	33.6
Cd	0.37 + 0.16i		2450

Absorption cross section for neutron wavelength of 1.8 Å. In the absence of a specified isotope, the values are for elements at their natural abundance for nuclei with complex scattering lengths. Table 1 gives a list of scattering lengths, incoherent scattering cross sections and absorption cross sections for elements occurring in polymers and some materials commonly used for shielding purposes. More complete lists can be found in the books by Kostorz [13] or Bacon [16]. Values for molecules or segments in polymer molecules are calculated as the sum over all elements present in the molecule, thus the scattering length of styrene is 2.33×10^{-12} cm. Scattering length densities are obtained by multiplying the molecular scattering length by the factor $N_A d/M$, where N_A is Avogadro's number, M the molecular weight of the molecular unit, d the density (or for polymers the reciprocal partial specific volume), using styrene again as an example then ϱ is calculated as 1.42×10^{10} cm cm^{-3}.

The most striking feature of Table 1 is the considerable difference in scattering length and incoherent cross section for the isotopes of hydrogen. It is this large difference in b values, and consequently ϱ, which has made small angle neutron scattering a powerful technique in polymer science. Values of ϱ for hydrogenous and deuterated analogues of some common monomer units and solvents are given in Table 2, a fuller list is given by Ullman. Referring to Eq. (14), we note that the prospects for observing the scattering arising from the single particle form factor is related to the magnitude of $(\varrho_p - \varrho_m)^2$, the contrast factor. Incorporation of a small amount of deuterated polymer into its hydrogenous analogue thus provides a means of observing the structural properties at a high total concentration of polymer, whilst the actual scattering species are dilute. This doping technique has been widely used for bulk homopolymers, networks, semi-crystalline polymers and concentrated solutions [2-4,17]. Furthermore, Table 2 shows that certain solvents have either a negative or much reduced scattering length density relative to the deuterated versions. Thus using a mixture of deuterated and hydrogenous solvents permits a gradual change in the scattering length density which can be used to match that of specific parts of the scattering particle. Figure 2 shows this process diagramatically for a core-shell

Table 2. Scattering length densities

Monomer/Solvent	$10^{-10}\varrho$ cm cm^{-3}	$10^{24}\sigma_{inc}$ cm^2
Ethylene	−0.34	319
Ethylene d_4	8.24	8
Styrene	1.47	638
Styrene d_8	6.30	16
Methyl Methacrylate	0.899	638
Methyl Methacrylate d_8	7.0	16
Isoprene	0.22	638
Isoprene d_8	5.49	16
Water	−0.56	159
Water d_2	6.34	4
Cychohexane	−0.55	956
Cychohexane d_{12}	5.76	24
Xylene	0.76	797
Xylene d_{10}	5.85	20

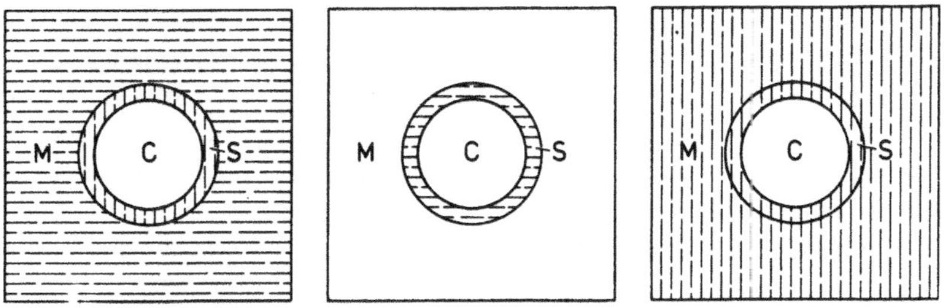

Fig. 2. Contrast variation method for a core (C) — shell (S) particle immersed in a matrix (M). Scattering length densities as indicated

particle immersed in a continuous medium. This technique of contrast variation is widely used in biological applications of neutron scattering [18] and the use of this technique in studying copolymers is discussed later.

5 Scattering Functions

The character of the single particle form factor, $F_p(Q)$, is dependent on the shape of the scattering particle and therefore many different form factors can be calculated. Calculation of $F_p(Q)$ from geometrical factors has been discussed by Guinier and Fournet [15] and the subject more recently reviewed by Porod [19]. Apart from these two sources, scattering functions have been presented by Kerker [20] and by Kratky [21], a particularly useful collection of such functions has been provided by Burchard [22]. Reproduced here are expressions for the most useful scattering functions in the small angle scattering study of copolymers either in solution or solid state.

1. Gaussian Coil [23]

$$(F_p(Q))^2 = 2/(Q^2\langle s^2\rangle)^2\,[\exp(-Q^2s^2) - 1 + Q^2s^2] \tag{15}$$

For small values of $Q\langle s^2\rangle$, the exponential can be expanded and the expression evaluated to give (limiting the expansion to terms in $(Q^2s^2)^3$:

$$(F_p(Q))^2 = 1 - Q^2\langle s^2\rangle/3$$

This evaluation can be used to obtain the radius of gyrations via the classical Zimm plot of I^{-1} as a function of Q^2 since:

$$I^{-1} \propto (1 - Q^2\langle s^2\rangle/3)^{-1}$$

$$\propto \left(1 + \frac{Q^2\langle s^2\rangle}{3}\right)$$

2. Solid Sphere [24)]

$$(F_p(Q))^2 = [(3/(QR)^3)(\sin(QR) - QR\cos(QR)]^2 \tag{16}$$

$$= 9\pi/2(QR)^3 J_{3/2}^2(QR) \tag{17}$$

Where R is the sphere radius and $J_{3/2}(X)$ is a Bessel function of the first kind of order 3/2 with argument X.

3. Solid Cylinder [25)]

$$(F_p(Q))^2 = \int_0^{\pi/2} \frac{\sin^2(QL\cos\theta)}{Q^2L^2\cos^2\theta} \frac{4J_1^2(QR\sin\theta)}{Q^2R^2\sin^2\theta} \sin\theta \, d\theta \tag{18}$$

Where 2L = cylinder length, 2R = cylinder diameter, θ the angle between the scattering vector and the cylinder axis. In the absence of any orientation effects then equation 18 simplifies to:

$$(F_p(Q))^2 = 4J_1^2(QR)/(QR)^2 \tag{19}$$

4. Lamellar Particle

In the absence of orientation and with the scattering vector normal to the lamellar particle surface, then:

$$(F_p(Q))^2 = \frac{\pi}{(QL)} J_{1/2}^2(QL/2) \tag{20}$$

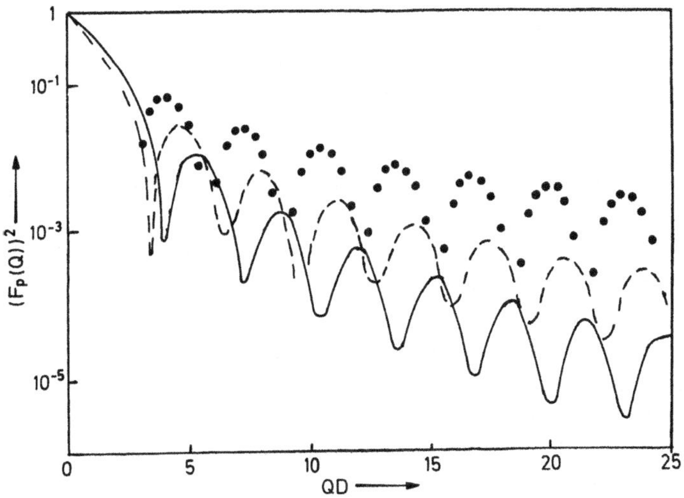

Fig. 3. Single particle form factor scattering for regular particles.
——— Sphere; --------- Cylinder; · · · · · · Lamella
D is characteristic particle dimension (radius or half thickness)

Equations (17), (19) and (20) acquire considerable importance in the analysis of small angle neutron scattering from block copolymers and their nature is displayed in Fig. 3. In this semi logarithmic plot the characteristic maxima are observed with increasing Q, however, the higher order maxima are severely damped and only observable in such semi-logarithmic plots. The maxima in such plots appear at characteristic values of QD (where D is the dimension parameter used in the argument to the Bessel Function), Table 3 sets out these values for the three particle types.

Table 3. Maxima in scattering functions for uniform spheres, cylinders and lamellae

Order	QR_s	QR_c	$QL/2$
0	0	0	0
1	5.77	4.69	3.98
2	9.10	8.24	7.39
3	12.32	11.36	10.51
4	15.52	14.56	13.64
5	18.69	17.6	16.69
6	21.85	20.67	19.89
7	25.01	23.86	23.1

Apart from the solid regular particles dealt with in 2, 3 and 4 above, other simple particle scattering laws include those for hollow spheres, ellipsoids and discs. These have been discussed and summarised in other publications [15, 20–22] and reference to these should be made for further details.

6 Experimental Methods

Small angle neutron diffractometers are available at relatively few research centres, whilst they differ in minor aspects, e.g. wavelength available, monochromatisation method, all diffractometers have so many common features that only one will be described. The major problem is the comparatively low flux of neutron sources in relation to the spatial resolution necessary for measurements at small scattering vector. These two factors make the use of two dimensional position sensitive detectors mandatory. Other factors in the construction of such instruments have been discussed by Schmatz et al. and by Schelten and Hendricks [27]. Most of the results for small angle neutron scattering from polymers have been obtained using the D11 diffractometer at the high flux reactor of the Institut Laue Langevin, Grenoble [28], France. Other instruments which have been used are the D17 diffractometer also at Grenoble, the small angle scattering instrument at AERE Harwell [29] and diffractometers at the University of Missouri [30], the National Bureau of Standards in Washington [31], and at Oak Ridge, Tennessee [32].

A schematic sketch of the D11 instrument is shown in Fig. 4. Thermal neutrons produced by the enriched uranium core of the reactor pass through a liquid deuterium cold source, thereby reducing their energy and increasing the flux at longer wave-

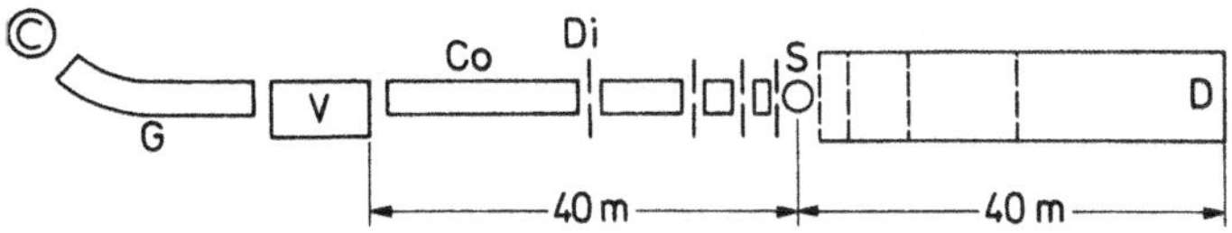

Fig. 4. Schematic diagram of the D11 small angle neutron diffractometer at the Institut Laue-Langevin Grenoble, France

lengths. A curved waveguide filters out fast neutrons and γ-rays and channels the neutron beam onto a helical slot velocity selector which produces 'monochromatic' neutrons whose wavelength is determined by the rotational velocity of the monochromator. In point of fact the neutrons have a distribution in wavelength about the mean value, but this is generally known and usually small. Monochromatic neutrons are guided on to the sample, the incident beam cross sectional area being defined by a diaphragm (usually cadmium) immediately before the sample. Scattered neutrons pass down a flight tube and are collected by the two dimensional detector. The detector may be placed at discrete distances, up to 40 metres, from the sample, the sample-detector distance defining the range of scattering vector and its resolution. Typically the detector is a 64 by 64 array of 1 cm^2 individual cells each of which is able to record incident neutrons, higher resolution is possible with 128 by 128 arrays of 0.5 cm^2 cells. Neutron counts registered on the detector are continually accumulated by a mini-computer, permitting on line inspection of the data. At the end of the measurement the data is stored on a local disc memory and also automatically transferred to a large mainframe computer. Apart from performing these data handling tasks, the local computer also controls instrumental features (wavelength selection, movement of sample changer, etc.) and acts as a 'watchdog' in that it can automatically stop and start measurements given control parameters as well as reporting on the current state of the sample (temperature, pressure, magnetic field strength, etc.). After collection of the data, it is usually regrouped as a radial distribution in intensity about the beam centre on the detector.

Subsequently, each spectrum is corrected for detector efficiency, normalised to unit time and unit thickness of sample, before subtracting background scattering, i.e. scattering due to solvent, container and the incoherent scattering from the specimen. The latter of these is generally most troublesome and great efforts are made to assess its contribution to the total scattering. Various methods have been adopted, usually dependent on the system being investigated. The aim of such data analysis programme suites is to produce final data which is machine independent and free from artefacts [33]. With care in the performance of the experiment and objective data analysis this goal can generally be achieved.

7 Experimental Results

Relatively few small angle neutron scattering studies have been performed on block copolymers in comparison to those on homopolymers, the majority of the published work is concerned with the solid state with little work reported on block copolymer solutions. Results on isotopic block copolymers, e.g. polystyrene molecules with part of the chain synthesised from deuterated styrene [34], will not be considered here, since these molecules exemplify the correlation hole effect discussed by de Gennes [10]. True block copolymers only are considered here, wherein the components are chemically distinct from each other.

7.1 Solid State of Pure Block Copolymers

The characteristic feature of block copolymers in the solid state is their microphase separated structure, with domains of the minor component dispersed in a matrix of the major component [35]. Domain symmetry is chiefly determined by copolymer composition, however, the route to the solid state (i.e. melt prepared or solvent cast) may influence this factor. The major aim of recent small angle scattering experiments on block copolymers has been the investigation of current statistical thermodynamic theories of domain formation. In this respect, it should also be noted that small angle X-ray scattering has been used to investigate block copolymers, the work of Hashimoto et al. [36–38] being particularly noteworthy.

7.2 Theories of Domain Formation

An exhaustive, critical review of the status concerning current statistical thermodynamic analysis of block copolymer domain formation will not be presented here. A precis only of the major theories is given and the predictions from each noted, fuller details are available in the original publications. Furthermore, whilst the earlier theories of Meier [39] and Williams [40–42] were important in stimulating interest and defining the questions to be addressed, they are not considered here since the more recent ideas encompass all the features of the earlier theories.

Perhaps the most detailed statistical thermodynamic theory of microdomain formation is that due to Helfand and Wasserman [43–47]. In common with other theories, the pivotal feature of the theory is the setting up of an equation for the free energy of formation of the microphase separated solid state. Three components of this equation pertain to domain parameters. Firstly domain growth is promoted by the desire to reduce the ratio of domain surface area to domain volume thereby reducing the interfacial free energy. Domain growth is opposed by the necessary localisation of the joint between blocks at the domain matrix boundary, thus preventing significant excursions of one block into a region occupied by the other. The third contribution also opposes domain growth by the necessity to preserve uniform density within a phase. Localisation of the joint would entail an increase in segment density near the interface if the block configuration were Gaussian distributed. Uniform density is maintained by chain expansion into less probable configurations, i.e. an entropic limitation. The minimisation of this free energy equation is simplified by the 'Narrow Interphase Approximation', which states that the distance (a_I) over which the density of one phase smoothly changes to that of the second, is such that $a_I/d_{int} \ll 1$, where d_{int} is the interdomain separation. Additionally, a_I is independent of copolymer molecular weight, composition or domain morphology being determined by local forces only. Minimisation of the free energy equation identifies the equilibrium domain size, D (sphere or cylinder radius or lamellar thickness), whilst the separation, d_{int}, is calculated assuming a close packed structure and sphericalised unit cells in conjunction with the interrelation between phase volumes and unit cell volumes. This theory allows D and d_{int} to be calculated as a function of molecular weight of copolymer in conjunction with other parameters such as the interfacial area per block joint and the number of molecules per unit domain area, it predicts the following scaling relations:

$$D \propto M_D^{0.67}$$

$$d_{int} \propto M^{0.67}$$

Where M_D is the molecular weight of the domain forming block and M the molecular weight of the copolymer.

The NIA theory has been criticised from two viewpoints. Leibler [48] points out the restrictive nature of assuming a constant, narrow interface at the domain boundary and that the NIA requires a specification of mesophase symmetry before the free energy equation can be minimised. By contrast, Leibler considers the onset of microphase separation from the homogeneous region, where the concept of an interphase is invalid. By analysing monomer density fluctuations using the random phase approximation of de Gennes [49], Leibler writes the free energy, near the microphase separation temperature, as a function of an order parameter and identifies copolymer composition and the product χN as the important terms, where χ is the Flory interaction parameter and N the degree of polymerisation of the block. Although this theory makes no quantitative predictions regarding domain separation and size, it does predict the mesophase symmetry, the wavelength of the density fluctuation of maximum growth ($\approx d_{int}$), the phase diagram at the microphase separation temperature and the occurence of transitions between the mesophases. The theory predicts that for copolymer weight fraction compositions which are not equal to 0.5, a body centred cubic arrangement of spheres or a hexagonally close packed array of cylinders is obtained on phase separation with the regime of spherical domain stability being extremely narrow.

Hong and Noolandi [50–53] have published a series of papers on the statistical thermodynamics of heterophase polymer systems, they point out that the NIA theory requires an ad hoc compressibility term added to the free energy equation due to the need to restrict density fluctuations. Their theory is essentially a mean field theory where the free energy equation consists of a Flory-Huggins term and a fluctuation term describing inhomogeneities in the copolymer. The fluctuation term is minimised by the identification of a fluctuation wavelength of maximum growth, which is a function of χ and copolymer composition, defining the periodicity and symmetry of the mesophase. Consequently, this theory has similarities to that of Leibler's, however, a dependence of domain separation on molecular weight is predicted such that:

$$d_{int} \propto M^{1/2}$$

In summary, Helfands NIA theory predicts molecular weight dependencies of domain size, separation and other parameters of the phase separated copolymers with the presumption of an interface of constant thickness at the domain boundary.

Such definitive forecasts do not result from the theories of Leibler or Hong and Noolandi. However, the latter theory describes in detail the phase diagram for copolymer — homopolymer mixtures and is therefore pertinent to the X-ray work of Roe [54, 55]. Similarly, Leibler's theory provides a detailed description of a microphase separation mechanism and thus is of value in the interpretation of experiments investigating this phenomenon. Small angle neutron scattering data reported to date has been mainly concerned with pure styrene-diene block copolymers which are fully microphase separated and thus examined D, d_{int} and interfacial layer thickness as a function of molecular weight and composition and therefore comparison has usually been made with the NIA theory of Helfand.

7.3 Pure Block Copolymers in the Solid State

Domain Size and Separation

The earliest published paper dealing with SANS from block copolymers appeared in 1982 [56] and provided an overview of the areas which have been of continuing interest since. Thus diffraction patterns for hydrogenous styrene-isoprene copolymers were analysed in terms of the long range organisation of the domains using an analogy to elementary cubic and simple close packed structures. For one copolymer with spherical domains, SANS data was obtained at higher Q in an effort to observe the single particle form factor scattering from the domains. Such scattering was observed but at a very weak level, however, the domain radius obtained in conjunction with the domain separation determined from the Bragg pattern enabled the calculation of the styrene volume fraction assuming a particular arrangement of the spheres. The best agreement was obtained for a face centred cubic arrangement of spheres. This paper also reported on a single determination of the interfacial layer thickness and a value for the radius of gyration in the spherical microdomain. Little comparison with theoretical predictions was made since the range of molecular weights for any one domain symmetry was small.

Cohen and co-workers [57–60] have produced a series of papers on styrene-butadiene copolymers with spherical butadiene microdomains where the butadiene phase was deuterated. Domain sizes were determined by fitting a theoretical curve, including a modification for an interfacial layer, to the data and including a distribution of domain size. They noted that domain radius obtained by SANS was consistently larger than that from transmission electron microscopy, a fact attributed to a systematic artefact of the staining process in preparing samples for electron microscopy [57]. However, this artefact has yet to be identified. This work was expanded in a later publication [58] on similar materials, for which a series of well resolved Bragg peaks were evident for $0.01 < Q < 0.03\,Å^{-1}$. However, they pointed out that the long range arrangement of the spherical butadiene microdomains could not be definitively established due to the modulation of the interference function by the single particle form factor. Volume fractions of butadiene were calculated using SANS domain radii for a simple cubic and a body centred cubic array of spheres, by comparison of these values with the experimental volume fraction they conclude that a body centred cubic array is formed. Whilst this agrees with the theoretical prediction of Leibler [48], a re-working of their SANS data on the basis of a face centred cubic lattice of spheres

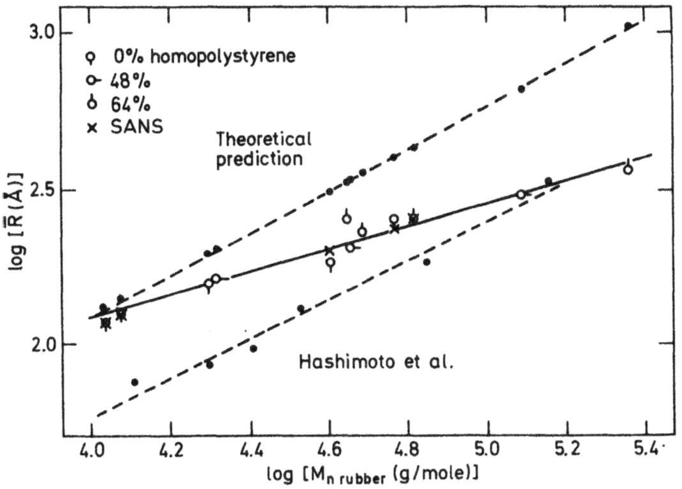

Fig. 5. Log-log plot of mean sphere size as a function of the molecular weight of the butadiene block of styrene-butadiene block copolymers. Upper dashed line from Helfand NIA theory: lower dashed line is a fit to Hashimoto's results. Ref. [59]

produces even better agreement with their quoted butadiene volume fraction. These workers have also investigated the influence of casting solvent on the structural parameters, using toluene, benzene and a mixture of tetra-hydrofuran and 2-butanone as solvents. No significant difference in either domain size or macrolattice plane spacing was observed, moreover the percentage standard deviation of domain size distribution was constant at circa 10%. SANS determined domain radii were combined with values obtained by transmission electron microscopy (the latter after correction by the factor noted above), and the results compared with small angle X-ray data of Hashimoto and the theoretical predictions of the NIA theory, Fig. 5 shows this comparison. From NIA theory it is expected that:

$$R_s \propto M_d^{0.67}$$

Where R_s = sphere radius and M_d is the molecular weight of the domain forming block, the results obtained by Cohen et al. [59] scale as:

$$R_s \propto M_d^{0.37}$$

This discrepancy was explained on the basis of the argument used by Hashimoto [37], i.e. adjustment of spherical domain size requires mass transport through the copolymer and the energy barrier to this will increase as the solvent evaporates during the casting process. Although not explicitly stated, this energy barrier presumably originates from a viscosity increase. Consequently, whilst the explanation of Cohen et al. concerning the increasing disagreement as the domain block molecular weight increase superficially concurrs with this idea, it ignores the increased viscosity of the polystyrene matrix. This is especially so for those specimens where polystyrene homopolymer has been added to adjust the butadiene weight fraction to that commensurate with spherical microdomain symmetry.

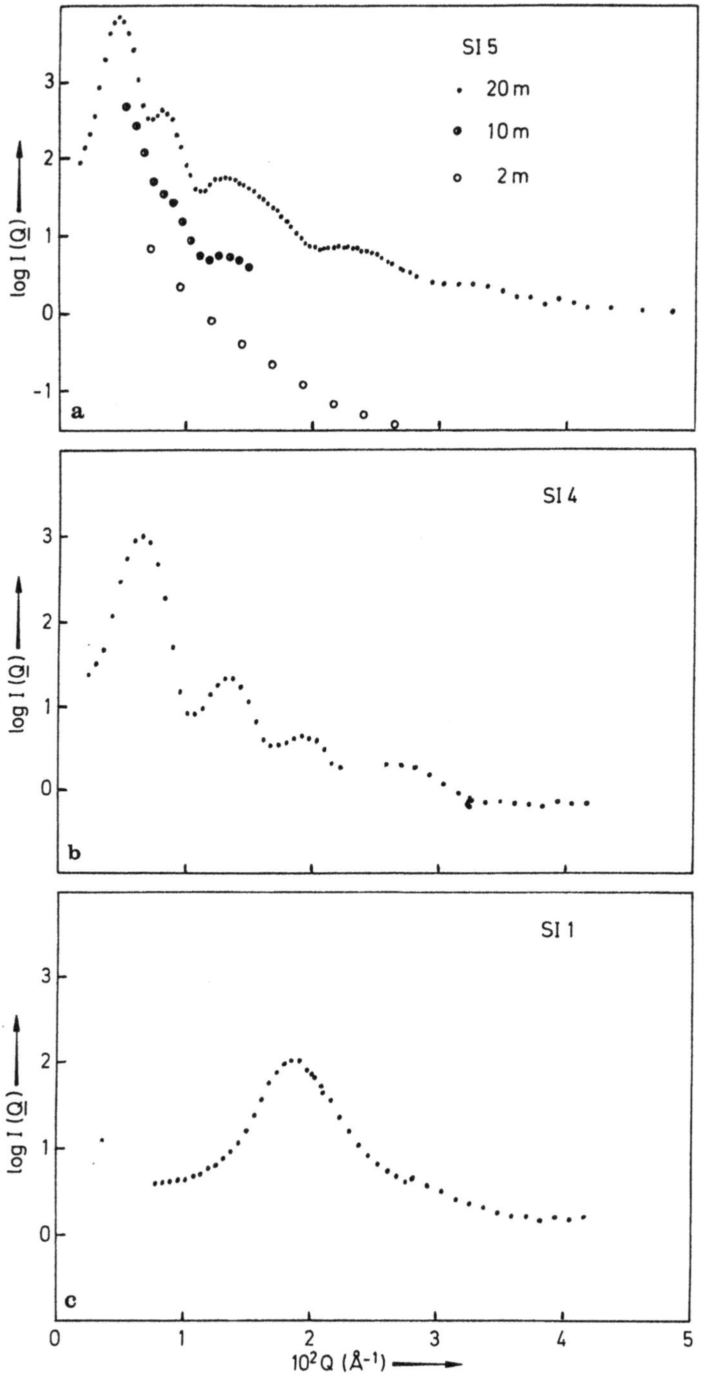

Fig. 6a–c. Diffraction envelopes for hydrogenous diblock copolymers of styrene and isoprene. (Ref. [61]). **a** Cylindrical isoprene domains; **b** Lamellar domains; **c** Spherical styrene domains; **a** illustrates the decrease in resolution on moving the detector

To date, the fullest report on SANS studies of styrene-diene block copolymers has been that of Richards and Thomason[61]. A range of styrene-isoprene di and triblock copolymers were prepared such that each domain symmetry was available over a range of molecular weight. Figure 6 shows diffraction patterns obtained for each domain morphology using fully hydrogenous samples. These were obtained by combining data collected over a series of discrete Q ranges, as specified in Fig. 6a. This latter figure also demonstrates the loss in resolution inherent in moving a detector with a fixed array of cells into differing ranges of scattering vector Q. Contour plots of the scattered intensity were generally isotropic about the incident neutron beam, however, as Fig. 7 shows, lamellar block copolymers show a preferential alignment

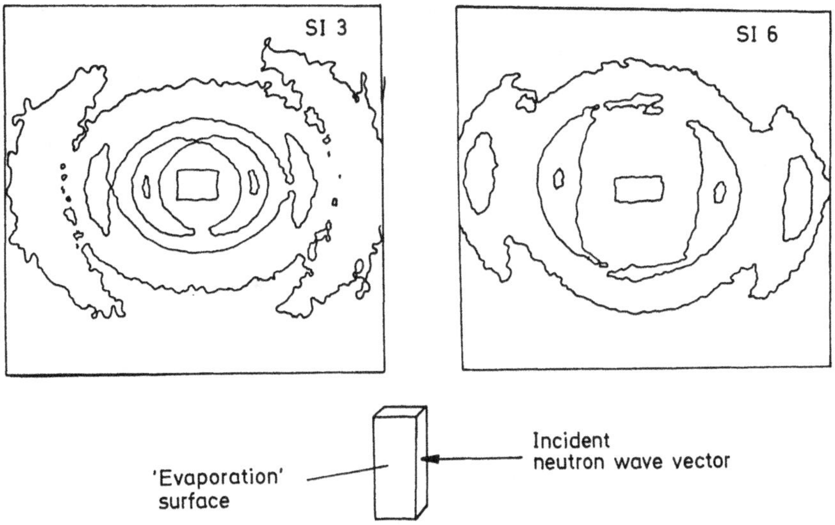

Fig. 7. Isointensity contour plots for lamellar diblock copolymers of styrene and isoprene. (Ref. [61])

of the sheets parallel to the evaporation surface. Notwithstanding this fact, the Q values for the Bragg peaks for lamellar specimens were identical whether the incident beam was parallel or perpendicular to the lamellar normals, clear evidence of disorder in the specimens. The diffraction pattern for block copolymers with spherical microdomain symmetry underlines the observation made by Cohen et al. namely that insufficient detail is obtained to define domain packing definitively. These Bragg peaks were interpreted as arising from a crystalline macrolattice formed by the domains and consequently interpreted using the equations relating lattice plane spacings to unit cell dimensions and the Miller indices of the Bragg peaks observed. Whilst this is reasonably straightforward for lamellar and cylindrical domain morphologies, confirmation of spherical domain long range order required calculation of volume fractions and comparison with values determined experimentally. This comparison suggested that a face centred cubic arrangement of spheres prevailed.

Calculation of volume fractions required values of domain radius or thickness, these wers determined from SANS data at intermediate Q, i.e. where $A(Q) = 1$,

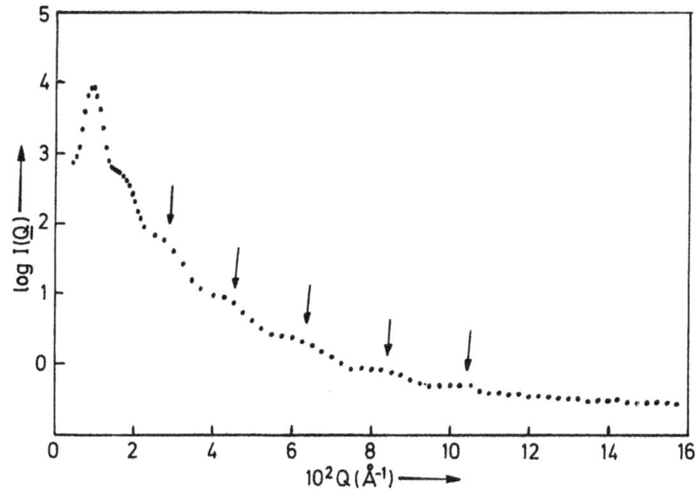

Fig. 8. Scattering envelope to intermediate Q values for a diblock copolymer of deuterated styrene and hydrogenous isoprene with cylindrical styrene domains (Ref. [61])

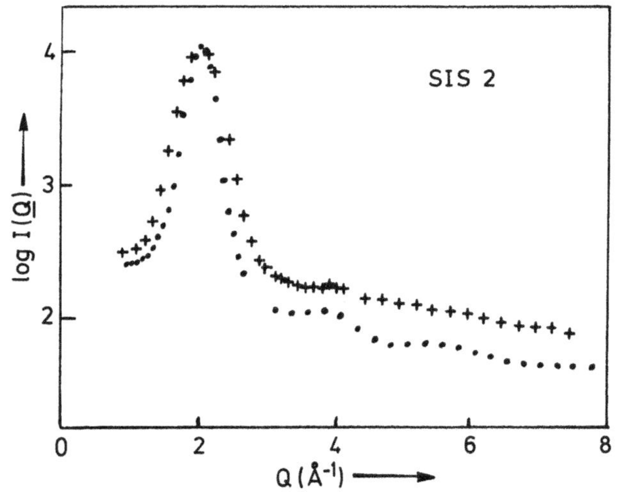

Fig. 9. Influence of thermal annealing on the diffraction pattern of a hydrogenous triblock copolymer with spherical styrene domains. + as prepared; ● after annealing at 150 °C under vacuum (Ref. [61])

using copolymers with a fully deuterated styrene phase. Such data are shown in Fig. 8, and a number of weak maxima are evident from which the domain size may be determined using the figures quoted in table 3. Evidently, these maxima are much more diffuse than the theoretical curves in Fig. 3. Domain size polydispersity is the cause of these diffuse maxima and this is discussed in more detail later. All the samples had been cast from toluene solutions, and bearing in mind the possibility of non-equilibrium structures being obtained, some samples were annealed at 150 °C. Figure 9 shows the effect of this annealing on the diffraction pattern for one sample.

A sharpening of Bragg peaks is observed but no shift in position, the sharpening was interpreted as a consequence of the growth of 'grains' within the sample wherein all unit cells were in register with each other.

Values of domain size and separation were discussed in terms of Helfand's NIA theory, tacitly assuming that the two components of the block copolymer constituted a symmetric polymer pair, i.e. equality of Kuhn statistical step lengths, b, and monomer molar density. Whilst this is approximately true for values of b for styrene and isoprene, the densities are quite different. However, calculations of domain size and domain separation as a function of molecular weight for both styrene and isoprene domains show that both types have the same molecular weight dependence and moreover the difference in the values for either styrene or isoprene domains is negligible. Figure 10

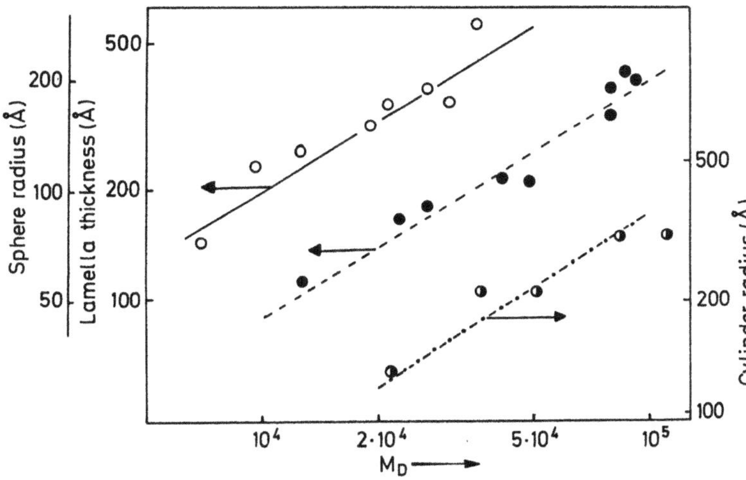

Fig. 10. Domain size variation with molecular weight of domain forming block, M for styrene-isoprene block copolymers. ○ Spherical domain; ● Lamellar domains; ◐ Cylindrical domains

shows a log-log plot of domain dimensions as a function of domain block molecular weight for the three domain symmetric encountered. The lines drawn through the data points are calculated from the NIA theory and agree reasonably well, furthermore the magnitude of both theoretical and experimental values are similar over the whole range of molecular weight investigated in contrast to the results of Cohen et al. [59] and Hashimoto et al. [36,37]. Least squares fits of straight lines to these data gave the following scaling relations:

$$R_s = 0.89 M_d^{0.52}$$

$$R_c = 0.55 M_d^{0.56}$$

$$L = 0.22 M_d^{0.66}$$

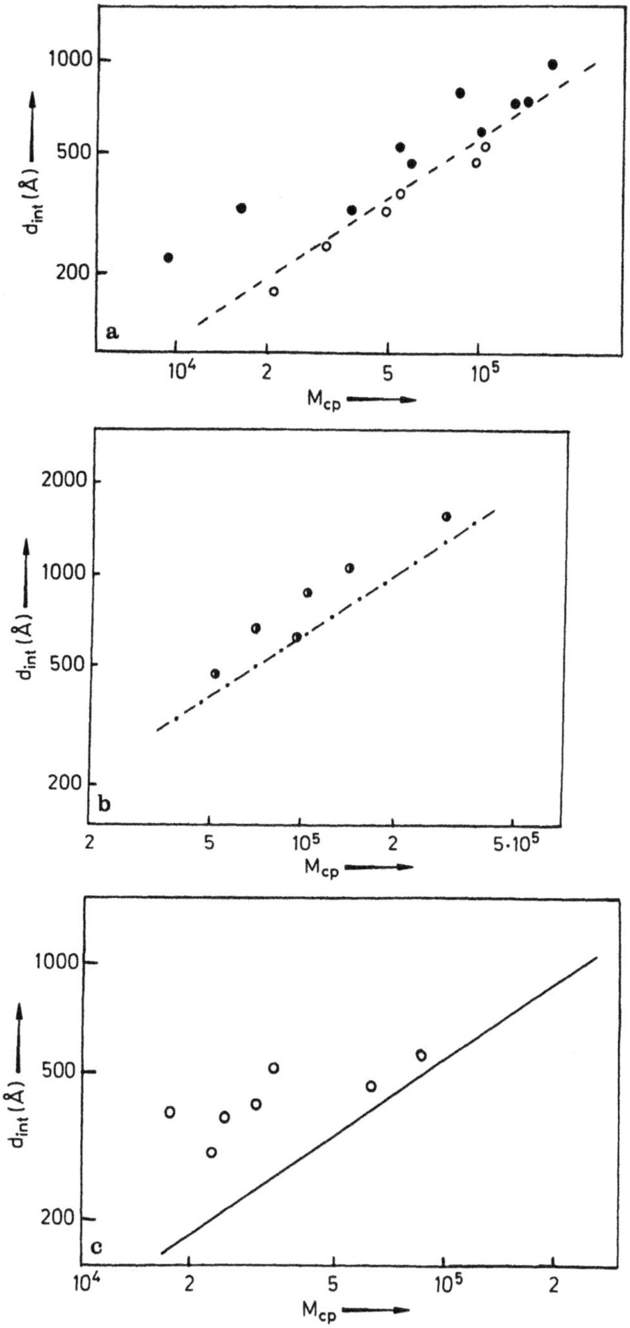

Fig. 11a–c. Domain separation variation with copolymer molecular weight for styrene-isoprene block copolymers. **a** lamellar domains; **b** cylindrical domains; **c** spherical domains. Lines calculated from Helfand NIA theory. ○ = data points from X-ray work of Hashimoto

Where R_c and L are the cylinder radius and lamellar thickness respectively and the units are Ångstroms. The deviation from the expected 0.67 exponent from NIA theory is greatest for spherical microdomains and this may reflect energy barriers to mass transport between domains preventing equilibrium being achieved, as discussed by Hashimoto [37]. This deviation becomes even more apparent when domain separations are examined. Since domain size and domain separation are connected via the copolymer composition, they should both display the same molecular weight exponent. Figure 11 shows the SANS determined domain separations as a function of copolymer molecular weight in a log-log plot, together with lines calculated from the NIA theory. Theoretical values of d_{int} were lower than experimental values for all domain symmetries, however, for lamellar and cylindrical domains the exponents in the scaling relations were in fair agreement with NIA theory, the actual equations being:

$$d_{int} = 0.32M^{0.67} \qquad \text{(cylinders)}$$

and

$$d_{int} = 0.997M^{0.56} \qquad \text{(lamellae)}$$

For spherical microdomains the relation obtained was:

$$d_{int} = 8.6M^{0.36}$$

in no agreement with theory. Possible sources of this disagreement were discussed and it was concluded that the spherical domains become locked at some non-equilibrium separation during the evaporation process, thereafter the approach to equilibrium becoming infinitely slow. Molecular weight polydispersity is thought not to have any influence since essentially equilibrium domain sizes were obtained.

Domain-Matrix Interface

A key factor in the NIA theory is the constant value of the interfacial layer thickness at the domain-matrix boundary. There has been much discussion on the existence and magnitude of an interfacial region where partial mixing of the two components takes place. Its existence has been established by positron annihalation studies [62], whilst its magnitude is still a source of interest. Williams [40-41] has proposed that this interface is unsymmetric and of considerable extent, in terms of volume fraction, whilst Krause [63] concludes that the interphase volume fraction is very small. Mechanical and dynamical mechanical properties have been discussed [64-67], in greater or lesser detail, in terms of the interphase volume fraction and Williams [68] has successfully modelled viscoelastic properties of block copolymers by specifically incorporating an interphase volume fraction. The validity of this approach has been commented on by Cohen et al. [67] who also review the importance of the interphase in the interpretation of mechanical properties. On the basis of a symmetric density variation across the interface, the value of a_I calculated from NIA theory is 14 Å. Determination of the interphase thickness turns on evaluation of the deviations of the scattered intensity from Porod's Law [69], a possibility first pointed out by Ruland [70]. Porod's

Law pertains to particles with sharp boundaries and random orientation, thus when $QD \gg 1$ for all dimensions, D, of the particle:

$$I(Q) = K\, 2\pi (A_p/V_p^2)\, Q^{-4}$$

Where A_p and V_p are the surface area and volume of the particle respectively. The existence of an interphase of smoothly changing density at the particle boundary can be modelled by convoluting the sharp boundaries with a smoothing function, the Fourier transform of which, $H(Q)$, will appear in the equation for the scattered intensity:

$$I(Q) = K\, 2\pi (A_p/V_p^2)\, Q^{-4} H^2(Q) \tag{21}$$

Application of this type of equation to small angle X-ray data has been discussed by Koberstein et al. [71] and Roe [72], both point out the crucial importance of accurate subtraction of the background. This is especially true in X-ray work where the electron density difference between phases is small and where the background increases (due to thermal density fluctuations) with increasing Q. For SANS from part deuterated copolymers, the change in scattering length density between two phases is very large as Fig. 12 shows. Furthermore Koberstein [73] has demonstrated that the incoherent background scattering is far greater than thermal density contributions and moreover it is flat over the full Q range generally used for SANS experiments, consequently SANS appears to offer the best possibility to determine a_I. Notwithstanding these advantages of SANS, determination of a_I still requires very careful measurements

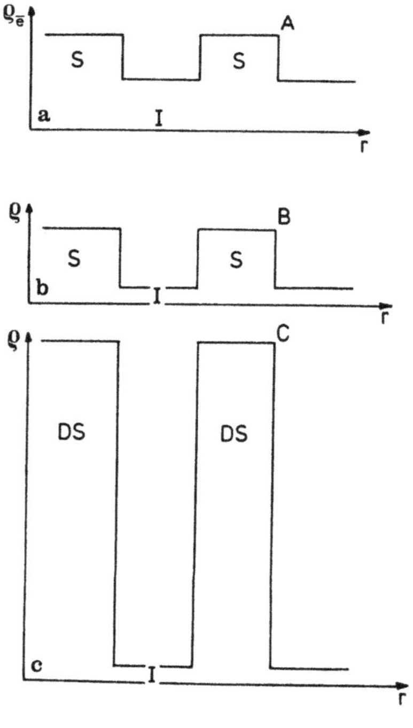

Fig. 12a–c. Schematic variation of scattering densities for styrene-isoprene block copolymers. **a** Electron density variation for small angle X-ray scattering; **b** Scattering length density variation for hydrogenous styrene-isoprene copolymers; **c** Scattering length density variation for deuterated styrene-hydrogenous isoprene block copolymers

since in the range of Q used, the signal to noise (background) is not high. Whilst the form of H(Q) in equation 21 can be chosen to be in exact agreement with the density profile obtained by Helfand, it is not convenient for determination of a_I, consequently it has been more usual for H(Q) to be represented by a Gaussian distribution:

$$H(Q) = \exp(-\sigma^2 Q^2/2)$$

which makes σ the second moment of the density distribution across the interface. The thickness associated with this distribution, t, is then $\sqrt{12}\sigma$ whilst a_I, the parameter defined by Helfand, is 0.637t. Incorporation of the Gaussian distribution of H(Q) results in the intensity expression:

$$I(Q) = K\, 2\pi(A_p/V_p^2)\, Q^{-4} \exp(-\sigma^2 Q^2)$$

Therefore:

$$\mathrm{Ln}(Q^4 I(Q)) = \mathrm{Ln}\,(K\, 2\pi A_p/V_p^2) - \sigma^2 Q^2$$

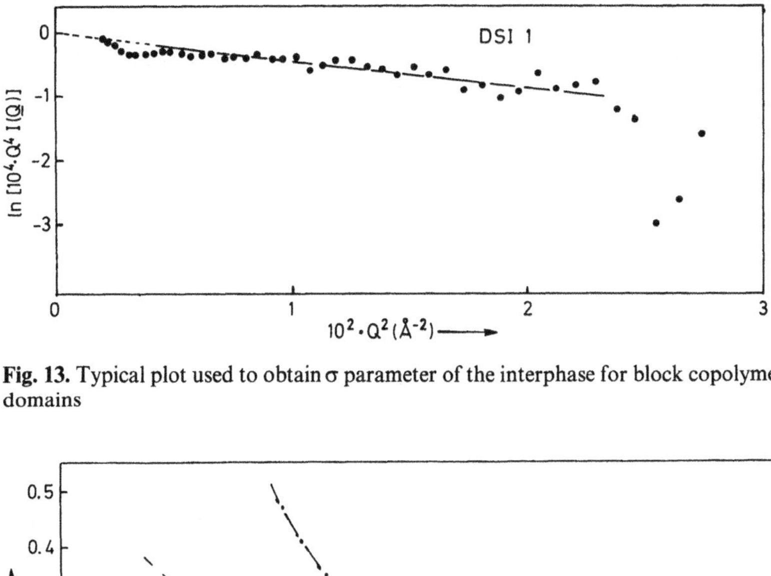

Fig. 13. Typical plot used to obtain σ parameter of the interphase for block copolymers with spherical domains

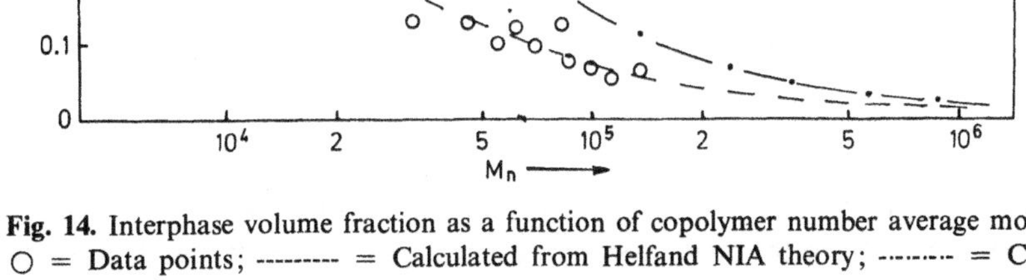

Fig. 14. Interphase volume fraction as a function of copolymer number average molecular weight. O = Data points; --------- = Calculated from Helfand NIA theory; -·-·-·-·- = Calculated from Meier's theory

A typical plot of data according to equation is shown in Fig. 13, and a straight line can easily be drawn through the data points. For lamellar and cylindrical microdomains, this process is subject to some correction to allow for orientation effects [74], nonetheless the values of a_I obtained are reasonably constant with the average value being 12.8 Å a value which gives support to the NIA theoretical value of 14 Å. This value for a_I was also obtained by Cohen et al. [59] for spherical butadiene microdomains cast from different solvents and by the X-ray work of Hashimoto et al. [36, 37], thus supporting the assumption that local interactions are dominant in determining the nature of the interphase. Interfacial volume fraction is more important for analysis of mechanical properties, Fig. 14 shows this volume fraction as a function of molecular weight, the best agreement being with Helfands NIA model.

Single Chain Scattering

The major application of SANS in polymer science has been the evaluation of the radius of gyration of polymer molecules in situations inaccessible to light or X-ray scattering. Generally the technique has been to 'dope' a hydrogenous polymer with a small percentage of its deuterated counterpart, the excess scattering between the two being interpreted via Eq. (15). Equation (15) can be written in a fuller form [56] which enables the weight average molecular weight of the scattering species to be obtained as well as the z average mean square radius of gyration. Thus at $Q = 0$, $\langle F_p(Q) \rangle^2 = 1$ and

$$\frac{1}{M} = cI_c(0)\, 4\pi t T_s \bar{v}^2 (\varrho_p - \varrho_m)^2 / (I(0)\,(1 - T_c)\, N_A)$$

c	= concentration of labelled species
$I_c(0)$, $I(0)$	= Extrapoluted intensity at $Q = 0$ for calibrant and sample respectively
T_c, T_s	= Neutron transmission of calibrant and sample respectively
t	= sample thickness
$\bar{v}$	= partial specific volume of labelled species
$\bar{M}_w$	= weight average molecular weight

The use of dilute 'solutions' of deuterated homopolymers in hydrogenous homopolymers results in the absence of any intermolecular terms in the final expression for the excess scattering, however this also means that the intensity is weak. Higher concentrations of deuterated material have been used [75, 76], up to 50% by weight, to increase signal to noise ratio. This technique does, however, require that both matrix and dopant have the same molecular weight *and* molecular weight distribution. For the determination of radius of gyration, ideally the Q range used should be such that $QR_g \lesssim 1$.

This requirement, in conjunction with typical block molecular weights, means that the desired Q range falls within that wherein the Bragg peaks from the long range domain order are dominant, consequently subtraction of the undoped background has to be carried out with care.

The straightforward approach of subtracting the scattered intensity of the hydrogenous copolymer from the same copolymer containing a percentage of part deuterated copolymer has been used in two or three cases. Richards and Thomason [56] obtained the radius of gyration and molecular weight of deuterated styrene blocks in a styrene-isoprene block copolymer with spherical styrene microdomains. The molecular weight obtained from SANS data agreed reasonably well with that obtained by gel permeation chromatography, thus indicating a random distribution of deuterated blocks throughout the domains. From the same data, the radius of gyration obtained was essentially that pertaining to the unperturbed state of the molecule.

Renewed interest in polymer blends has provoked a re-examination of the scattering laws for multicomponent systems, not only with a view to obtaining thermodynamic parameters, but also to assess the possibilities of removing the interference from the other components. This question has been particularly addressed by Jahshan and Summerfield [77] and Koberstein [73], they show that for a two phase system containing a proportion of deuterated material, essentially a three component polymer, then Eq. (14) should be written as:

$$I(Q) = I_0 \frac{N}{V} \Delta\Omega \left[\frac{V_p^2 N_p}{N} (X\varrho_p + \varrho_p(1-X) - \varrho_m)^2 (F_p(Q))^2 A(Q) + X(1-X)(\varrho_p^D - \varrho_p)^2 S_s(Q) c_D + \frac{\sigma_{inc}}{4\pi} \right]$$

Where ϱ_p^D is the scattering length density of the deuterated component of the domain forming block, whilst ϱ_p is the scattering length density of its hydrogenous counterpart; X is the molefraction of deuterated copolymer; $S_s(Q)$ is the scattering law of the deuterated block in the copolymer, usually the expression for a Gaussian

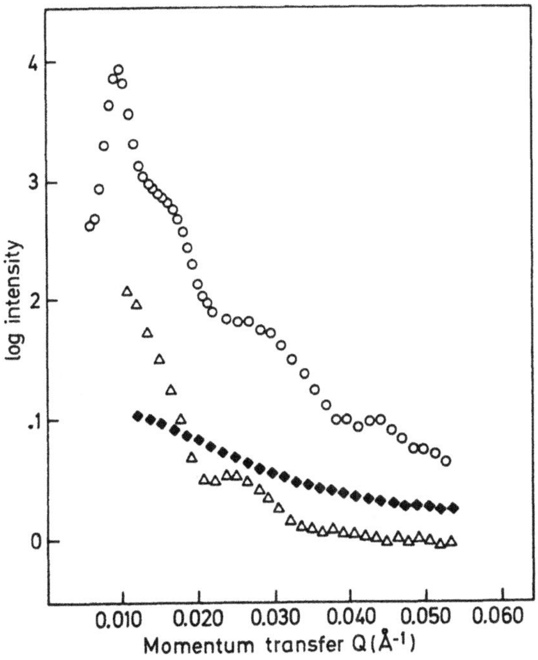

Fig. 15. Small angle neutron scattering data for SB_d3 (○), SB7 (△) and a mixture of the two ( ) where the interference function scattering has been contrast matched out. Ref. [60]

coil is used, and c_D is the concentration of labelled chains per unit volume. Inspection of the equation indicates that if X can be manipulated sufficiently to make $(X\varrho_p^D + \varrho_p(1 - X) - \varrho_m) = 0$, then only the self scattering law, $S_s(Q)$, will be observed. This approach has been successfully used by Cohen et al. in their work on styrene-butadiene copolymers [60]. Their experiments were made on specimens wherein 16% of the butadiene was deuterated, these being obtained by mixing a fully hydrogenous copolymer with a copolymer with fully deuterated butadiene blocks. Figure 15 shows a semi log plot of their data, the hydrogenous/deuterated mixture is notable for the absence of any Bragg peaks, Table 4 gives their results from SANS and molecular characterisation. For the combination $SB_d1/SB1$ the agreement between SANS and g.p.c. molecular weights is good whilst the radius of gyration of the butadiene block is essentially unperturbed. Such is not the case for the combination $SB_d3/SB7$ the SANS molecular weight being considerably larger than the g.p.c. value. Whilst this may be indicative of clustering of deuterated butadiene blocks in a few domains, it is noteworthy that the compositions of SB_d3 and SB7 are considerably different, therefore the microphase structure of the two copolymers cannot be identical, a circumstance which violates one of the conditions defined by Koberstein [73] as necessary for the use of equation.

Table 4. Radii of gyration (R_g) and molecular weights of deuterobutadiene blocks in styrene-butadiene block copolymers [60]

Sample	$10^{-3}M_w^B$	W_B	R_g (Å)	$10^{-3}M_w^B$ (SANS)
SB1	11	0.12		
SB7	59	0.09_6		
SB_d1	13	0.136		
SB_d3	54	0.362		
$SB_d1/SB1$			44	19
$SB_d3/SB7$			133	179

Regretably, this technique cannot be used for determining the radius of gyration in the styrene domains. It is impossible to adjust $(X\varrho_p^D + \varrho_p(1 - X) - \varrho_m)$ to zero for styrene-butadiene or styrene-isoprene block copolymers when the deuterated component is the styrene block.

Hadziioannou et al. [78] have reported on chain dimensions in lamellar diblock copolymers of styrene and isoprene oriented to a 'single crystal' form by melt shearing. The dimensions of the styrene blocks normal to the lamellar surface were determined by the doping technique and with the incident beam normal to the lamellar planes and using only the scattering where contamination from interference function could be neglected. This naturally increases the statistical error since only a part of the detector is used. The mean square distance of the deuterated styrene blocks normal to the surface was in between values calculated for freely interpenetrating Gaussian coils and for chains confined to separated cylindrical tubes.

Apart from these studies of single block dimensions in domains, there have only been two other published attempts. Han et al. [79] makes a brief reference to results obtained for lamellar styrene-isoprene copolymers. Since no data were given, no further comment will be made. Cooper et al. [80] in the same publication report on an attempt to measure the radius of gyration of polytetramethylene oxide (PTMO) units in a polyurethane block copolymer. The technique advocated by Koberstein [73], and discussed earlier, was used to counteract the interference scattering from the two phases, however it was noted that there was evidence for residual interference scattering. For the PTMO molecular weight, 1445, a radius of gyration of 16.4 Å was obtained, this should be compared with an unperturbed value of circa 15 Å calculated from the characteristic ratio, C_∞, of polytetrahydrofuran.

Computer Modelling of Small Angle Neutron Scattering from Block Copolymers

Whilst data and parameters pertaining to the copolymer structure can be extracted directly from experimental results, the modelling of the observed scattering envelope can provide a deeper insight by noting the influence of adjusting model parameters. Hashimoto et al. [31] have attempted this for the single particle form factor scattering obtained from X-rays. By this means a distribution in domain size was evaluated and the interphase scattering was incorporated directly. These formulae were used by Cohen et al. [58, 59] in their SANS work on styrene-butadiene copolymers. This treatment has been extended by Richards and Thomason [82] to include instrumental factors, e.g. wavelength distribution and detector resolution and the effects of domain orientation for anisometric domains (lamellae and cylinders). Additionally, the incoherent background contribution to the signal was calculated in a formal manner. For spherical domains the inclusion of a Gaussian distribution of domain size with a standard deviation of 0.1, in conjunction with incoherent background scattering has a drastic influence on the possibility of observing single domain scattering as Fig. 16 shows. The advantages of deuterating one component are evident. Domain orientation was modelled using a modification of an equation afforded by Fournet [25]

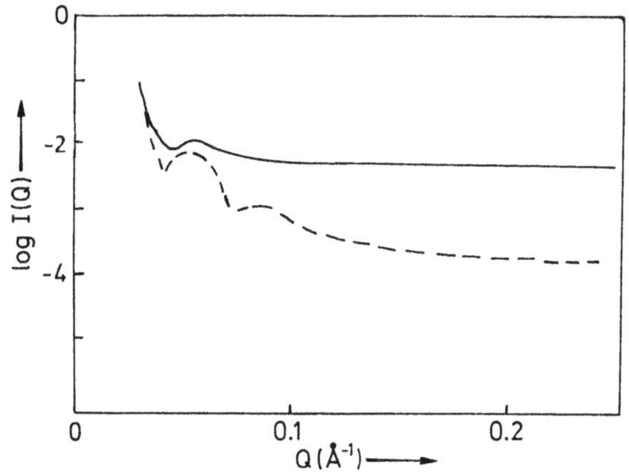

Fig. 16. Calculated single particle form factors for spherical styrene domains of radius 100 Å. ——— = Hydrogenous domains in a hydrogenous isoprene matrix; -------- = Deuterated domains in a hydrogenous isoprene matrix. Domains have a Gaussian distribution of radius with a percentage standard deviation of 10%

for cylindrical particles, which essentially factorises the scattering into axial and cross section components. If anisometric particles are not oriented such that their cross section is within 90° ± 20° to the scattering vector, Q, then the scattered intensity rapidly drops to the level of background scattering, and no information regarding single domain parameters can be extracted. Inclusion of the appropriate parameters does, however, enable experimental data to be reproduced.

Interference function scattering was modelled by using a paracrystalline lattice model as a basis, with the adoption of the equations for X-ray scattering from low molecular weight crystalline materials. Thus the Bragg peaks essentially result from a 'powder' diffraction pattern from the block copolymers. Reasonable agreement

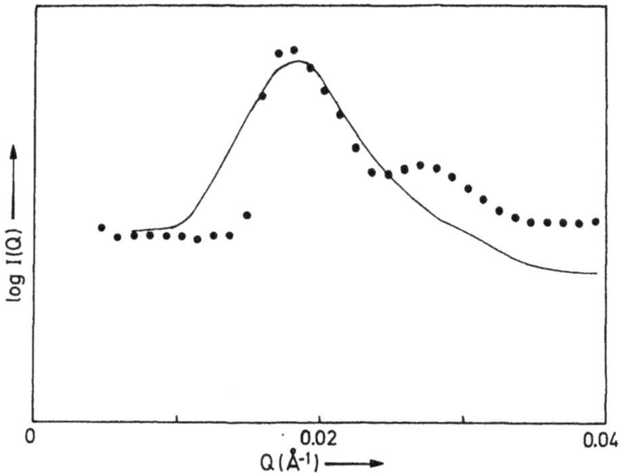

Fig. 17. Comparison of experimental neutron scattering data (———) with calculated interference function scattering (. . . . . .) for spheres on a paracrystalline face centred cubic lattice

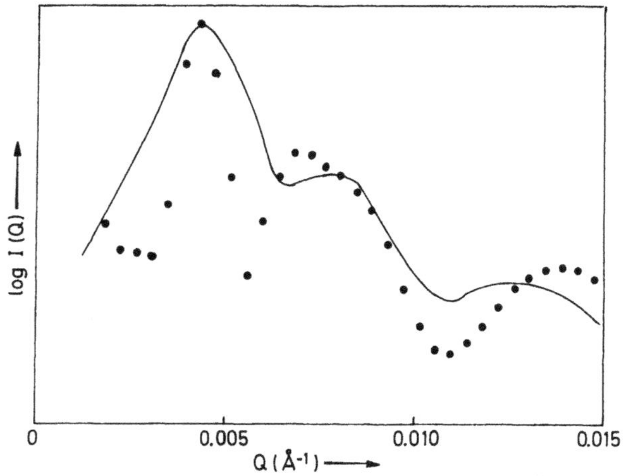

Fig. 18. Calculated interference function scattering (. . . . . .) compared with experimental data (———) for cylinders on a paracrystalline hexagonally close packed lattice

with experimental data is obtained if spherical domains are modelled using a face centered cubic lattice (Fig. 17) whilst cylindrical domains appear to be hexagonally close packed (Fig. 18). Although the agreement using such simple models is gratifying it is clear that these cannot be completely accurate since modelled Bragg peaks are somewhat narrower and for higher order peaks of larger amplitude. Perhaps a statistical approach using correlation functions may provide a better insight, such attempts are yet to be made.

7.4 Copolymers in Solution

The configuration of block copolymers in solution has attracted experimental and theoretical interest for many years [83-86]. Two models are generally discussed for molecularly dissolved copolymers; a segregated configuration where the chemically disimilar blocks occupy distinct regions in space, minimising contacts, and secondly a mutually interpenetrating random coil model. In the segregated structure it is expected that each block should have the configuration of the respective homopolymer whilst the interpenetrating model leads to expanded configurations.

Bushuk and Benoit [87] analysed the light scattering behaviour of copolymers many years ago, they demonstrated that the intensity of scattered light depended on the contrast between copolymer and solvent. Additionally, Leng and Benoit [88] showed that the angular variation of the scattered intensity also depended on the contrast. In consequence the molecular weight and radius of gyration obtained by light scattering are apparent values (unless obtained under special conditions, vide infra) and have to be corrected to true values. Fuller discussions of the application of light scattering to copolymers in solution have been provided by Benoit and Froelich [89] and in the papers of Kratochvil [90, 91]. The equations developed for light scattering can be applied directly to small angle neutron scattering with adaptation for the differing contrast factors [92]. Thus the scattered intensity can be written as:

$$I(Q) = K^* c M_{App}(\varrho_c - \varrho_s)\, S_c(Q)$$

Where M_{App} is the apparent molecular weight of the copolymer with scattering length density ϱ_c dissolved in a solvent of scattering length density ϱ_s and the scattering law is $S_c(Q)$. Consider a block copolymer with type A blocks and type B blocks, where W_A is the weight fraction of component A, then,

$$\varrho_c = W_A \varrho_A + (1 - W_A)\, \varrho_B$$

where ϱ_A and ϱ_B are the scattering length densities of the copolymer components. To simplify subsequent formulae, contrast factors are denoted by K_i, thus,

$$K_A = \varrho_A - \varrho_S; \qquad K_B = \varrho_B - \varrho_S \qquad \text{and} \qquad K_C = \varrho_C - \varrho_S$$

The apparent molecular weight is then given by:

$$M_{App} = \frac{K_A K_B}{K_C^2} M_C + (K_A(K_A - K_B)/K_C^2)\, W_A M_A + (K_B(K_B - K_A))/K_C^2 W_B$$

Whilst the scattering law is:

$$S_C(Q) = \frac{1}{K_C^2}[W_A^2 K_A^2 S_A(Q) + (1 - W_A)^2 K_B^2 S_B(Q) + 2W_A(1 - W_A) K_A K_B S_{AB}(Q)]$$

Inspection of these two equations reveals the advantages of contrast variation. Thus if the solvent can be chosen (or adjusted) such that either K_A or K_B is zero then the equations may be greatly simplified, hence if K_A is zero, then:

$$M_{App} = \frac{K_B^2}{(1 - W_A) K_C^2} M_B$$

and

$$S_C(Q) = \frac{(1 - W_A)^2 K_B^2 S_B(Q)}{K_C^2}$$

$$= S_B(Q) \text{ since } K_C = (1 - W_A) K_B$$

If the contrast factors and the copolymer compositions are known, then the molecular weight of the block can be calculated directly. It should be noted however, that determination of the true molecular weight of the copolymer requires measurements to be made in at least three solvents with different K_C, K_A and K_B values, followed by the solution of simultaneous equations. The limitation of light scattering in such an analysis is associated with the thermodynamic changes inherent in using different solvents to obtain a sufficiently large change in optical contrast factor. This restriction does not appear in small angle neutron scattering since a wide range in solvent scattering length density without consequent thermodynamic changes can be obtained by mixing hydrogenous and deuterated analogues of the desired solvent, e.g. cyclohexane, $C_6H_{12}\varrho = -0.27 \times 10^{10}$ cm cm^{-3}; $C_6D_{12}\varrho = 6.71 \times 10^{10}$ cm cm^{-3}.

Assuming that the scattering law for each block and the whole copolymer molecule can be represented by the Debye formula (Eq. (15)), then the apparent radius of gyration of the copolymer is given by:

$$\langle s_{App} \rangle = Y^2 \langle s_A^2 \rangle + (1 - Y) \langle s^2 \rangle_B + Y(1 - Y) \langle s_{AB}^2 \rangle$$

where $Y = W_A K_A / K_C$

$$\langle s_{AB}^2 \rangle = \langle s_A^2 \rangle + \langle s_B^2 \rangle + \langle L^2 \rangle$$

Where $\langle s_A^2 \rangle$ and $\langle s_B^2 \rangle$ are the mean square radii of gyration of the A and B blocks and $\langle L^2 \rangle$ is the mean square separation between the centres of mass of the A and B blocks. Replacing for $\langle s_{AB}^2 \rangle$ in equation gives:

$$\langle s_{App}^2 \rangle = Y \langle s_A^2 \rangle + (1 - Y) \langle s_B^2 \rangle + Y(1 - Y) \langle L^2 \rangle$$

This equation shows that the apparent mean square radius of gyration of a copolymer when plotted as a function of Y (i.e. measured in solvent of different scattering length density), should have a parabolic shape with downward curvature. Ionescu et al. [92] showed that this curve should be little altered in the presence of excluded volume, they also demonstrated that the curve can become severely distorted if the copolymers are polydispersed in molecular weight. Figure 19 shows the schematic variation in

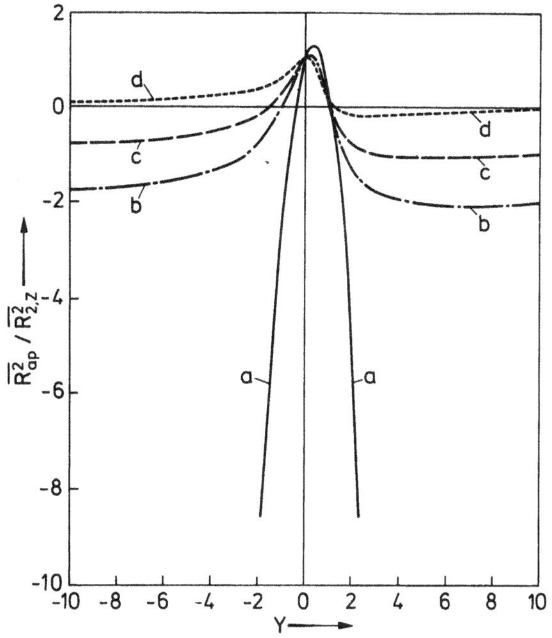

		λ_1	λ_2	$(\overline{R^2_{ap}}/\overline{R^2_{2,z}})_{Y\to\infty}$
———	a	1	1	
—·—	b	1.33	1.33	1.923
– – –	c	1.5	1.5	0.90
········	d	2	2	0

Fig. 19. Influence of polydispersity on the relative variation of R^2_{app} ($= \langle s^2 \rangle_{app}$) with Y from Numerical calculation for a block copolymer with a Schulz-Zimm distribution. $\lambda_i = \bar{M}_{w,i}/\bar{M}_{n,i}$. Ref. [92]

the shape of the parabola for a block copolymer with a Schulz-Zimm distribution, where the λ parameter is the ratio $\bar{M}_w/\bar{M}_n$ for each block. Ionescu et al. [93] also examined this analysis experimentally by SANS measurements on two styrene-isoprene block copolymers wherein the styrene was fully deuterated. The solvents used were cyclohexane and toluene, contrast variation being obtained by different combinations of the deuterated and hydrogenous isomers. For the copolymer with a narrow distribution of molecular weights, the variation of radius of gyration was in excellent agreement with the theoretical parabola. The second copolymer, being somewhat polydisperse, was reasonably well fitted by the modified equation of the parabola, using $\bar{M}_w/\bar{M}_n$ ratios obtained from gel permeation chromatography. No evidence for a segregated configuration in either toluene or cyclohexone was forthcoming from this study.

Han and Mozer [94] have combined SANS with light scattering to evaluate the configuration of a styrene-methyl methacrylate diblock copolymer when dissolved in toluene. Here again the styrene block was fully deuterated and SANS experiments were made in deuterated toluene thereby providing the radius of gyration of the methyl methacrylate block, since the deuteropolystyrene has essentially zero contrast in this solvent. The radius of gyration of the deuterostyrene block was obtained from light scattering in hydrogenous toluene, and also from SANS in deuterotoluene. In view of the non-negligible scattering from the deuterostyrene block in deuterotoluene, then the results obtained for the methacrylate block must be treated with caution. From the radii of gyration obtained a partially segregated core and shell model was proposed with the styrene block forming the shell. Notwithstanding, the apparently excellent agreement between light scattering and SANS values of the radius of gyration for the deuterostyrene block, some comments regarding these results are appropriate. Firstly, if the reported radius of gyration is correct, then an expansion factor of circa 2.4 for the styrene block is indicated and this seems very large in comparison to the stated block molecular weight of 90×10^3. Additionally the minimum Q at which measurements were made was 0.011 Å, with most of the range being considerably larger than this. Consequently the necessary Guinier condition, $Q \leqq 1/R_g$, was not met. Lastly, the molecular weight of the styrene block is such that the particle scattering function in light scattering is approximately 1 and therefore, according to Kratochvil [95], the prospects for measuring the particle size are slight.

Styrene-methyl methacrylate diblock copolymers have also been studied in a selective solvent [96], i.e. a solvent which is thermodynamically better for one or other block. For these experiments, p-xylene was used as the solvent since it had been previously reported [83] as a theta solvent for polymethyl methacrylate at circa 40 °C. Separate SANS measurements were made on a copolymer with a deuterated styrene block dissolved in deuterated xylene and in hydrogenous xylene at temperatures of 30 °C and 40 °C. Deutero xylene has the same scattering length density as deuterostyrene, consequently the scattered intensity is solely attributable to the methyl methacrylate block when the copolymer is dissolved in this solvent. At the two temperatures used conventional Zimm plots were obtained, however the methyl methacrylate block molecular weight obtained at 30 °C was twice that at 40 °C indicating the formation of a bimolecular multimer at 30 °C. From the radii of gyration obtained at both temperatures it appears that the methyl methacrylate block in the multimer

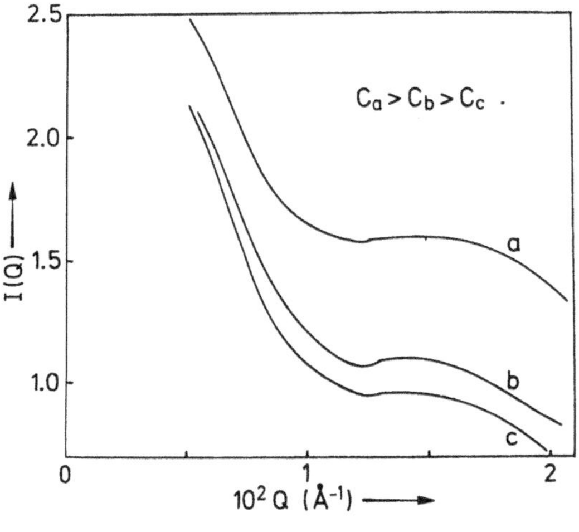

Fig. 20. Experimental scattering curves for solutions of a deuterostyrene-methyl methacrylate block copolymer in p-xylene

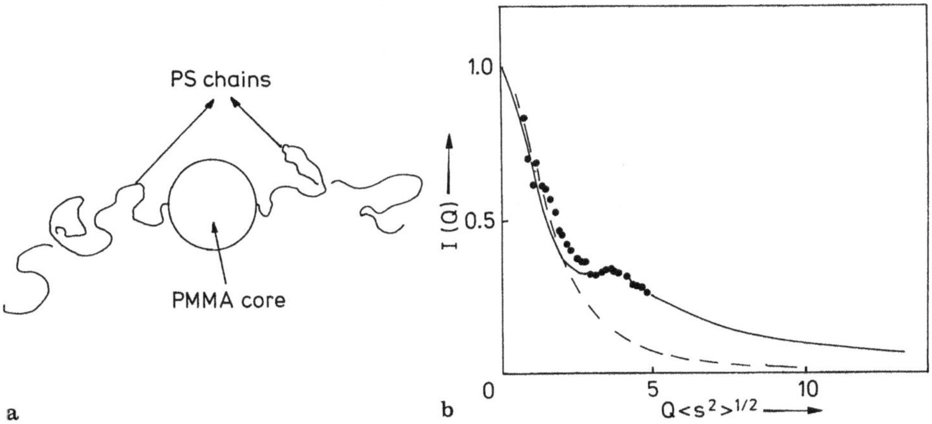

a b

Fig. 21a. Schematic sketch of model used in calculation of scattering law for styrene-methyl methacrylate diblock copolymer in solution; **b** Comparison of rotational isomeric state calculation (———) with data (●). Dashed line is the scattering for a Gaussian chain

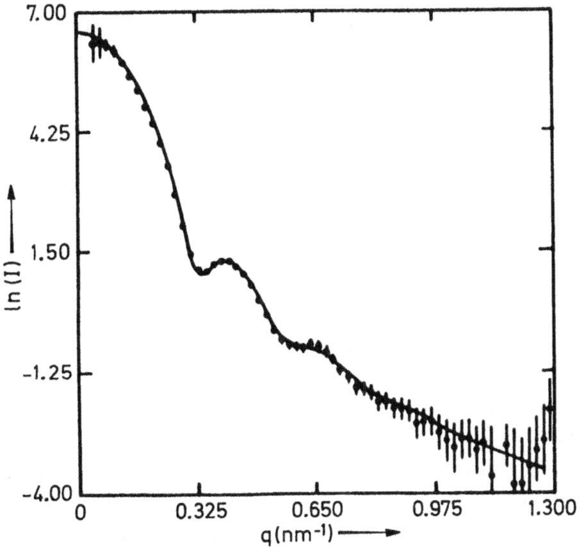

Fig. 22. Scattering curve for a deuterostyrene-butadiene diblock copolymer at 5% concentration in a polybutodiene matrix. ● Data ——— Model fit. Ref. [97]

had a compact structure, however little could be inferred regarding the multimer structure. Greater insight into the structure was obtained from SANS measurements at 30 °C using hydrogenous xylene as the solvent, wherein the deuterated styrene blocks are the observable particles. Figure 20 shows the scattering obtained, a distinct maximum appearing at higher Q values. The fact that this maximum did not move to different Q values as the copolymer concentration was changed, showed that it did not arise from interference between neutrons scattered from different deutero styrene blocks. The model proposed was that of a central spherical core of polymethy methacrylate with the deuterostyrene blocks somehow disposed at the periphery.

Modelling of the observed scattering using hollow sphere models with thin shell or finite shell thicknesses was attempted, but fits were only obtained for unrealistically large central core radii (in relation to the radius of gyration measured). A model that produced an excellent fit to the scattering function was a central sphere with two chains attached to the surface. The scattering law was calculated by assuming the central sphere was invisible and using a Monte Carlo method in conjunction with a rotational isometric state description of the chains attached to the surface. A sketch of the model and the agreement between theory and experiment is shown in Fig. 21.

A special case of a copolymer in solution has been reported by Selb et al. [97]. They used SANS to study a deutero styrene-butadiene diblock copolymer dissolved in butadiene homopolymer of low molecular weight ($1600 < M_w < 6500$). The scattering patterns at low concentration ($\leqq 5\%$) showed a series of smeared maxima which were interpreted as arising from spherical deutero styrene micelles. At concentrations greater than 5%, Bragg scattering due to the interference function became evident at low Q values. Scattering curves were simulated by convoluting the single particle form factor for a sphere with a log normal distribution function, a comparison of fit and experiment is shown in Fig. 22. From these experiments it was concluded that as the molecular weight of the polybutadiene matrix for the styrene block increased the degree of association of the deuterostyrene blocks increased with consequent increase in the radius of the spherical micelle. Surprisingly, a decrase in spherical micelle radius was noted as the copolymer concentration was increased.

7.5 Deformation of Block Copolymers in the Solid State

Electron microscopy and small angle X-ray scattering have been used to investigate the structure of block copolymers samples prepared in flow fields [98-100]. However, in some cases the domain morphology is not that pertaining to the equilibrium situation, and furthermore the data can provide no information on the mechanism by which the final structure is obtained.

Richards and Thomason [101] have reported some preliminary results on styrene-isoprene copolymers in uniaxial extension. Although a large range of extension ratio was investigated, only a limited range of Q was available, Fig. 23 shows a series of contour plots for the specimen investigated. The most striking feature is the increasing fragmentation of the Debye-Scherrer rings as the extension ratio of the copolymer is increased. This was interpreted as an orientation effect due to alignment of the 'grains' in the copolymer. Mullin and Richards [102] have extended this work by examining a variety of styrene-isoprene copolymers over a wider range in Q. The change in position of the main diffraction maximum with bulk extension ratio is shown in Fig. 24 whilst Fig. 25 shows the bulk extension ratio compared with the extension ratio calculated from the displacement of the Bragg peak. Evidently, the domain separation does not transform affinely with bulk extension ratio and this may be supporting evidence for the grain alignment mechanism tentatively proposed by Richards and Thomason. A more complete description of the structural changes devolving from deformation must refer to the paracrystalline nature of the long range order. Fronk and Wilke [103] have described such a model for the small angle X-ray scattering of oriented semi-crystalline polymers. The three dimensional intensity patterns obtained showed general resemblances to those obtained for block copoly-

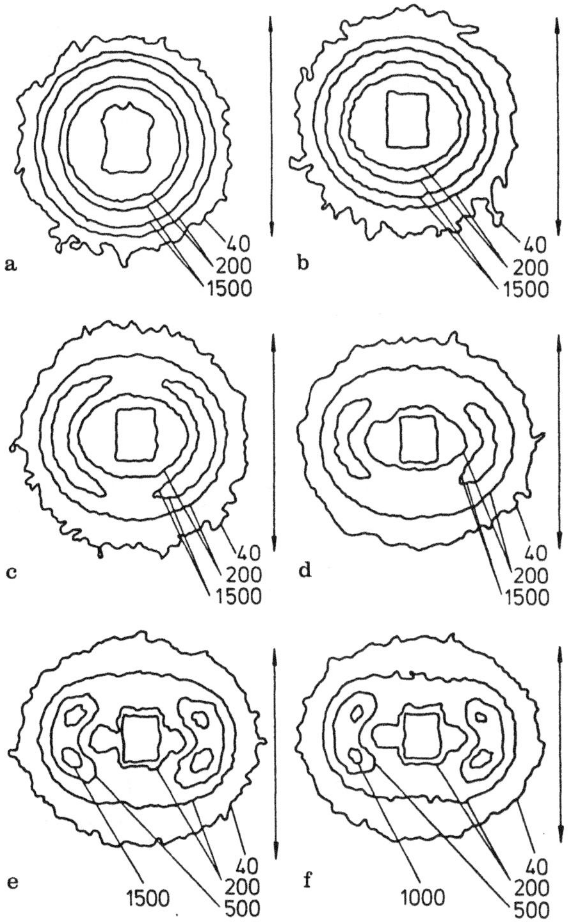

Fig. 23a–f. Isointensity contour plots, for a triblock copolymer styrene and isoprene with spherical styrene domains, as a function of extension ratio. **a** 1.0; **b** 1.1; **c** 1.2; **d** 1.5; **e** 2.0; **f** 3.0. Direction of stretch indicated by the arrow

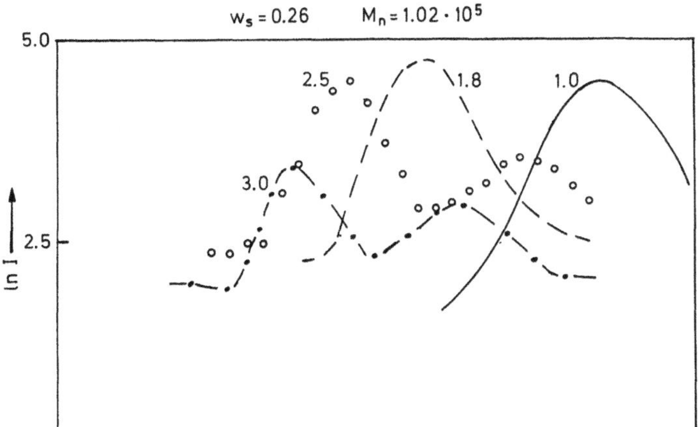

Fig. 24. Typical dependence of Bragg Peak position with extension ratio for a triblock copolymer of styrene and isoprene with spherical styrene domains. Bulk extension ratio as indicated in each peak

mers, i.e. increased fragmentation as the orientation (extension ratio) increased. The model utilises a three dimensional para-crystalline macrolattice [104], and although it has as yet only been applied to lamellar crystallites, the treatment appears to be applicable to other lattices by substituting the appropriate lattice factor.

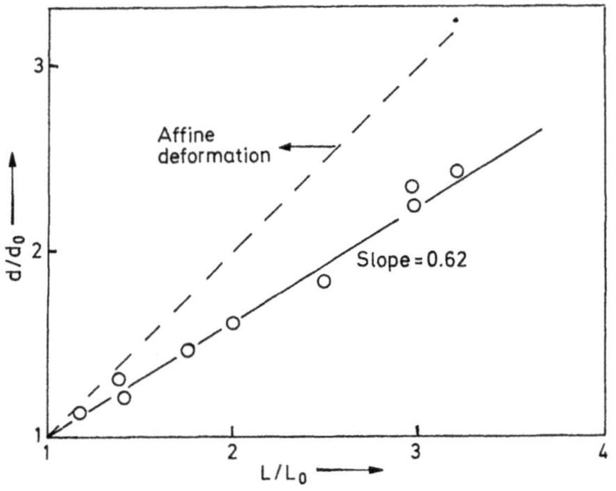

Fig. 25. Lattice plane extension ratio as a function of applied bulk extension ratio for a styrene-isoprene triblock copolymer with spherical styrene domains

8 Conclusions

Evidently small angle neutron scattering can provide much information on the structure of block copolymers in both the solid state and in solution. Appropriate deuteration enables particular features to be 'highlighted' and thus considerably simplifying the extraction of meaningful physical parameters. Only a few systems have been studied in any detail, but the results obtained demonstrate admirably the power of the technique, e.g. the evaluation of structural features over a length scale ranging from ~20 Å to ~2000 Å. Furthermore, both supramolecular and intramolecular aspects of the structure can be examined in favourable circumstances. With the iminent operation of new, high intensity pulsed neutron sources, an exciting range of time resolved experiments becomes feasible, e.g. stress relaxation and microphase separation experiments in real time, which proffer the prospect of greater insight into the structural organisation of block copolymers.

9 References

1. Annual Report of the Institut Laue-Langevin, Grenoble, France
2. Richards, R. W.: Chapter 5 in 'Polymer Characterisation — 1' ed. J. V. Dawkins. Applied Science Publishers. Barking 1978
3. Higgins, J. S.: in 'Treatise on Materials Science and Technology Vol. 15'. Ed. G. Kostorz, Academic Press. London 1979
4. Maconnachie, A., Richards, R. W.: Polymer *19*, 739 (1978)
5. Ballard, D. G. H.: Ch. 11 in 'Macromolecular Chemistry, Volume 1'. Specialist Periodical Report. Royal Society of Chemistry 1980

6. Ballard, D. G. H., Janke, E.: Chapter 11 in 'Macromolecular Chemistry Volume 2'. Specialist Periodical Report. Royal Society of Chemistry 1982
7. Ullman, R.: Ann. Rev. Mater. Sci. *10*, 261, 1980
8. Higgins, J. S.: Chapter **000** in 'Macromolecular Chemistry, Volume 3'. Specialist Periodical Report. Royal Society of Chemistry 1984
9. 'Organisation of Macromolecules in the Condensed Phase'. Farad. Disc. of R.S.C. No. 68 (1979)
10. de Gennes, P. G.: 'Scaling Concepts in Polymer Physics'. Cornell University Press. 1979
11. Marshall, W., Lovesey, S. W.: 'Theory of Thermal Neutron Scattering'. Clarendon Press. Oxford 1971
12. Turchin, V. E.: 'Slow Neutrons', Israel Program for Scientific Translations, Jerusalem, 1965
13. Kostorz, G.: 'Treatise on Materials Science and Technology. Volume 15. Neutron Scattering'. Academic Press. New York. 1979
14. Enderby, J. E.: Ann. Rev. Phys. Chem. *34*, 155 (1983)
15. Guinier, A., Fournet, G.: 'Small Angle Scattering of X-rays', Wiley, New York 1955
16. Bacon, G. E.: 'Neutron Diffraction'. Clarendon Press, Oxford 1967
17. Richards, R. W.: in Developments in Polymer Characterisation — ed. J. V. Dawkins. Applied Science Publishers 1984 (to be published)
18. Zaccai, G., Jacrot, B.: Ann. Rev. Biophys. Bioeng. *12*, 139 (1983)
19. Porod, G.: Chapter 2 in 'Small Angle X-ray Scattering'. Ed. O. Glatter and O. Kratky. LAcademic Press, London 1982
20. Kerker, M.: 'The Scattering of Light and Other Electromagnetic Radiation'. Academic Press, London 197
21. Kratky, O.: Prog. Biophys. and Mol. Biol. *13*, 105 (1963)
22. Burchard, W.: Chapter 10 in 'Applied Fibre Science, Volume 1' ed. F. Happey. Academic Press, 1978, London
23. Debye, P.: Ann. Physik. *46*, 809 (1915); J. Phys. Colloid. Chem. *51*, 98 (1947)
24. Rayleigh, Lord: Proc. Roy. Soc. *A90*, 219 (1914)
25. Fournet, G.: Bull. Soc. Franc. Min. et Crist. *74*, 39 (1951)
26. Schmatz, W., Springer, T., Schelten, J., Ibel, K.: J. Appl. Crystallogr. *7*, 96 (1978)
27. Schelten, J., Hendricks, R. B.: J. Appl. Crystallogr. *11*, 297 (1978)
28. 'Neutron Research Facilities at the ILL High Flux Reactor'. Office of the Scientific Secretary, Institut Laue Langevin. 1983
29. 'Neutron Beam Instruments at Harwell'. HMSO. 1978
30. Mildner, D. F. R., Berliner, R., Pringle, O. A., King, J. S.: J. Appl. Cryst. *14*, 370 (1981)
31. Han, C. C., Mozer, B.: Macromolecules *10*, 44 (1977)
32. National Center for Small Angle Scattering Research Users' Guide. Oak Ridge National Laboratory. Tennessee 1980
33. Ghosh, R. E.: 'A Computing Guide for Small Angle Scattering Experiments'. Institut Laue-Longevin 81GH 29T
34. Duplessix, R., Cotton, J. P., Benoit, H., Picot, C.: Polymer *20*, 1181 (1979)
35. Gallot, B. R. M.: Adv. Polym. Sci. *29*, 85 (1978)
36. Hashimoto, T., Shibayama, M., Kawai, H.: Macromolecules *13*, 1237 (1980)
37. Hashimoto, M., Shibayama, M., Kawai, H.: Macromolecules *13*, 1660 (1980)
38. Fujimura, M., Hashimoto, H., Kurahashi, K., Hashimoto, T., Kawai, H.: Macromolecules *14*, 1196 (1981)
39. Meier, D. J.: J. Polym. Sci. *C26*, 81 (1969); Polym. Prepr. *11*, 400 (1970); Polym. Prepr. *15*, 171 (1974)
40. Leary, D. F., Williams, M. C.: J. Polym. Sci. *B8*, 335, 1970
41. Leary, D. F., Williams, M. C.: J. Polym. Sci. Polym. Phys. Edn. *11*, 345 (1973)
42. Henderson, C. P., Williams, M. C.: J. Polym. Sci. Polym. Lett. Edn. *17*, 257 (1979)
43. Helfand, E.: Macromolecules *8*, 552 (1975)
44. Helfand, E., Wasserman, Z. R.: Macromolecules *9*, 879 (1976)
45. Helfand, E., Wasserman, Z. R.: Macromolecules *11*, 960 (1978)
46. Helfand, E., Wasserman, Z. R.: Macromolecules *13*, 994 (1980)
47. Helfand, E., Wasserman, Z. R.: Chapter in 'Developments in Block Copolymers — 1' ed. I. Goodman. Applied Science Publishers, Barking, Essex
48. Leibler, L.: Macromolecules *13*, 1602 (1980)

49. de Gennes, P. G.: J. Phys. (Paris) *31*, 235 (1970)
50. Hong, K. M., Noolandi, J.: Macromolecules *14*, 727 (1981)
51. Hong, K. M., Noolandi, J.: Macromolecules *14*, 1229 (1981)
52. Noolandi, J., Hong, K. M.: Macromolecules *15*, 482 (1982)
53. Hong, K. M., Noolandi, J.: Macromolecules *16*, 1083 (1983)
54. Zin, W. C., Roe, R. J.: Macromolecules *17*, 183 (1984)
55. Roe, R. J., Zin, W. C.: Macromolecules *17*, 189 (1984)
56. Richards, R. W., Thomason, J. L.: Polymer *24*, 581 (1981)
57. Berney, C. V., Cohen, R. E., Bates, F. S.: Polymer *23*, 1222 (1983)
58. Bates, F. S., Cohen, R. E., Berney, C. V.: Macromolecules *15*, 589 (1982)
59. Bates, F. S., Berney, C. V., Cohen, R. E.: Macromolecules *16*, 1101 (1983)
60. Bates, F. S., Berney, C. V., Cohen, R. E., Wignall, G. D.: Polymer *24*, 519 (1983)
61. Richards, R. W., Thomason, J. L.: Macromolecules *16*, 982 (1983)
62. Djermouni, B., Ache, H. J.: Macromolecules *13*, 168 (1980)
63. Krause, S.: Macromolecules *11*, 1288 (1978)
64. Shen, M., Kaelble, D. H.: J. Polym. Sci. Polym. Phys. Ed. *8*, 149 (1970)
65. Krause, G., Rollmann, K. W.: J. Polym. Sci. Polym. Phys. Ed. *14*, 1133 (1976)
66. Chen, Y.-D. M., Cohen, R. E.: J. Appl. Polym. Sci. *21*, 629 (1977)
67. Bates, F. S., Cohen, R. E., Argon, A. S.: Macromolecules *16*, 1108 (1983)
68. Diament, J., Soong, D. S., Williams, M. C.: in 'Contemporary Topics in Polymer Science Vol. 4'. Ed. W. J. Bailey. Plenum, New York 1982
69. Porod, G.: Kolloid. Z. *124*, 83 (1951), *125*, 51, 108 (1952)
70. Ruland, W.: J. Appl. Crystallogr. *4*, 70 (1971)
71. Koberstein, J. T., Morra, B., Stein, R. S.: J. Appl. Crystallogr. *13*, 34 (1980)
72. Roe, R. J.: J. Appl. Crystallogr. *15*, 182 (1982)
73. Koberstein, J. T.: J. Polym. Sci. Polym. Phys. Edn. *20*, 593 (1982)
74. Richards, R. W., Thomason, J. L.: Polymer *24*, 1089 (1983)
75. Akcasu, A. Z., Summerfield, G. C., Jahshan, S. N., Han, C. C., Kim, C. Y., Yu, H.: J. Polym. Sci. Polym. Phys. Ed. *18*, 863 (1980)
76. Tangari, C., Summerfield, G. C., King, J. S., Berliner, R., Mildner, D. F. R.: Macromolecules *13*, 1546 (1980)
77. Jahshan, S. N., Summerfield, G. C.: J. Polym. Sci. Polym. Phys. Ed. *18*, 1859 (1980)
78. Hadziioannou, G., Picot, C., Skoulios, A., Ionescu, M. L., Mathis, A., Duplessix, R., Gallot, Y., Lingelser, J. P.: Macromolecules *15*, 263 (1982)
79. Amis, E. J., Glinka, C. J., Han, C. C., Hasewaga, H., Hashimoto, T., Lodge, T. P., Matsushita, Y.: Polym. Prepr. Am. Chem. Soc. *24*, 215 (1983)
80. Miller, J. A., Cooper, S. L., Han, C. C., Pruckmayr, G.: Poly. Prepr. Am. Chem. Soc. *24*, 398 (1983)
81. Hashimoto, T., Fujimura, M., Kawai, H.: Macromolecules *13*, 1660 (1980)
82. Richards, R. W., Thomason, J. L.: Submitted to Macromolecules
83. Kotaka, T., Tanaka, T., Ohnuma, H., Murakami, Y., Inagaki, H.: Polym. J. *1*, 245 (1970)
84. Inagaki, H., Miyamoto, T.: Makromol. Chem. *87*, 166 (1965)
85. Tanaka, T., Kotaka, T., Ban, K., Hattori, M., Inagaki, H.: Macromolecules *10*, 961 (1977)
86. Cowie, J. M. G.: Chapter 1 in 'Developments in Block Copolymers — 1' edited by I. Goodman. Applied Science Publishers, Barking 1982
87. Bushuk, W., Benoit, H.: Con. J. Chem. *36*, 1616 (1958)
88. Leng, M., Benoit, H.: J. Polym. Sci. *57*, 263 (1962)
89. Benoit, H., Froelich, D.: Chapter 11 in 'Light Scattering from Polymer Solutions', ed. M. B. Huglin. Academic Press. N.Y. 1972
90. Tuzar, Z., Kratochvil, P., Strakova, D.: Eur. Polym. J. *6*, 1113 (1970)
91. Kratochvil, P., Sedlacek, B., Strakova, D., Tuzar, Z.: Makromol. Chem. *148*, 271 (1971)
92. Ionescu, L., Picot, C., Duval, M., Duplessix, R., Benoit, H., Cotton, J. P.: J. Polym. Sci. Polym. Phys. Ed. *19*, 1019 (1981)
93. Ionescu, L., Picot, C., Duplessix, R., Duval, M., Benoit, H., Cotton, J. P., Lingelser, J. P., Gallot, Y.: J. Polym. Sci. Polym. Phys. Edn. *19*, 1035 (1981)
94. Han, C. C., Mozer, B.: Macromolecules *10*, 44 (1977)

95. Kratochvil, P.: Chapter 7 in 'Light Scattering from Polymer Solutions' ed. M. B. Huglin. Academic Press. N.Y. 1972
96. Edwards, C. J. C., Richards, R. W., Strepto, R. F. T.: Unpublished results
97. Selb, J., Marie, P., Rameaui, A., Duplessix, R., Gallot, Y.: Polym. Bull. *10*, 444 (1983)
98. Folkes, M. J., Keller, A.: in 'The Physics of Glassy Polymers' ed. R. N. Haward. Applied Science Publishers. London 1973
99. Hadzioannou, G., Skoulios, A.: Macromolecules *15* (1982)
100. Inoue, T., Moritani, M., Hashimoto, T., Kawai, H.: Macromolecules *4*, 500 (1971)
101. Richards, R. W., Thomason, J. L.: Polymer *24*, 275 (1983)
102. Mullin, J. T., Richards, R. W.: Work in progress
103. Fronk, W., Wilke, W.: Colloid and Polymer Sci. *261*, 1010 (1983)
104. Wilke, W.: Acta Cryst. *A39*, 864 (1983)

Editor: H.-J. Cantow
Received November 23, 1984

Formation of Living Propagating Polvmer Radicals in Microspheres and Their Use in the Synthesis of Block Copolymers

Tsuneyuki Sato and Takayuki Otsu
Department of Applied Chemistry, Faculty of Engineering, Osaka City University, Sumiyoshi-ku, Osaka 558, Japan

When acrylamide derivatives such as N-methylacrylamide (NMAAm) and N-methylmethacrylamide (NMMAm) are polymerized in adequate solvents by radical initiators, the resulting polymers are dispersed as microspheres which contain living propagating radicals. In the polymerization of NMMAm in benzene with AIBN, the solvent-cage escaping cyanopropyl radicals are converted to living poly-(NMMAm) radicals in high yields, ca. 90%. These living polymer radicals react easily with other vinyl monomers at room temperature to give stable propagating radicals of the second monomers, resulting in block copolymers. These reaction processes were investigated by ESR.

1 Introduction

In the usual radical polymerization, the propagating polymer radicals are generally highly reactive, and hence short-lived. However, some polymerization systems have been reported to involve long-lived (living) propagating radicals.

The heterogeneous polymerization is a good method for the formation of living propagating radicals. When certain vinyl monomers are polymerized in poor solvents, the propagating radicals are buried alive within the resulting polymers. A typical example is the bulk polymerization of acrylonitrile (AN). The photo-sensitized polymerization of AN with di-tert-butyl peroxide(DBPO) was reported to form living poly(AN) radicals in a yield of 12.7%, based on the initiating tert-butoxy radicals [1]. The poly(AN) radicals reacted with other vinyl monomers to produce block copolymers [2, 3]. Vinyl chloride was also converted to its living polymer radicals, which could be similarly used to prepare block copolymers [3].

Recently, Seymour et al. have applied the heterogenous polymerization method to the synthesis of many kinds of block copolymers [4-6]. They have emphasized that the solubility parameters are useful to find the optimum conditions for block copolymerization [7].

γ-Ray irradiation of solid monomers such as acrylamide derivatives [8] and acrylic acid [9] yielded the corresponding living polymer radicals. Such polymer radicals were allowed to react with second monomers to give block copolymers [10].

Emulsion polymerization also represents a method to obtain surviving polymer radicals, which were isolated in micelles [11]. The living radicals in micelles were available for block copolymer synthesis [12, 13].

A matrix polymerization was reported to proceed through a living radical mechanism, where methacrylic acid(MAA) was polymerized in water in the presence of chitosan(CTS) [14]. The poly(MAA) radicals were considered to survive due to com-

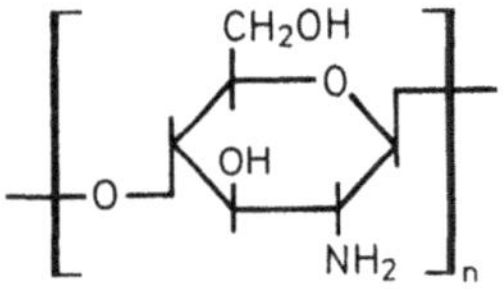

(CTS)

plexation with CTS. They reacted with styrene (St) to produce block copolymers [15].

It is interesting that inclusion polymerizations gave also living polymer radicals [16]. The structures of the propagating radicals formed in the polymerization of pentadiene derivatives in a deoxycholic acid canal were studied by ESR [17].

Considerable attention has recently been paid to plasma-initiated polymerization of vinyl monomers, in which long-lived propagating radicals of ultra-high molecular weight were formed even in homogeneous polymerization systems [18-21]. Some block copolymers were prepared by using plasma-initiated polymerization.

It is also noteworthy that the radical polymerization of methyl methacrylate(MMA) in phosphoric acid resulted in living polymers in spite of its homogeneous nature [22].

Stabilization of polymer radicals by complexing with metal ions was considered in the polymerization of MMA with the chromium(II) acetate/benzoyl peroxide system [23,24], although a recent report explained this living nature on the basis of a viscosity effect [25]. This system was applied to the synthesis of an AN-MMA block copolymer [26]. The polymerization of MMA with the cobaltocene/bis(ethyl acetoacetato)copper(II) system also showed a living character and was used in preparing a MMA-St block copolymer [27].

Vinyl polymerization with initiator-transfer reagent-terminator (iniferter) is interesting as a living polymerization model, in which a reversible combination of polymer radicals with stable primary radicals occurs continuously during the polymerization [28-30]. The resulting telechelic polymers initiated the polymerization of other monomers to give block copolymers in good yields [31-33].

Recently, we have found that acrylamide derivatives such as N-methylacrylamide-(NMAAm) and N-methylmethacrylamide (NMMAm) were polymerized by radical initiators in adequate solvents to form polymer microspheres, which contained the very stable propagating radicals of the amide monomers in high concentrations. Furthermore, the living polymer radicals were found to react readily with other vinyl monomers at room temperature, yielding block copolymers. We have also investigated these reactions by means of ESR. This article reviews our recent work on the formation of living propagating radicals, their reactions with vinyl monomers, and their use in block copolymer synthesis.

2 Formation of Living Propagating Polymer Radicals in Microspheres [34]

When NMAAm was polymerized in benzene at room temperature by using di-tert-butyl peroxalate(DBPOX) as the initiator, the polymerization mixture afforded the ESR spectrum shown in Fig. 1a. This three line spectrum is characteristic of the propagating radicals of acrylic monomers [35] and assignable to the poly(NMAAm) radical(III). The 1:2:1 triplet is due to the fact that the coupling constant (ca. 25G) of the α-hydrogen is similar to that of one of the two β-hydrogens and that the other β-

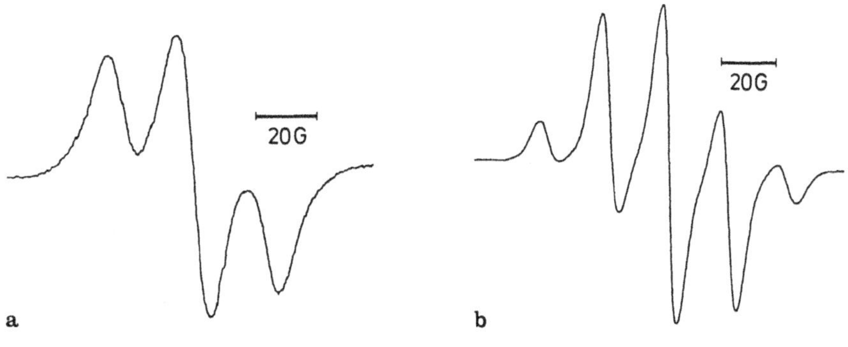

Fig. 1a. ESR spectrum of the NMAAm/DBPOX/benzene system after reaction for 5 h at room temperature; [NMAAm] = 3,87 mol/l, [DBPOX] = 1.78×10^{-2} mol/l. **b.** ESR spectrum of the NMMAm/DBPOX/benzene system after reaction for 55 h at room temperature; [NMMAm] = 3.27 mol/l, [DBPOX] = 4.27×10^{-2} mol/l

hydrogen has a much smaller constant. Such non-equivalence of β-hydrogens results from restricted rotation around the C—C bond in the polymer radicals [36]. The poly(NMAAm) radical is formed according to the following scheme;

$$\text{t-BuO}-\text{OCO}-\text{COO}-\text{O}-\text{t-Bu} \rightarrow 2\,\text{t-BuO}^{\cdot} + 2\,CO_2 \qquad (1)$$
(DBPOX)

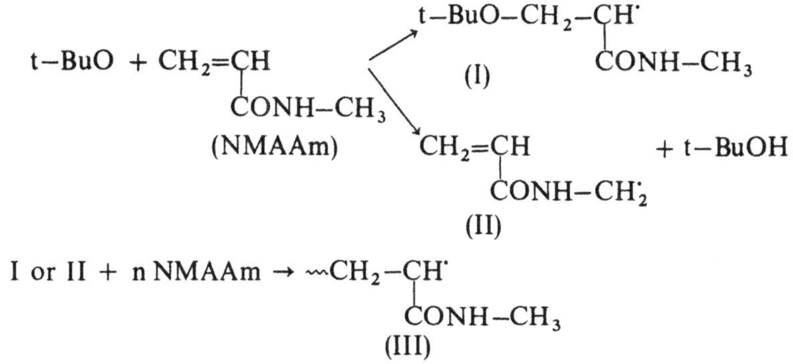

(2)
(3)
(4)

DBPOX, a low temperature initiator, decomposes easily at room temperature into t-butoxy radicals and carbon dioxide (Eq. (1)) [37]. The t-butoxy radical undergoes both addition to the monomer (Eq. (2)) and hydrogen-abstraction from the N-methyl group (Eq. (3)) [38]. Radicals I and II attack the monomer and propagate yielding poly(NMAAm) radicals (Eq. (4)). The stable poly(NMAAm) radical was similarly formed in the photo-sensitized polymerization with azo-bis-isobutyronitrile(AIBN) or di-tert-butyl peroxide(DBPO). In the latter system, the concentration of living radical III reached 1×10^{-3} mol/l [39].

Spectrum (b) in Fig. 1 was observed when NMMAm was polymerized in benzene by DBPOX. The five line spectrum is due to the propagating radical of NMMAm(IV), in which the three hydrogens of the methyl group and one of the β-hydrogens interact similarly with the unpaired electron [36]. The living poly(NMMAm) was also

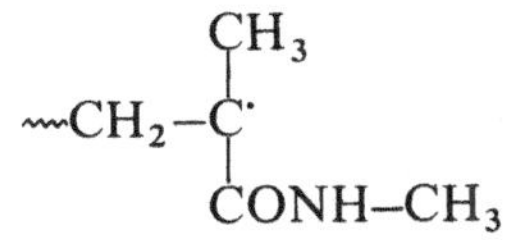

(IV)

obtained in dioxane and carbon tetrachloride, where the polymerization proceeded heterogenously. No polymer radical was observed in a homogeneous polymerization in ethanol, a good solvent for poly(NMMAm). The poly(NMMAm) radical in benzene survived in a degassed sealed tube over several months at room temperature. When exposed to air, however, it reacted readily with oxygen to give a peroxy radical.

Acrylamide (AAm) and methacrylamide (MAm) were likewise converted into the living propagating radicals V and VI by their photo-sensitized polymerization with DBPO in dioxane [39]. ESR spectra of polymer radicals V and VI are shown as (a) and

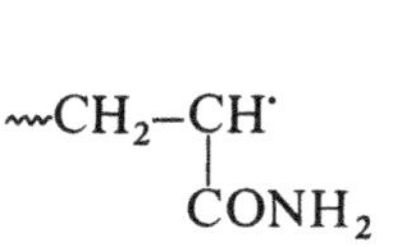
(V)

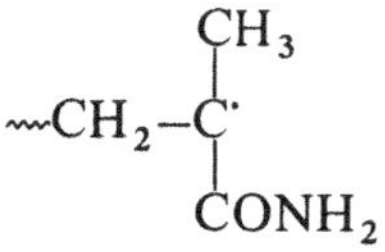
(VI)

(b) in Fig. 2, respectively. Spectrum (a) due to the poly(AAm) radical has two shoulder peaks at both sides of the main three lines absorption, while the poly(NMAAm) radical does not show such shoulders. A similar observation of shoulder peaks has been reported for the polymer radical formed in a solid state polymerization of AAm [40]. These shoulders may be due to a different conformation of the propagating radical.

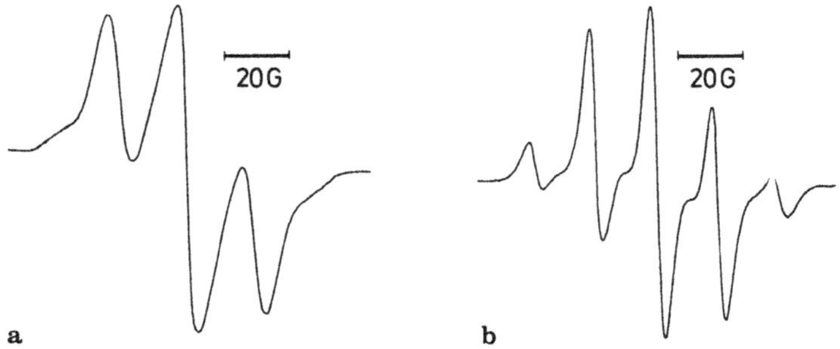

Fig. 2a. ESR spectrum of the AAm/DBPO/dioxane system after irradiation for 3.7 h at room temperature; [AAm] = 2.8 mol/l, [DBPO] = 0.43 mol/l. **b.** ESR spectrum of the MAm/DBPO/dioxane system after irradiation for 4.3 h at room temperature; [MAm] = 2.4 mol/l, [DBPO] = 0.43 mol/l

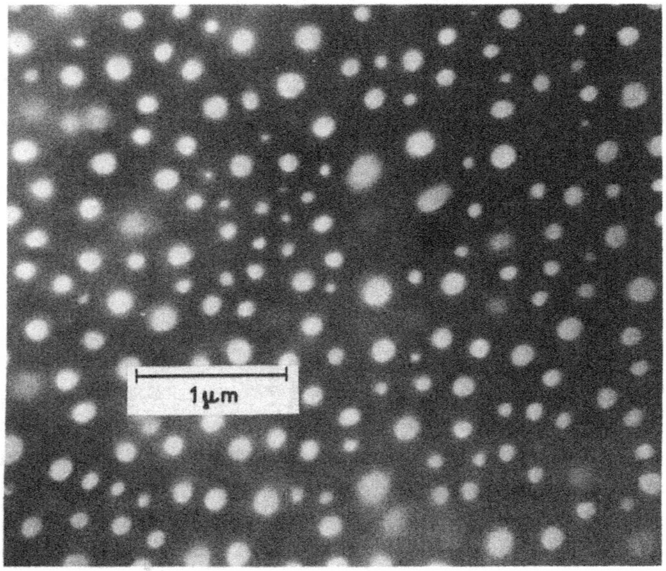

Fig. 3. Transmission electron micrograph of poly(NMMAm) in a poly(p-MeSt) matrix

The five line spectrum (b) observed in the MAm polymerization is closely similar to that of the poly(NMMAm) radical (Fig. 1b), although the former is more sharp compared with the latter.

Figure 3 shows a transmission electron micrograph of poly(NMMAm) in a poly(p-methylstyrene)(p-MeSt) matrix. The sample was prepared by an AIBN-initiated polymerization of a p-MeSt solution containing dispersed poly(NMMAm). Poly-(NMMAm) formed in a photo-sensitized polymerization of NMMAm with DBPO in benzene was used for this electron microscopic examination.

As it can be seen from Fig. 3, the poly(NMMAm) microscpheres of 1000–3000 Å diameter are dispersed in the poly(p-MeSt) matrix. Thus, on polymerization, the amide monomers are converted to polymer microspheres which are dispersed in the polymerization system [41]. These microspheres are assumed to contain many living propagating polymer radicals.

The microspheres of poly(NMAAm) and poly(NMMAm) are swollen in aromatic solvents such as St, anisole, and benzonitrile. Interestingly, these solutions containing swollen microspheres showed a remarkable thermochromism under sunlight [41]. The color changed from blue to violet, to yellowish red and finally to yellow with increasing temperature, although the temperature range of color change depended on the system used. The coloration is explained as a result of a Christiansen filter effect exerted by swollen microspheres and aromatic solvents [42].

As mentioned above, some amide monomers are easily converted into their living propagating radicals when they undergo photo-sensitized polymerization with DBPO in benzene or dioxane at room temperature. Therefore, it is of interest to examine the post-effect in the photo-sensitized polymerization of the amide monomers. Fig. 4 shows the results obtained in the post-polymerizations of AAm in dioxane and NMMAm in benzene at room temperature [39]. As it can be seen from the figure, a small post-effect was observed in the AAm polymerization. On the other hand, no

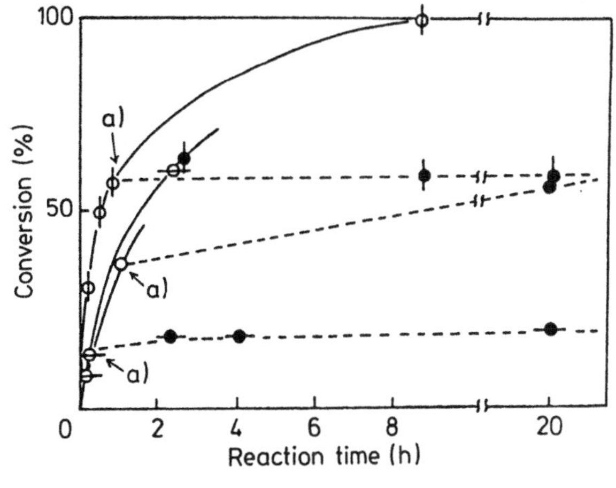

Fig. 4. Post-polymerization of AAm and NMMAm at room temperature; a) irradiation was stopped. Open circles: under irradiation; filled circles: after stopping irradiation; ○: [AAm] = 3.52 mol/l, [DBPO] = 0.217 mol/l, in dioxane; ⦶: [NMMAm] = 1.96 mol/l, [DBPO] = 0.217 mol/l, in benzene; Am] = 4.90 mol/l, [DBPO] = 0.217 mol/l, in benzene; ⊖: [NMMAm] = 4.90 mol/l, [DBPO] = 0.217 mol/l, in benzene

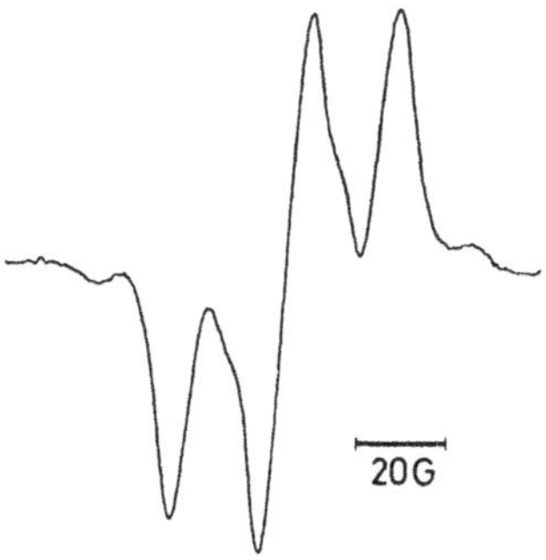

Fig. 5. ESR spectrum of the poly(NMMAm) radical/MA system after reaction for 4 h at room temperature; a benzene solution of NMMAm (2.45 mol/l) and DBPO (0.542 mol/l) was irradiated for 2 h at room temperature before MA was added

post-polymerization took place under the present conditions in the NMMAm polymerization, although almost complete polymerization was observed when irradiation was continued (uppermost curve in Fig. 4).

However, the living poly(NMMAm) radical reacted readily with other vinyl monomers at room temperature [34]. Figure 5 shows an ESR spectrum of the reaction mixture obtained when the poly(NMMAm) radical (IV) was allowed to react with methyl acrylate (MA). This spectrum is assigned to a mixture of the propagating radicals from MA (three line spectrum) and a small amount of poly(NMMAm) radicals. This indicates that a large portion of the poly(NMMAm) radicals was converted into the stable propagating radical of MA (VII) (Eq. (5)).

$$IV + n\,CH_2=\underset{\displaystyle COOCH_3}{\underset{|}{CH}} \rightarrow \sim\!CH_2-\underset{\displaystyle COOCH_3}{\underset{|}{CH^{\cdot}}} \qquad (5)$$

(VII)

Next, polymerizations of vinyl monomers were attempted by using poly(NMMAm) radicals as initiator. The poly(NMMAm) radicals were prepared by polymerization in benzene with DBPOX at room temperature for 2 days. The polymerization system was heated for additional 3 h at 50 °C to decompose the unreacted initiator. To this solution, a second monomer was added and then allowed to stand without stirring at room temperature for a day. The results are summarized in Table 1. Thus, the living poly(NMMAm) radicals can readily initiate vinyl polymerization.

Table 1. Polymerization of vinyl monomers with poly(NMMAm) radicals in benzene at room temperature[a]

Second monomer	Reaction time (h)	Yield (%)[b]
MA	20	ca. 100
AN	24	41
MMA	24	36

[a] [NMMAm] = 3.27 mol/l, [DBPOX] = 2.5×10^{-2} mol/l, [second monomer] = 62.5 vol % with respect to whole solution;
[b] Calculated based on second monomer

3 Kinetic Study on the Polymerization of NMMAm with Formation of Living Propagating Radicals in Microspheres [43)]

3.1 Polymerization of NMMAm with AIBN in Benzene

The polymerization was carried out in a sealed tube which was degassed by the freezing and thawing method. The polymerization proceeded heterogeneously. The resulting polymer was isolated by pouring the polymerization mixture into a large excess of ethyl acetate (AcOEt).

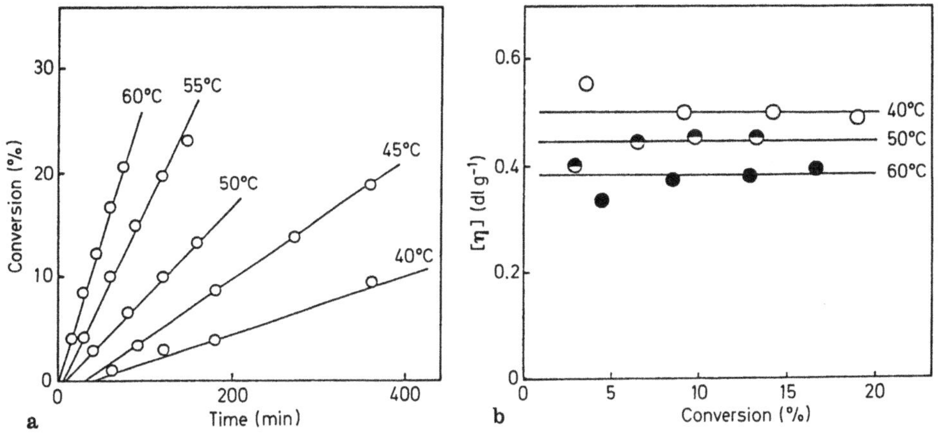

Fig. 6. Time-conversion curves (**a**) and [η] of poly(NMMAm) (**b**) in the polymerization of NMMAm with AIBN in benzene; [NMMAm] = 3.68 mol/l, [AIBN] = 1.01×10^{-2} mol/l

Figure 6a shows the time-conversion curves obtained in the temperature range from 40 °C to 60 °C. A short induction period was observed. It was separately confirmed that the polymerization was not inhibited, but retarded in this period. The induction period seems to be characteristic of the heterogeneous polymerization in benzene since it was not observed in the homogeneous one in methanol. The reason for the induction period will be discussed in Sect. 3.4.

The intrinsic viscosity ([η], at 30 °C in methanol) of the resulting poly(NMMAm) was almost independent of the monomer conversion (Fig. 6b), although it decreased with increasing temperature. This indicates that a stationary state for the active centres of the polymerization was reached under the present conditions.[1]

1 The NMMAm monomer is, for the most part, converted into living propagating radicals; apparently a unimolecular termination by polymer radical occlusion occurs. Presumably, only a small portion of the living polymer radicals function as active centers for the polymerization, while the others are dormant in the microspheres. Otherwise, MW and hence [η] of the poly(NMMAm) would increase with the conversion.

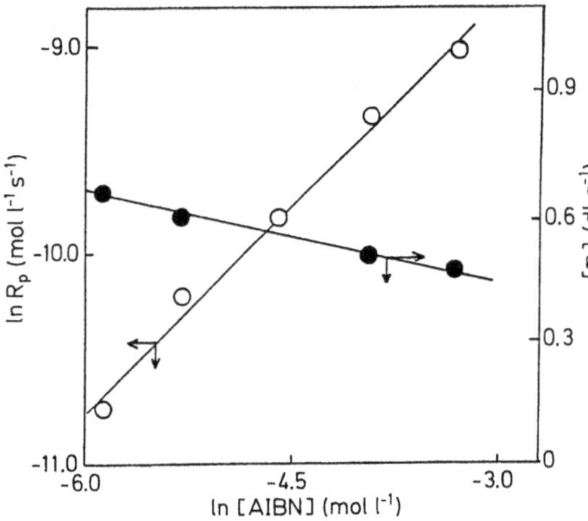

Fig. 7. Effect of [AIBN] on the polymerization rate (R_p) (○) and [η] of poly(NMMAm) (●) at 50 °C in benzene; [NMMAm] = 3.68 mol/l

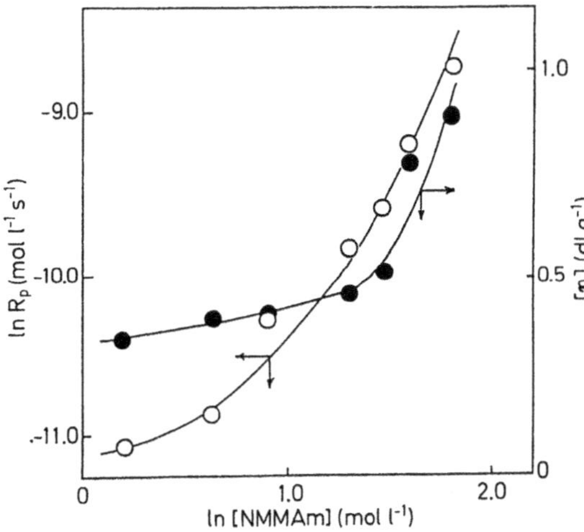

Fig. 8. Effect of [NMMAm] on the polymerization rate (R_p) (○) and [η] of poly(NMMAm] (●) at 50 °C in benzene; [AIBN] = 1.01×10^{-2} mol/l

The overall activation energy of the polymerization was calculated to be 96.3 kJ/mol, which is somewhat higher than those (ca. 85 kJ/mol)[44] observed for conventional homogeneous polymerizations.

Figure 7 shows the relationship between polymerization rate (R_p) and AIBN concentration, at 50 °C, where the monomer concentration was kept constant at 3.68 mol/l. R_p was proportional to the 0.68th order of the initiator concentration. A similar dependence (0.63th order) was observed when the NMMAm concentration was 2.45 mol/l. Such higher initiator order, compared with the 0.5th order in the usual radical polymerization, has also been reported in the heterogeneous polymeriza-

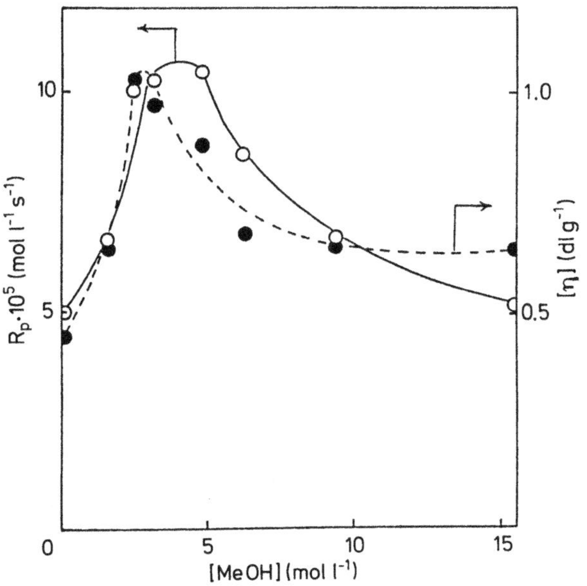

Fig. 9. Effect of methanol on the polymerization rate (R_p) (○) and (η) of poly(NMMAm) (●) at 50 °C in benzene; [NMMAm] = 3.68 mol/l, [AIBN] = 1.01×10^{-2} mol/l

tion of AN [45–47]. The effect of the initiator concentration on [η] of poly(NMMAm) is also shown in Fig. 7. The molecular weight of the polymer decreased with increasing initiator concentration.

As shown in Fig. 8, the present polymerization exhibited no definite relationship between R_p and the monomer concentration. The order of R_p with regard to the monomer concentration increased with increasing NMMAm concentration. Similar results were observed in the polymerization of AN in benzene [46]. [η] of the resulting polymer increased with the NMMAm concentration in a similar manner (Fig. 8). This anomalous behavior seems to originate from a different affinity of the polymer to monomer and benzene as reaction medium, which influenced the relative rate ratio of propagation to termination. In order to check the assumption, the effect of methanol on the polymerization was investigated at 50 °C. Methanol is a good solvent for poly(NMMAm). The results obtained are shown in Fig. 9. R_p exhibited a maximum with increasing methanol concentration. The molecular weight of the polymer formed showed a similar behavior. These results will be discussed in Sect. 3.4.

3.2 Determination of the Formation Rate of Living Propagating Radicals by Means of ESR

NMMAm is polymerized with AIBN in benzene to give living propagating radicals according to the following scheme:

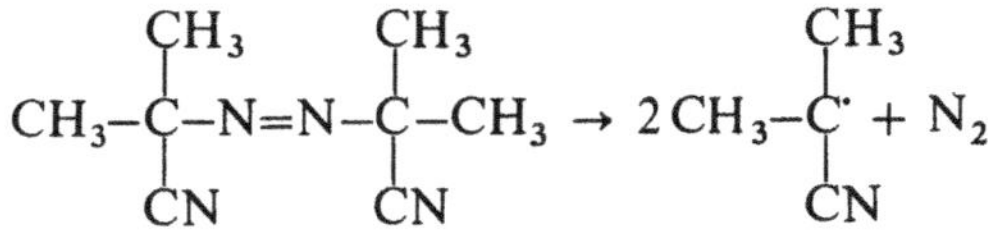

(AIBN) (VIII) (6)

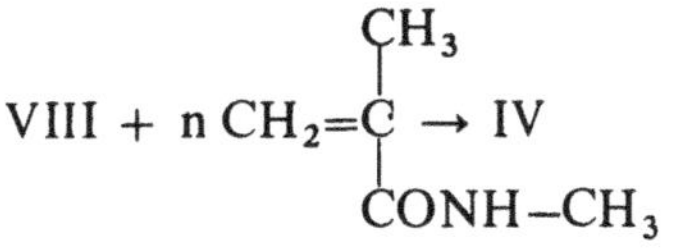

(NMMAm) (7)

The formation of living radical IV in benzene was investigated by ESR. Fig. 10a shows the ESR spectrum change of the polymerization mixture with time at 40 °C; the initial concentrations of NMMAm and AIBN were 2.45 and 1.21×10^{-2} mol/l, respectively. These first-derivative spectra were integrated to evaluate the concentration of polymer radicals, using 1,3,5-triphenylverdazyl(TPV), a stable radical, as a reference. The concentration of polymer radicals thus determined at 44, 50 and 55 °C was plotted against the reaction time in Fig. 10b. Straight lines were observed at each temperature, with a short induction period, similar to that observed for the polymerization of NMMAm in benzene (Fig. 6a). From the slopes of the straight

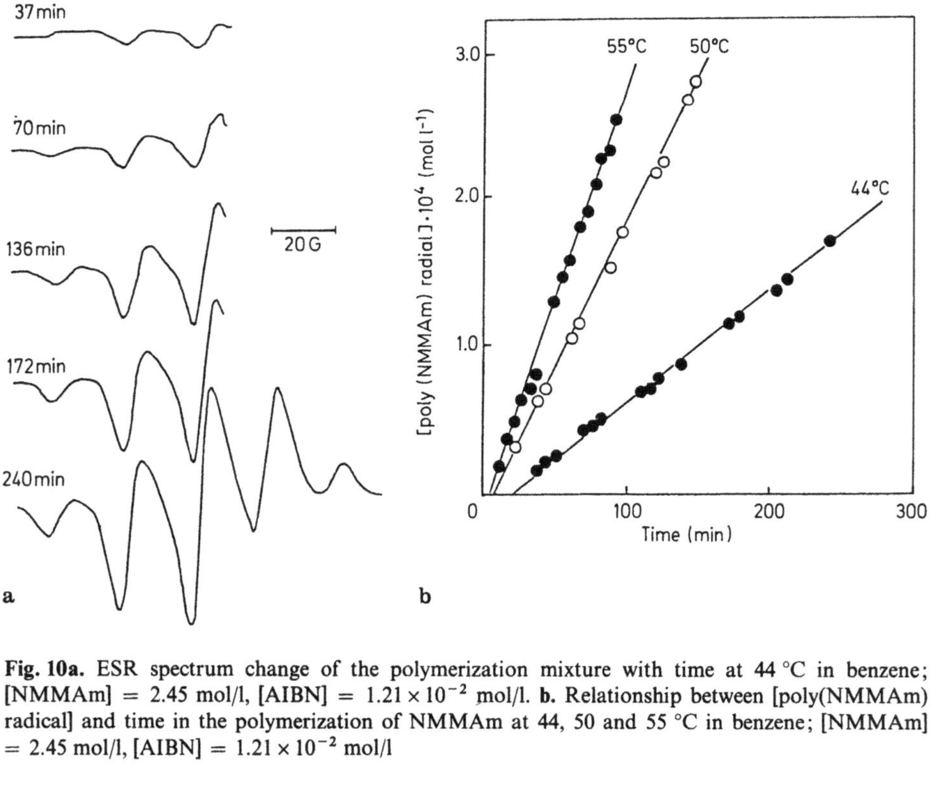

Fig. 10a. ESR spectrum change of the polymerization mixture with time at 44 °C in benzene; [NMMAm] = 2.45 mol/l, [AIBN] = 1.21×10^{-2} mol/l. **b.** Relationship between [poly(NMMAm) radical] and time in the polymerization of NMMAm at 44, 50 and 55 °C in benzene; [NMMAm] = 2.45 mol/l, [AIBN] = 1.21×10^{-2} mol/l

Table 2. Rate of living radical formation (R_r) in the polymerization of NMMAm with AIBN in benzene[a]

Temp. (°C)	$R_r \times 10^8$ (mol l^{-1} sec^{-1})	$R_d \times 10^8$ [b] (mol l^{-1} sec^{-1})	f[c]
44	1.29	1.13	0.57
50	3.36	2.83	0.59
55	4.85	5.80	0.42

[a] [NMMAm] = 2.45 mol/l, [AIBN] = 1.21×10^{-2} mol/l;
[b] R_d (decomposition rate of AIBN) was calculated with Eq. (8);
[c] $f = R_r/2R_d$

lines, the rate of living radical formation was determined. The decomposition rate of AIBN and the efficiency of living radical formation were calculated by using the following rate constant equation (Eq. (8))[48];

$$k_d\ (sec^{-1}) = 1.58 \times 10^{15} \exp(-128.9\ kJ/RT) \tag{8}$$

The values obtained are presented in Table 2.

In order to evaluate the initiator efficiency under the present conditions, the de-

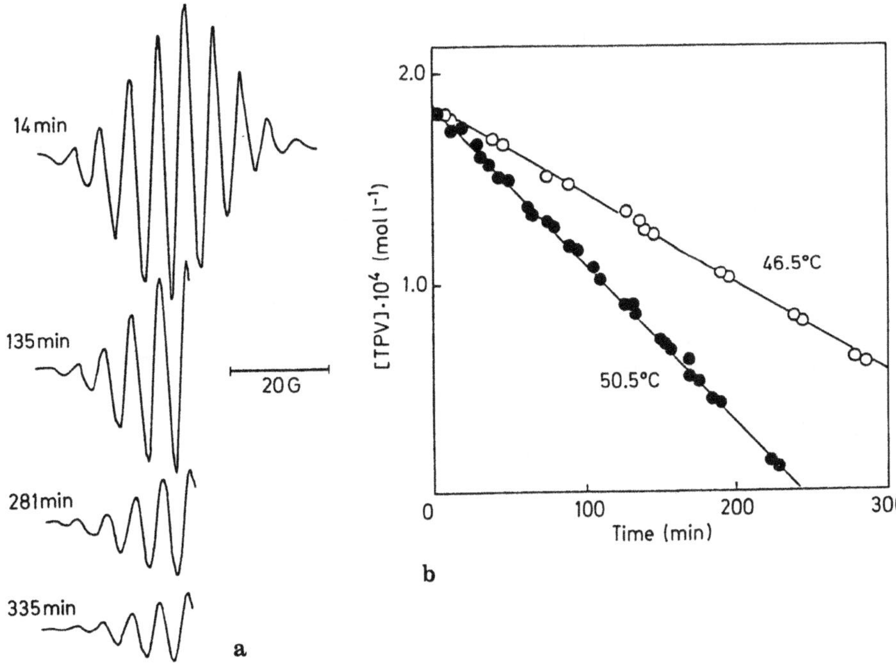

Fig. 11a. ESR spectrum change of TPV with time in the polymerization of NMMAm with AIBN at 46.5 °C in benzene; [NMMAm] = 2.45 mol/l, [AIBN] = 4.02×10^{-3} mol/l. **b.** Relationship between time and [TPV] in the polymerization of NMMAm with AIBN at 46.5 and 50.5 °C in benzene; [NMMAm] = 2.45 mol/l, [AIBN] = 4.02×10^{-3} mol/l

composition of AIBN was carried out in the presence of TPV, with the concentrations of NMMAm and AIBN 2.45 and 4.02×10^{-2} mol/l, respectively. As shown in Fig. 11a, the ESR spectrum due to TPV decreased in intensity with time, being accompanied by a fading of the green color of TPV in the reaction mixture. TPV was consumed by the reaction with cyanopropyl(VIII) or NMMAm propagating radicals. Fig. 11b shows plots of the TPV concentration versus time at 46.5 and 50.5 °C. From the plots, the rate of TPV disappearance was estimated and its values are shown in Table 3,

Table 3. Rate of disappearance (R_v) of 1,3,5-triphenylverdazyl (TPV) and initiator efficiency of AIBN[a]

Temp. (°C)	$R_v \times 10^9$ ($mol\ l^{-1}\ sec^{-1}$)	$R_d \times 10^9$ [b] ($mol\ l^{-1}\ sec^{-1}$)	f[c]
46.5	7.22	5.59	0.65
50.5	12.70	10.10	0.63

[a] [NMMAm] = 2.45 mol/l, [AIBN] = 4.02×10^{-3} mol/l, [TPV] = 1.86×10^{-4} mol/l;
[b] R_d (decomposition rate of AIBN) was calculated with Eq. (8);
[c] $f = R_v/2R_d$

together with the calculated decomposition rate and initiator efficiency of AIBN at each temperature. The initiator efficiency of AIBN has formerly been determined by using TPV as inhibitor at 30 °C to be 0.54 for MMA and 0.68 for St [49], which are similar to our values (0.63 and 0.65).

Comparison of the efficiency of living radical formation with that of the initiator leads to the noticeable conclusion that the cyanopropyl radicals escaping the solvent cage were converted into living poly(NMMAm) radicals in very high yields [89% (44 °C), 92% (50 °C) and 66% (55 °C), based on the escaping cyanopropyl radicals]. This finding points to a unimolecular termination by burying the living radicals within the resulting poly(NMMAm) microspheres.

3.3 Electron Microscopic Study of the Resulting Polymer Microspheres

Figure 12a shows an electron micrograph of the polymer particles formed in the polymerization of NMMAm (3.68 mol/l) with AIBN (1.01×10^{-2} mol/l) at 50 °C for 160 min in benzene. A cluster of several polymer particles of 0.8–1.8 μm diameter was observed. Such a wide size distribution of the particles suggests that an agglomeration of the primary particles occurred during polymerization. Furthermore, the particles had rough surfaces, implying that the propagation proceeded at isolated points on polymer surfaces.

Figure 12b shows an electron micrograph of the polymer obtained at a higher monomer concentration (6.12 mol/l). Aggregation of the particles progressed further so that the particles were no more spherical. Addition of methanol, a good solvent for poly(NMMAm), gave similar results. The diameter of the polymer microspheres decreased with increasing temperature.

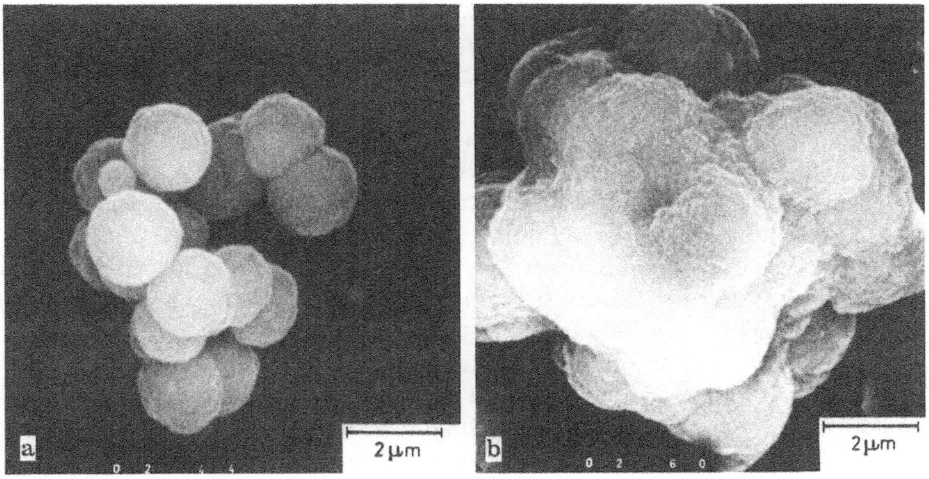

Fig. 12. Scanning electron micrograph of the polymer formed in the polymerization of NMMAm with AIBN at 50 °C in benzene; **a.** [NMMAm] = 3.68 mol/l, [AIBN] = 1.01×10^{-2} mol/l, 160 min. **b.** [NMMAm] = 6.12 mol/l, [AIBN] = 1.01×10^{-2} mol/l, 90 min

3.4 Polymerization Mechanism

The rate of polymerization, R_p, of NMMAm with AIBN in benzene can be expressed by the following equation:

$$R_p = k[AIBN]^n[NMMAm]^m$$

with n = 0.63–0.68, and m = 1–2.5. One might have anticipated that the order of 0.63–0.68 with regard to the initiator concentration would be the consequence of co-occurrence of normal bimolecular termination and unimolecular termination due to propagating radical occlusion [45–47]. However, this is not the case in our system. Most of the cage-escaping cyanopropyl radicals were converted into stable propagating radicals buried in the resulting polymer microspheres, indicating that under the present conditions unimolecular termination due to polymer radical occlusion prevails.

On the other hand, even a complete unimolecular termination by occlusion of polymer radicals may be expected to result in an apparent initiator order lower than unity, because the rate of occlusion depends on that of polymer formation, R_p. An increase of the initiator concentration causes an increase of R_p and consequently of the occlusion rate. That is, the rate of unimolecular termination due to polymer radical occlusion increase with the initiator concentration. In fact, the molecular weight of poly(NMMAm) decreased with increasing initiator concentration, as shown in Fig. 7. First order in the initiator concentration should be observed only if the lifetime of the active centres were independent of the initiator concentration. The R_p dependence on the monomer concentration was not linear but increased with increasing NMMAm concentration. Since poly(NMMAm) is considered to have a higher affinity to the monomer than to benzene (nonpolar solvent), swelling of the polymer microspheres is accelerated by increased NMMAm concentration. This suppresses termination due to radical occlusion. As a result, both R_p and [η] of the resulting polymer increase similarly with the NMMAm concentration, as shown in Fig. 8. Highly swollen polymer microspheres are expected to have a marked tendency to aggregation (Fig. 12b).

Addition of a relatively small amount (< ca. 3 mol/l) of methanol, a good solvent for poly(NMMAm), led to similar results. However, the presence of larger amounts of methanol accelerates bimolecular termination between polymer radicals, and hence causes a decrease of R_p and of the polymer molecular weight. Consequently, maxima in R_p and [η] of the polymer were observed when the methanol concentration was changed (Fig. 9). N,N-Dimethylformamide, a good solvent for poly(AN), shows a similar behavior in the AN polymerization in benzene [46].

As shown in Figs. 6 and 10b, short induction periods were observed at 50 °C, both in the polymerization and the formation of living propagating radicals. We suggest that in the early stages of the polymerization several oligomer radicals of NMMAm associate to form a nucleus, where a relatively high concentration of radicals accelerates the reaction between these radicals, and therefore retards the polymerization. This is considered to be responsible for the induction periods.

Nuclei thus formed grow to primary polymer particles, followed by agglomeration to larger polymer particles. Oligomer radicals formed later are adsorbed on the surfaces of polymer particles and then propagate until they are occluded by newly formed polymer chains. The interaction between propagating radicals and polymer particles through hydrogen-bonding is so strong that the adsorbed polymer radicals cannot move and the propagation proceeds in isolated regions. As a result, the formation efficiency of living propagating radicals is very high and the surface of polymer microspheres is rough (Fig. 12a).

Following the above considerations, the polymerization of NMMAm with AIBN in benzene is envisaged as shown schematically in Fig. 13.

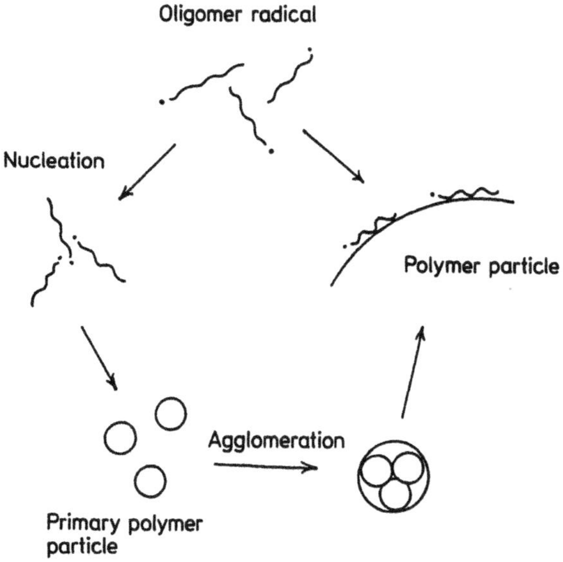

Fig. 13. Schematic picture of the polymerization mechanism

4 ESR Study of the Reactions of Living Poly(NMAAm) and Poly(NMMAm) Radicals with Vinyl Monomers [39]

The living propagating radicals of NMAAm and NMMAm react readily with other vinyl monomers at room temperature. These reactions lead to easy formation of stable propagating radicals of the second monomers, and hence are expected to be useful for the clarification of the propagation mechanism in radical polymerization. It is also of great interest to obtain information concerning the propagating terminal radicals in radical copolymerization, especially in the alternating copolymerization which propagates via change transfer complexes between acceptor and donor monomers.

Thus, the reactions of the polymer radicals with vinyl monomers or binary monomer mixtures were investigated by means of ESR.

A benzene solution of NMAAm, or NMMAm, and DBPO as photoinitiator was irradiated by a 100 W high pressure mercury lamp at room temperature in a degassed sealed ESR tube. To this reaction mixture, a second monomer or a binary monomer mixture was added by using a break-seal technique. After a given time, the ESR spectrum of the reaction mixture was recorded at room temperature.

4.1 Reaction of the Poly(NMAAm) Radical with Vinyl Monomers

Figure 14 shows the ESR spectrum change of the reaction mixture of propagating radical III and MMA. The three line spectrum of the poly(NMAAm) radical changed with time, becoming finally that of the propagating radical (IX) of MMA. A similar

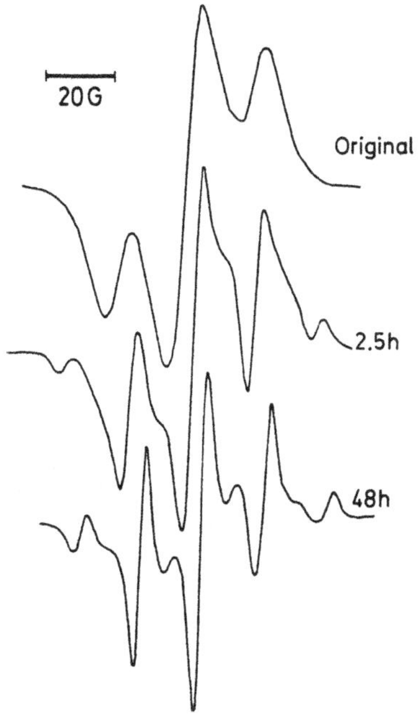

Fig. 14. ESR spectrum change of the poly(NMAAm) radical/MMA system; a benzene solution (0.25 ml) of NMAAm (2.3 mol/l) and DBPO (0.43 mol/l) was irradiated for 2 h at room temperature and then a mixture of MMA (0.25 ml) and benzene (0.2 ml) was added

nine line spectrum of the poly(MMA) radical has been reported by other workers[36, 50]. This indicates that the poly(NMAAm) radicals reacted slowly with MMA to be converted to poly(MMA) radicals (Eq. (9)).

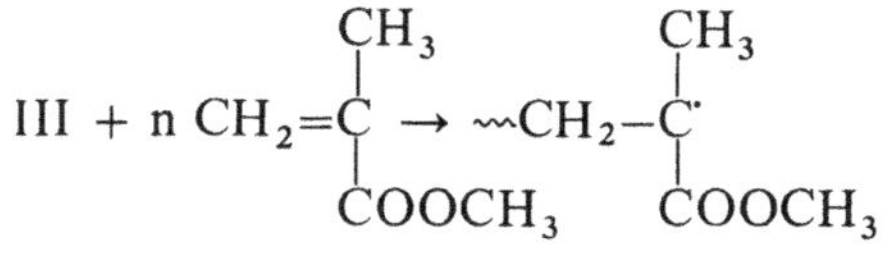

(MMA) (IX) (9)

Figure 15a shows an ESR spectrum observed in the reaction of poly(NMAAm) radicals with isopropenyl methyl ketone (IPMK). This spectrum is similar to that of the poly(MMA) radical and has therefore been assigned to the propagating polymer radical (X) of IPMK.

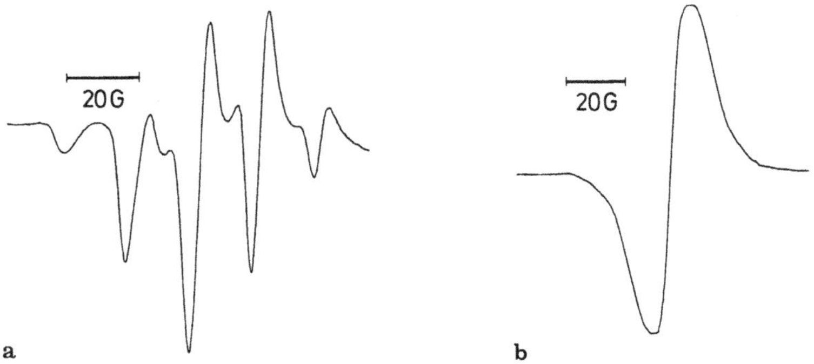

a b

Fig. 15a. ESR spectrum of the poly(NMAAm) radical/IPMK system after reaction for 2 days at room temperature; a mixture of IPMK (0.2 ml) and benzene (0.4 ml) was added. **b.** ESR spectrum of the poly(NMAAm) radical/PA system after reaction for 8 days at room temperature; 0.8 ml of PA was added

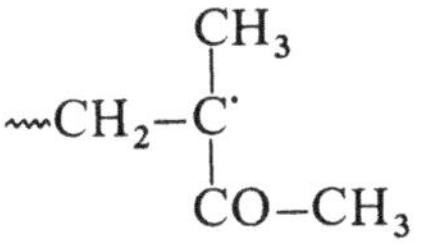

(X)

The nine line spectrum of the poly(MMA) radical has been explained as an overlap of a five line and a nine line spectrum which are due to a deformed conformation (a in Fig. 16) and a symmetrical one (b in Fig. 16) concerning the β-hydrogens in the poly(MMA) radical, respectively[36]. Comparison of the spectra of polymer radicals IX and X indicates that the relative population of the deformed conformation to the symmetrical one is similar in poly(MMA) and poly(IPMK) radicals.

On the other hand, as shown in Figs. 1b and 2b, the propagating polymer radicals of NMMAm and MAm show five line spectra, suggesting that these polymer radicals exist mainly in the deformed conformation.

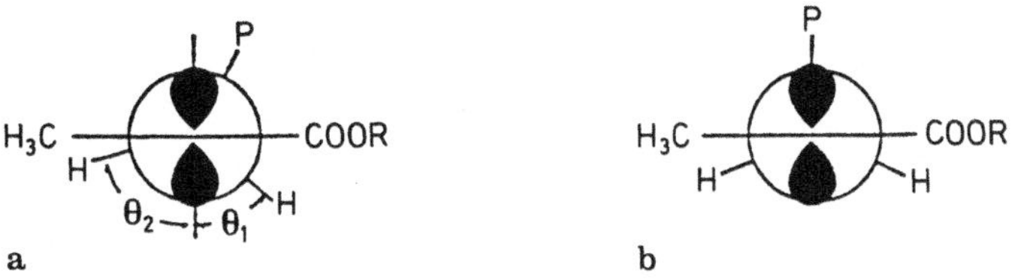

Fig. 16. Conformations of the poly(MMA) radical: **a.** deformed conformation ($\theta_1 = 45°$, $\theta_2 = 75°$), **b.** symmetrical conformation ($\theta_1 = \theta_2 = 60°$); P = polymer chain

Figure 15b shows the ESR spectrum of the reaction system of poly(NMAAm) radicals and phenylacetylene (PA). The broad singlet spectrum observed may be assigned to polymer radical XI from PA. The coupling constant of the β-hydrogen might be too small to overcome broadening factors such as interaction of the unpaired electron with the phenyl hydrogens and restriction of segmental motion of the polymer radical.

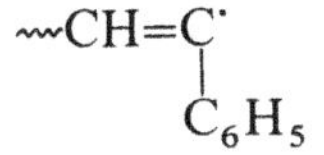
(XI)

It is very difficult to obtain ESR spectra of the propagating polymer radicals of 1,1-diphenylethylene (DPE) and α-methylstyrene (α-MeSt) in the usual radical polymerization, because these monomers have little homopolymerizability. However, the present method can yield easily the living propagating radicals of such monomers at room temperature.

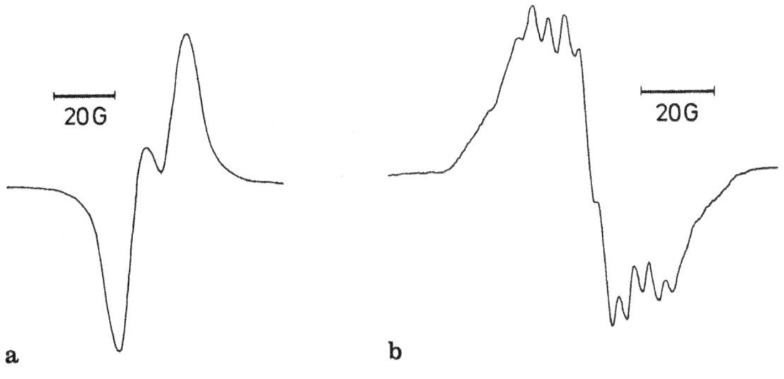

Fig. 17a. ESR spectrum of the poly(NMAAm) radical/DPE system after reaction for 5 days at room temperature, recorded at 50 °C; a mixture of DPE (0.2 ml) and benzene (0.4 ml) was added. **b.** ESR spectrum of the poly(MMAAm) radical/α-MeSt system after reaction for a day at room temperature; 0.5 ml of α-MeSt was added

Figure 17a shows the ESR spectrum of the reaction mixture of poly(NMAAm) radicals with DPE, recorded at 50 °C. The doublet spectrum observed is assigned to the propagating radical (XII) of DPE formed by addition of the monomer to polymer radical III (Eq. (10)). Polymer radical XII has mainly a deformed

$$\mathrm{III} + CH_2{=}\underset{\underset{C_6H_5}{|}}{\overset{\overset{C_6H_5}{|}}{C}} \rightarrow \sim\!\!\sim CH_2{-}\underset{\underset{C_6H_5}{|}}{\overset{\overset{C_6H_5}{|}}{C^{\cdot}}} \qquad (10)$$

(DPE) (XII)

conformation, since the doublet of this spectrum is considered to be due to the interaction of the unpaired electron with only one of the two β-hydrogens in the polymer radical.

Figure 17b shows the ESR spectrum observed in the reaction of poly(NMAAm) radicals with α-MeSt (Eq. (11)). This fine-splitted spectrum is quite different from those of the propagating radicals of other α-methyl-substituted monomers such as MMA, IPMK, MAm and NMMAm. This suggests that polymer radical XIII has a different conformation. However, the poly(MMA) radical has been reported to show a spectrum of 13 lines in methyltetrahydrofuran at very low temperature, which was interpreted as a consequence of a slight distortion of the symmetrical conformation of the β-hydrogens [36].

$$\mathrm{III} + CH_2{=}\underset{\underset{C_6H_5}{|}}{\overset{\overset{CH_3}{|}}{C}} \rightarrow \sim\!\!\sim CH_2{-}\underset{\underset{C_6H_5}{|}}{\overset{\overset{CH_3}{|}}{C^{\cdot}}} \qquad (11)$$

(α-MeSt) (XIII)

4.2 Reaction of the Poly(NMMAm) Radical with Vinyl Monomers

Figure 18 shows the ESR spectrum change of the system of poly(NMMAm) radical (IV) and AN (Eq. (12)). As shown in the figure, the five line spectrum of IV was gradually converted into the three line spectrum of the propagating radical (XIV) of AN, and then the triplet changed to a broad singlet. A similar three line spectrum has been observed on γ-ray irradiation of AN at low temperature, where AN oligomer was assumed to be formed [51], while the occluded polymer radical formed in the usual polymerization of AN shows a broad singlet spectrum [1]. The change in the spectrum of the reaction mixture of the polymer radical IV and AN seems to originate from the change in the environment around the radical center of poly(AN) radical XIV, depending on the number of AN monomeric units incorporated in the polymer radical.

$$\mathrm{IV} + CH_2{=}\underset{\underset{CN}{|}}{CH} \rightarrow \sim\!\!\sim CH_2{-}\underset{\underset{CN}{|}}{CH^{\cdot}} \qquad (12)$$

(AN) (XIV)

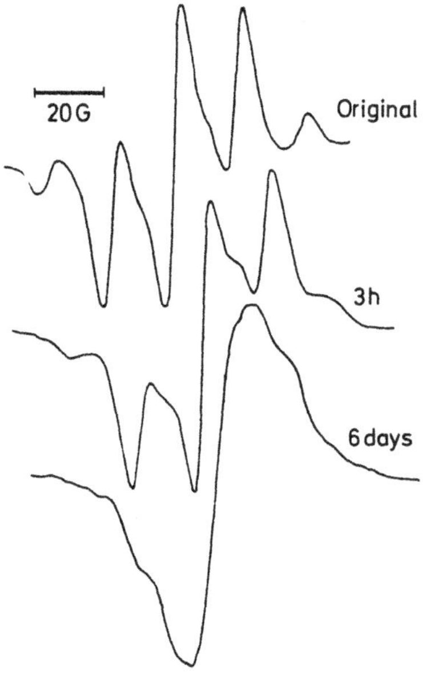

Fig. 18. ESR spectrum change of the poly(NMMAm) radical/AN system; a benzene solution (0.25 ml) of NMMAm (2.0 mol/l) and DBPO (0.43 mol/l) was irradiated for 2 h at room temperature and then a mixture of AN (0.17 ml) and benzene (0.33 ml) was added

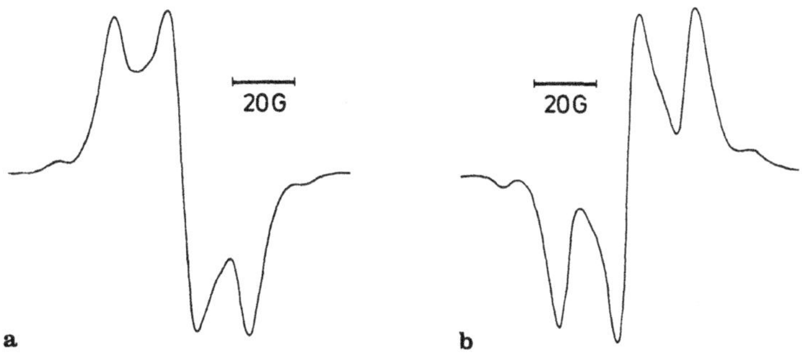

Fig. 19a. ESR spectrum of the poly(NMMAm) radical/St system after reaction for 22 h at room temperature; 0.5 ml of St was added; **b.** ESR spectrum of the poly(NMMAm) radical/NVP system after reaction for 46 h at room temperature; a mixture of NVP (0.17 mol) and benzene (0.33 ml) was added

Figure 19 shows ESR spectra obtained in the reaction of the poly(NMMAm) radical with St (a) and N-vinyl-2-pyrrolidone (NVP) (b). The propagating radicals of St (XV) and NVP (XVI) were observed as three line spectra, analogously to those of NMAAm and MA, all four being α-mono-substituted monomers [35].

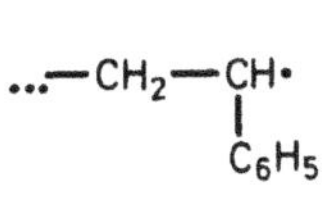

(XV)

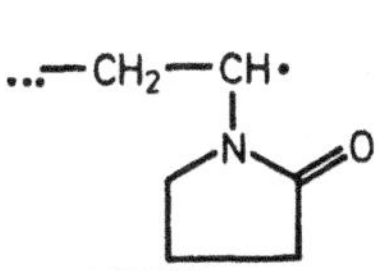

(XVI)

4.3 Reactions of the Poly(NMMAm) and Poly(NMAAm) Radicals with Binary Monomer Mixtures [50]

Figure 20 shows ESR spectra of the reaction mixtures obtained when the poly-(NMMAm) radical was allowed to react with a MMA/St mixture varying the relative concentrations of MMA and St. The spectrum change corresponded well to the change in the relative concentration of the monomer mixture added. The nine line spectrum due to poly(MMA) radical was mainly observed at the higher MMA concentration, while the poly(St) radical prevailed at the lower MMA concentration. This result conforms to that known in the usual radical copolymerization of MMA with St.

On the other hand, as shown in Fig. 21, the spectrum change with the relative concentration [MMA]/[St] in the poly(NMAAm) radical/MMA/St system was found to be fairly different from that in the poly(NMMAm) radical/MMA/St system

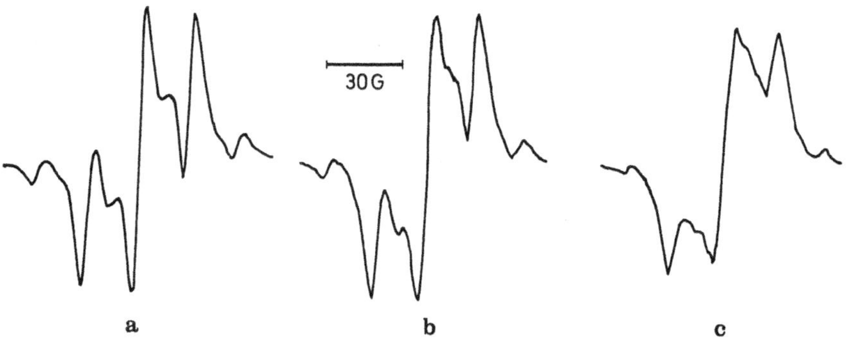

Fig. 20. ESR spectra of the poly(NMMAm) radical/MMA/St system at room temperature; **a.** a mixture (0.3 ml) of [MMA]/[St] = 2.1 was added and allowed to react for 7 days; **b.** a mixture (0.5 ml) of [MMA]/[St] = 1.1 was added and allowed to react for 10 days; **c.** a mixture (0.5 ml) of [MMA]/[St] = 0.27 was allowed to react for 14 days

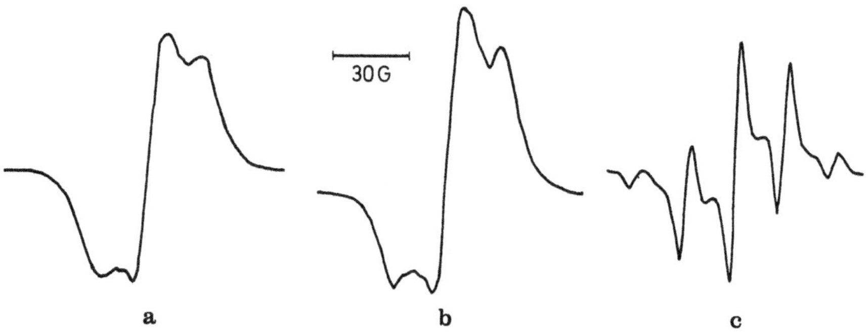

Fig. 21. ESR spectra of the poly(NMAAm) radical/MMA/St system at room temperature; **a.** a mixture (0.5 ml) of [MMA]/[St] = 0.27 was added and allowed to react for 7 days; **b.** a mixture (0.5 ml) of [MMA]/[St] = 4.3 was added and allowed to react for 7 days; **c.** a mixture (0.5 ml) of [MMA]/[St] = 8 was added and allowed to react for 7 days

(Fig. 20). The poly(St) radical was predominant even at the molar ratio of [MMA]/[St] = 4.3 in the poly(MMAAm) radical system, while the poly(MMA) radical was mainly observed at the relative concentration of [MMA]/[St] = 2.1 in the poly(NMMAm) radical system.

This difference might be due to higher selectivity of the poly(NMAAm) microspheres in the incorporation of monomers in comparison with the poly(NMMAm) microspheres. The structure of the poly(NMAAm) microspheres is considered to be more compact owing to more effective hydrogen bonding as compared with poly-(NMMAm), which might prevent more efficiently the diffusion of MMA into the microspheres.

The reaction of the poly(NMAAm) radical with the cyclohexyl methacrylate-(CHMA)/St mixture was investigated, because CHMA was anticipated to have less affinity to the poly(NMAAm) microspheres than MMA. As can be seen from Fig. 22, the preference of St over CHMA in the incorporation into the poly(NMAAm) microspheres was greatly enhanced compared with the MMA/St system. The propagating radical (XVII) of CHMA was present only in small amounts, even at [CHMA]/[St] = 8 (Fig. 22b). An authentic sample of the poly(CHMA) radical was prepared by the reaction of poly(NMAAm) radical with CHMA and its spectrum is shown in Fig. 22c.

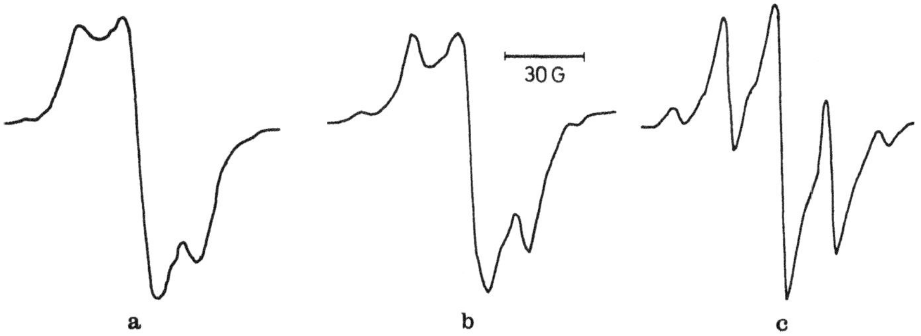

Fig. 22. ESR spectra of the poly(NMAAm) radical/CHMA/St system at room temperature; **a.** a mixture (0.5 ml) of [CHMA]/[St] = 2.6 was added and allowed to react for 5 days; **b.** a mixture (0.5 ml) of [CHMA]/[St] = 8 was added and allowed to react for 7 days; **c.** a mixture of CHMA (0.2 ml) and benzene (0.3 ml) was added and allowed to react for 5 days

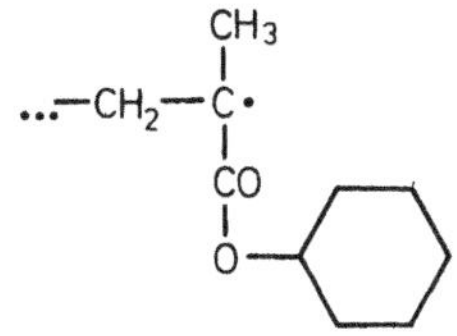

(XVII)

Next, some alternating copolymerization systems were investigated by using the poly(NMMAm) radical. The system poly(NMMAm) radical/diethyl fumarate(DEF)/ α-MeSt was examined first. In order to obtain the spectrum of the propagating

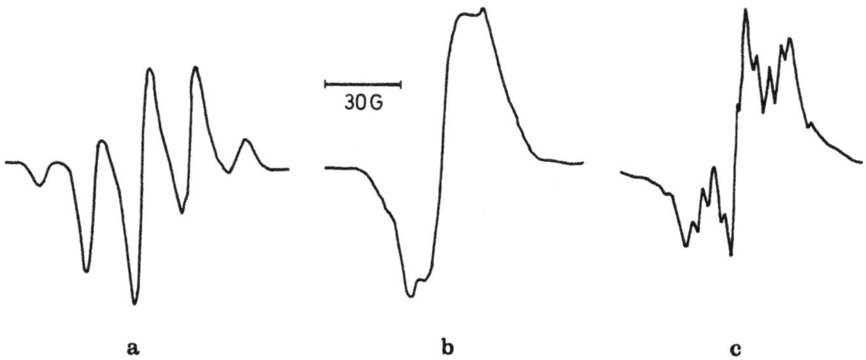

Fig. 23. ESR spectra of the poly(NMMAm) radical/DEF/α-MeSt system; **a.** 0.5 ml of DEF was added and allowed to react for 3 days at room temperature; **b.** the reaction mixture of (a) was further heated for 16 days at 70 °C; **c.** a benzene solution (0.7 ml) of DEF and α-MeSt([DEF]/[α-MeSt] = 1.2) was added and allowed to react for 10 days at room temperature and for 4 h at 70 °C

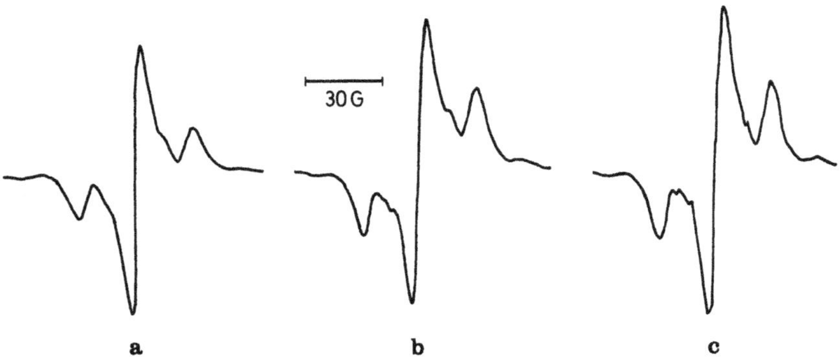

Fig. 24. ESR spectra of the poly(NMMAm) radical/MAn/α-MeSt system at room temperature; **a.** a solution of MAn (0.1 g) in benzene (0.6 ml) was added and allowed to react for 8 days; **b.** a solution of α-MeSt (0.12 g) and MAn (0.1 g) ([α-MeSt]/[MAn] = 1) in benzene (0.5 ml) was added and allowed to react for 9 days; **c.** a solution of α-MeSt (0.24 g) and MAn (0.05 g) ([α-MeSt]/[MAn] = 4) in benzene (0.4 ml) was added and allowed to react for 10 days

radical of DEF, poly(NMMAm) radicals were allowed to react with DEF. As shown in Fig. 23, only the five line spectrum (a) of the poly(NMMAm) radical was observed even after reaction for 3 days at room temperature. Further reaction at 70 °C for 16 days led ultimately to the observation of spectrum (b), which is assigned to the propagating radical (XVIII) of DEF. The poor reactivity of DEF is assumed to originate from the steric effect of the α- and β-carboethoxy groups, which lowers the reactivity

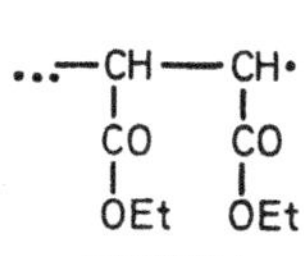

(XVIII)

of DEF toward the polymer radical and might hinder the incorporation of DEF into the poly(NMMAm) microspheres.

Spectrum (c) in Fig. 23 was obtained when a nearly equimolar mixture of α-MeSt and DEF was added to poly(NMMAm) radicals. The propagating radical α-MeSt (Fig. 17b) appears to be prevailing. It might be difficult for DEF to get to the active sites in the poly(NMMAm) microspheres in the presence of α-MeSt which has a high affinity to the microspheres. It is also possible that the poly(α-MeSt) radical is predominant as the terminal radical in this alternating copolymerization system.

Figure 24 shows the results observed for the reaction of the poly(NMMAm) radical with an α-MeSt/maleic anhydride(MAn) mixture, also a typical alternating copolymerization system. The propagating polymer radical (XIX) of MAn was prepared by the reaction of the poly(NMMAm) radical with MAn and its spectrum is shown in Fig. 24a. The triplet is considered to be due to the α- and β-hydrogens in radical XIX.

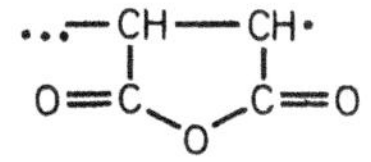

(XIX)

When an equimolar mixture of both monomers was used, the propagating radical from MAn was exclusively formed, with no observation of the complicated hyperfine splitted spectrum due to the poly(α-MeSt) radical (Fig. 17b). A very weak signal of the poly(α-MeSt) radical might be present, overlapped by the spectrum of radical XIX, at higher α-MeSt concentration ([α-MeSt]/[MAn] = 4), as shown in Fig. 24c. These results indicate that the poly(MAn) radical is the predominant terminal propagating site in the α-MeSt/MAn copolymerization system [2)].

a b

Fig. 25. ESR spectra of the poly(NMMAm) radical/MAn/DPE system at room temperature; **a.** a solution of MAn (0.05 g) and DPE (0.41 g) ([DPE]/[MAn] = 4.4) in benzene (0.2 ml) was added and allowed to react for 20 days; **b.** a solution of MAn (0.1 g) and DPE (0.2 g) ([DPE]/[MAn] = 1.1) in benzene (0.4 ml) was added and allowed to react for 3 days

2) Much attention has been paid to the propagating terminal radicals in alternating copolymerization via charge-transfer complexes of the two copolymerizing monomers (see e.g. Ref. [35)]). We believe that our method is very useful for identifying these radicals

ESR spectra of the poly(NMMAm) radical/MAn/DPE system are shown in Fig. 25. The propagating radical of DPE results in the doublet shown in Fig. 17a, that of MAn in a triplet (Fig. 24a). As can be seen from Fig. 25, both propagating radicals were observed in comparable concentrations, though the poly(MAn) radical was somewhat prevailing. The signal of the poly(DPE) radical intensified with increased DPE concentration.

5 Synthesis of Block Copolymers by the Reaction of Living Poly(NMAAm) Radical with Vinyl Monomers [53)]

As described above, when NMAAm or NMMAm is photo-polymerized in benzene at room temperature, the amide monomer is converted to polymer microspheres containing its living propagating radicals in high concentrations. These living radicals react easily with other vinyl monomers. These reactions can be applied to the synthesis of block copolymers.

A benzene solution of NMAAm (3.73 mol/l) and DBPO (0.175 mol/l) was irradiated at room temperature for 3 h by using a high pressure mercury lamp (100 W) to yield poly(NMAAm) radicals. A separate experiment revealed that the polymerization proceeded quantitatively and the resulting poly(NMAAm) had a weight-average molecular weight of 1.85×10^5 [54)].

To this polymerization mixture a benzene solution of a second monomer was added, by the break-seal method, to undergo block copolymerization. The resulting polymer was fractionated into three parts by Soxhlet extraction for 25 h. AcOEt was used first as the extracting solvent for homopolymers of the second monomers [benzene for poly(butyl acrylate)], which are considered to be produced mainly by chain transfer reactions. Then, homopoly(NMAAm) was extracted with methanol (MeOH). The insoluble residue was taken as block copolymer.

5.1 Reaction of the Poly(NMAAm) Radical with MA

Table 4 shows the results obtained when the poly(NMAAm) radical was allowed to react with MA for five days at room temperature. The added MA was polymerized

Table 4. Reaction of the poly(NMAAm) radical[a] with MA for 5 days at room temperature

Run	[MA][b]	Yield (%)		Fraction (wt %)		
	(mol/l)	Total	MA	AcOEt-soluble	MeOH-soluble	Insoluble
MA-1	2.78	90.7	81.0	14.8	18.9	66.2
MA-2	5.55	89.7	84.3	14.1	10.6	75.3
MA-3	8.33	93.5	91.3	11.0	5.6	83.3
MA-4	11.11	ca. 100	ca. 100	9.5	3.7	86.8

[a] A benzene solution (1.55 ml) of NMAAm (3.75 mol/l) and BBPO (0.175 mol/l) was irradiated for 3 h;

[b] The MA concentration of the benzene solution (2 ml) added to the prepolymerization mixture

Table 5. Effects of the reaction time and temperature on the reaction of poly(NMAAm) radical[a] with MA[b]

Run	Temp. (°C)	Time (h)	Yield (%) Total	Yield (%) MA	Fraction (wt %) AcOEt-soluble	Fraction (wt %) MeOH-soluble	Fraction (wt %) Insoluble
MA-5	50	10	42.5	27.6	12.9	9.0	79.1
MA-6	50	15	50.8	38.0	17.4	6.1	76.5
MA-7	50	25	54.3	42.4	11.0	5.3	83.7
MA-8	50	40	61.7	51.7	15.8	4.7	79.6
MA-9	40	15	70.8	63.2	10.5	3.6	85.9

[a] A benzene solution (1.55 ml) of NMAAm (3.75 mol/l) and DBPO (0.175 mol/l) was irradiated for 3 h;
[b] Undiluted MA (2 ml) was added to the prepolymerization mixture

in increasing yield with increasing concentration. When undiluted MA was used (run MA-4), its conversion reached almost 100%. The fraction of block copolymer (insoluble in both AcOEt and MeOH) was more than 60% of the total polymer in all runs and increased with the MA concentration. From the result of nitrogen analysis, the AcOEt-soluble part was found to contain little NMAAm, thus indicating that it was almost pure homopolymer from MA. On the other hand, the NMAAm content of the MeOH-soluble part was 70–75%, which suggests that it consisted of homopoly(NMAAm) and block copolymer. As shown below, this was also substantiated by its IR spectrum.

It is noteworthy that the amount of the MeOH-soluble part was only 20–30% of the NMAAm used and that this part contained not only homopoly(NMAAm) but also block copolymer. Thus, the NMAAm was, for the most part, incorporated in living poly(NMAAm) radicals which reacted effectively with MA.

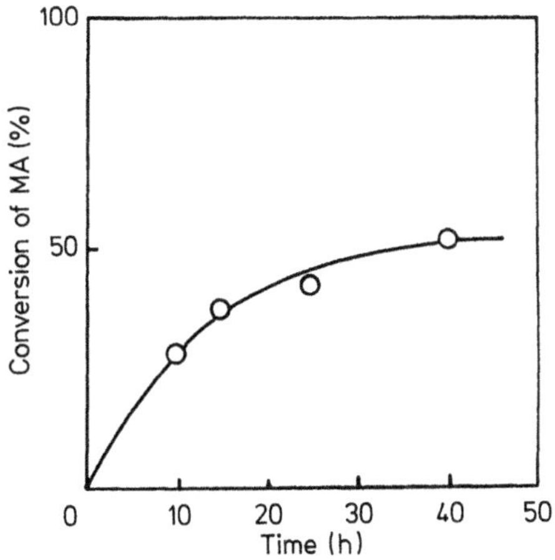

Fig. 26. Relationship between time and MA conversion in the reation of poly(NMAAm) radicals with MA at 50 °C; a benzene solution (1.55 ml) of NMAAm (3.73 mol/l) and DBPO (0.175 mol/l) was irradiated for 3 h before MA (2 ml) was added

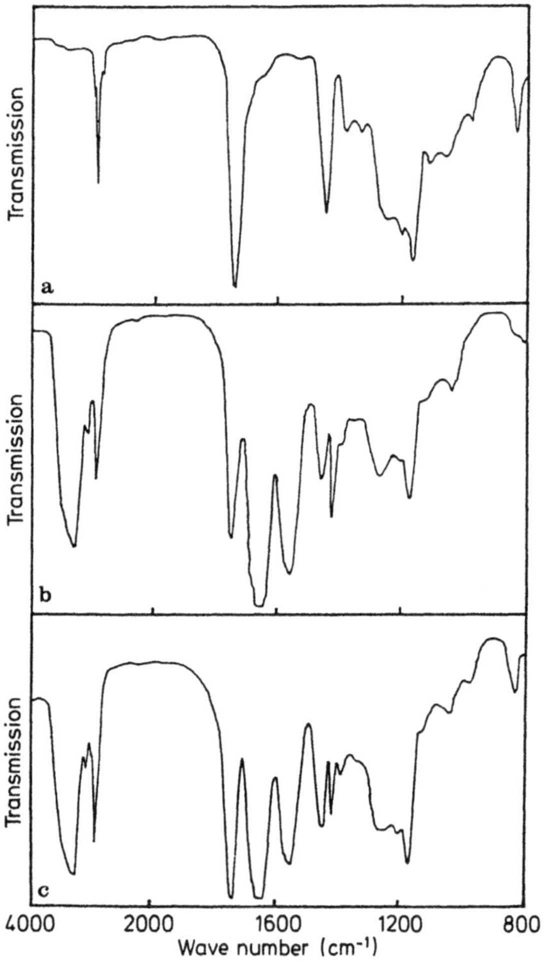

Fig. 27. IR spectra of the fractionated polymers in a NMAAm/MA system; **a.** AcOEt-soluble part, **b.** MeOH-soluble part, **c.** insoluble part

Table 5 shows the effects of temperature and time on the reaction of the poly-(NMAAm) radicals with MA. Although the MA conversion increased with increasing reaction time at 50 °C, this system tended toward dead-end polymerization as shown in Fig. 26. Furthermore, the conversion of MA at 40 °C (run MA-9) was higher than that at 50 °C. These results show that living propagating radicals were deactivated, probably by chain transfer and bimolecular termination, at the higher temperature. Block copolymer constituted 75–83% of the total resulting polymer at 40–50 °C.

Figure 27 shows IR spectra of the three fractions. The spectrum of the AcOEt-soluble fraction was identical with that of poly(MA). The MeOH-soluble part showed an absorption band at 1740 cm^{-1} due to ester-carbonyl groups, indicating that it contained block copolymer carrying short chains of poly(MA). These results are compatible with the nitrogen analysis. As expected, the solvent-insoluble part showed absorption bands observed in the homopolymers of NMAAm and MA, thus confirming that this part is a block copolymer.

5.2 Reaction of the Poly(NMAAm) Radical with Some Other Vinyl Monomers

Some other vinyl monomers were also allowed to react with the poly(NMAAm) radical at 40 °C for 15 h to prepare the corresponding block copolymers: ethyl acrylate (EA), butyl acrylate (BA), benzyl acrylate (BzA), MMA, CHMA and St.

Table 6. Reaction of the poly(NMAAm) radical[a] with some vinyl monomers[b] for 15 h at 40 °C

Second monomer	Yield for second monomer (%)	Fraction (wt %)		
		AcOEt-soluble	MeOH-soluble	Insoluble
MA	63.2	10.4	3.6	85.9
EA	62.8	9.7	2.1[c]	88.1
BA	36.0	8.9[d]	7.1	84.0
BzA	58.0	31.1	13.2	55.7
MMA	46.9	13.4	23.8	62.8
CHMA	26.1	9.6	38.6	51.8
St	11.8	6.6	24.4	69.0

[a] A benzene solution (1.55 ml) of NMAAm (3.75 mol/l) and DBPO (0.175 mol/l) was irradiated for 3 h;
[b] Undiluted second monomer (2 ml) was added to the prepolymerization mixture;
[c] Soluble in water;
[d] Soluble in benzene

The results obtained are summarized in Table 6. A longer alkyl chain of the acrylate monomer caused a decrease in its conversion. Block copolymers formed in these alkyl acrylate monomer systems constituted about 85% of the total polymers.

BzA was polymerized in a higher yield than BA, in spite of the greater bulkiness of the benzyl group. This seems to be the result of a high affinity of the aromatic group to poly(NMAAm), which causes BzA monomer to proceed easily to an active center. Poly(NMAAm) swells considerably in aromatic solvents such as benzene, acetophenone and methyl benzoate, but not in the aliphatics such as acetone, acetonitrile and AcOEt. However, the block copolymer yield in the BzA system was lower than with other acrylates. This is attributed to the participation of labile benzylic hydrogen of BzA in chain transfer reactions.

Methacrylate monomers showed relatively low conversions in comparison with the acrylates. In particular, the low reactivity of CHMA seems to result from its small affinity to the poly(NMAAm) microspheres, which causes slow diffucion of the monomer to the active reaction sites. Aliphatic hydrocarbons such as n-hexane and cyclohexane are not even miscible with NMAAm monomer. Consequently the CHMA system produced a larger MeOH-soluble part.

Styrene exhibited lower reactivity than the acrylate monomers. Since poly(NMAAm) swells in St, an aromatic solvent, the segmental movement of the polymer becomes easier and might lead to acceleration of bimolecular termination.

The results of nitrogen analysis and IR spectra revealed that the AcOEt-soluble (benzene-soluble for the BA system) parts were homopolymers of the second monomers, whereas the MeOH-soluble (water-soluble for the EA system) parts contained both homopoly(NMAAm) and block copolymers.

5.3 Thermogravimetric (TG) Study of Block Copolymers

The thermal degradation behavior of the block copolymers obtained was studied thermogravimetrically. As a typical example, TG and differential thermogravimetric (DTG) curves of the block copolymer from the poly(NMAAm)/EA system are shown in Fig. 28, together with those of poly(NMAAm) and poly(EA). The block copolymer decomposed at a higher temperature than poly(NMAAm), despite the fact that the former contains a poly(NMAAm) block. Similar behavior was observed in other copolymers. This phenomenon may be rationalized as follows. As already mentioned, the prepolymerization of NMAAm in benzene yields microspheres of poly(NMAAm)

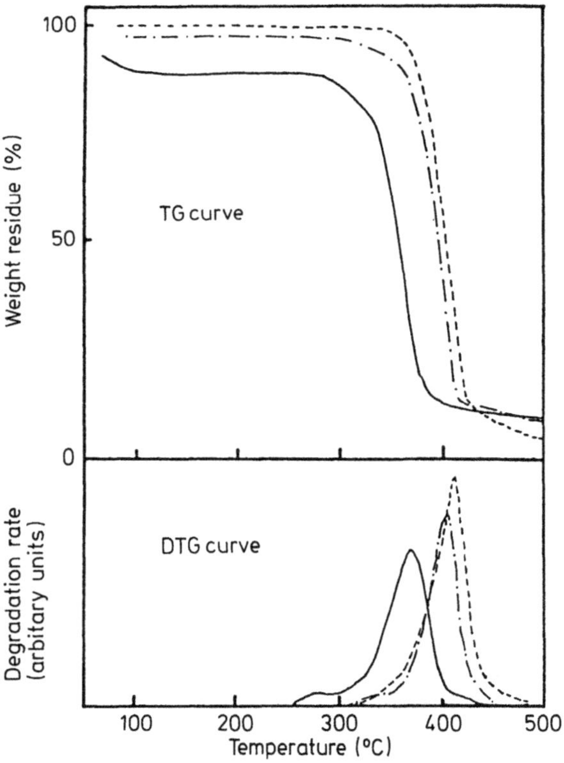

Fig. 28. TG and DTG curves of poly(NMAAm) (———), poly(EA) (---------) and NMAAm-EA block copolymer (-·-·-·-·-)

which contain living propagating radicals. When the poly(NMAAm) radicals react with the second monomer to produce a block copolymer, the poly(NMAAm) microspheres are surrounded by the second polymer. Because the second polymer is more thermostable, it protects the inner poly(NMAAm) block from thermal degradation.

5.4 Reaction of the Poly(NMAAm) Radical with MMA/St or CHMA/St Mixtures [52)]

As described above, an ESR study of the reaction of the poly(NMAAm) radical with a MMA/St (MS) or CHMA/St (CS) mixture led to the conclusion that St is incorporated into the poly(NMAAm) microspheres in preference to the methacrylate monomers, to react with the polymer radical. To confirm this conclusion, block copolymers were prepared by the reaction of poly(NMAAm) radicals with MS or CS mixtures, and the relative contents of methacrylate and St in the resulting polymers were determined. The poly(NMAAm) radical was allowed to react with monomer mixtures of different relative concentration for 15 h at 40 °C. The resulting polymer was fractionated into three parts (i.e., AcOEt-soluble, MeOH-soluble and insoluble parts). The AcOEt-soluble part consisted of random copolymer of methacrylate and St, while the MeOH-soluble part consisted of homopoly(NMAAm) and block copolymer. The solvent-insoluble part was block copolymer. The results obtained are summarized in Table 7.

Table 7. Reaction of the poly(NMAAm) radical[a] with MMA/St or CHMA/St mixture[b] for 15 at 40 °C

Run[c]	[St]/([St]+[methacrylate]) in comonomer	Conversion[d] (%)	Fraction		
			AcOEt-soluble	MeOH-soluble	Insoluble
MS-1	0.24	11.6	1.5	45.9	52.6
MS-2	0.48	12.3	1.5	39.2	59.3
MS-3	0.74	7.8	0.8	53.6	45.7
CS-1	0.34	11.2	2.0	57.9	40.1
CS-2	0.60	7.8	1.2	55.3	43.4
CS-3	0.82	9.2	1.4	33.9	58.7

[a] A benzene solution (1.55 ml) of NMAAm (3.75 mol/l) and DBPO (0.175 mol/l) was irradiated for 3 h;
[b] 2 ml of monomer mixture was added;
[c] MS and CS indicate the systems of MMA/St and CHMA/St, respectively;
[d] Based on the monomer mixture added

Table 8 lists the ratio [St]/([St] + [methacrylate]) in the fractions as determined by pyrolysis gas chromatography. The MeOH-soluble parts in all runs showed considerably higher St content in comparison to the random copolymer formed from the monomer feed of the same composition. On the other hand, the AcOEt-soluble and insoluble parts showed a St content nearly equal to that of the random copolymer. The tendency to higher St content in the MeOH-soluble part was more marked in the

Table 8. Compositions of block copolymers determined by pyrolysis gas chromatography[a]

Run	Fraction	[St]/([St]+[methacrylate])		
		In comonomer	In block copolymer	In random copolymer
MS-1	AcOEt-soluble	0.24	0.32	0.31
	MeOH-soluble		0.44	
	Insoluble		0.33	
MS-2	AcOEt-soluble	0.48	0.47	0.51
	MeOH-soluble		0.66	
	Insoluble		0.50	
MS-3	AcOEt-soluble	0.74	0.69	0.70
	MeOH-soluble		0.74	
	Insoluble		0.69	
CS-1	AcOEt-soluble	0.34	0.30	0.34
	MeOH-soluble		0.64	
	Insoluble		0.34	
CS-2	AcOEt-soluble	0.60	0.52	0.57
	MeOH-soluble		0.68	
	Insoluble		0.60	
CS-3	AcOEt-soluble	0.82	0.77	0.77
	MeOH-soluble		0.89	
	Insoluble		0.78	

[a] Pyrolysis conditions. MS series: pyrolysis temperature 500 °C, column (PEG-6000) temperature 80 °C. CS series: pyrolysis temperature 460 °C, column (PEG-6000) temperature 60–150 °C (heating rate 4 °C/min)

CS systems than in the MS systems. The block copolymers in the MeOH-soluble fraction are assumed to carry shorter chains of methacrylate/St random copolymer and longer ones of poly(NMAAm). These block copolymers are probably formed in the early stage of the reaction of the poly(NMAAm) radicals with the monomer mixture. In the early stage St is preferentially incorporated into the poly(NMAAm) microspheres, leading to the formation of block copolymer of higher St content. On the other hand, in the later stage the propagation might occur on or near the surfaces of the microspheres, resulting in random copolymerization. These results are compatible with those of the ESR (Figs. 21 and 22) and TG (Section 5.3) studies.

6 Synthesis of Block Copolymers by the Reaction of Living Poly(N-phenylmethacrylamide) Radicals with Vinyl Monomers [55]

Figure 29a shows an ESR spectrum of the polymerization mixture obtained when N-phenylmethacrylamide (NPMAm) was photo-polymerized at room temperature in benzene by using DBPO as sensitizer. The five line spectrum is characteristic of the propagating polymer radicals stemming from methacrylamide derivatives.

Thus, similarly to NMAAm and NMMAm, NPMAm was readily converted to its living propagating radical (XX) in the heterogeneous polymerization in benzene

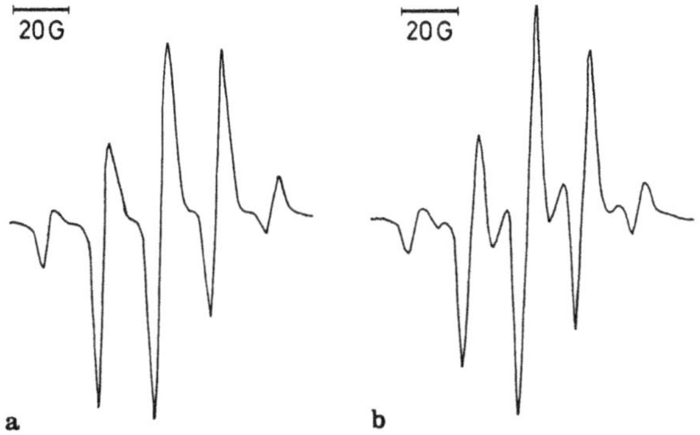

Fig. 29. ESR spectrum of the poly(NPMAm) radical/MMA system; **a.** a benzene solution (0.3 ml) of NPMAm (0.55 mol/l) and DBPO (0.1 mol/l) was irradiated for 3 h; **b.** to the above polymerization mixture a benzene solution (0.5 ml) of MMA (5.6 mol/l) was added and allowed to react for 30 min at room temperature

(Eq. (13)). Furthermore, this polymerization proceeded almost quantitatively (polymer yield: 94%).

$$\text{t-BuO}\cdot + n\,CH_2{=}\underset{\underset{NH-C_6H_5}{|}}{\underset{CO}{\underset{|}{\overset{\overset{CH_3}{|}}{C}}}} \rightarrow \sim\!CH_2{-}\underset{\underset{NH-C_6H_5}{|}}{\underset{CO}{\underset{|}{\overset{\overset{CH_3}{|}}{C}}}}\cdot \tag{13}$$

(NPMAm) (XX)

The spectrum (b) in Fig. 29 was observed when the living poly(NPMAm) radical was allowed to react with MMA at room temperature. This nine line spectrum is assigned to the propagating radical of MMA. This indicates that the poly(NPMAm) radical added easily to MMA to yield stable poly(MMA) radical (IX) under the present conditions (Eq. (14)).

$$XX + n\,MMA \rightarrow \sim\!CH_2{-}\underset{\underset{COOCH_3}{|}}{\overset{\overset{CH_3}{|}}{C}}\cdot \tag{14}$$

(IX)

The synthesis of block copolymers was attempted by using the reaction of living poly(NPMAm) radicals with vinyl monomers. Table 9 shows the results obtained when the poly(NPMAm) radical was allowed to react with some vinyl monomers. The second monomers were MMA, MA, AN, St and vinyl acetate (VAc).

MMA and MA were polymerized with moderate yields. However, AN, St and VAc scarcely reacted under the present conditions, judging from the yields and IR spectra of the resulting polymers. VAc, as a non-conjugative monomer, may be assumed to have only a slight reactivity toward the poly(NPMAm) radical [56];

Table 9. Reaction of the poly(NPMAm) radical with some vinyl monomers[a]

Second monomer (M_2)	Amount of M_2 (g)	Total yield (g)	Total yield (%)[b]	Recovered NPMAm (g)
MMA	1.88	1.17	37	0.02
MA	1.91	0.87	30	0.10
AN	1.61	0.31	~0	0.09
St	1.80	0.31	~0	0.09
VAc	1.86	0.25	~0	0.15

[a] A solution of NPMAm (0.4 g) and DBPO (0.05 ml) in benzene (2 ml) was irradiated at room temperature for 5 h, and then the second monomer was added and allowed to react at 40 °C for 30 h;
[b] Based on the second monomer

Table 10. Reaction of the poly(NPMAm) radical with MMA[a]

Run	Temp. (°C)	Time (h)	Total yield (g)	Total yield (%)[b]	Recovered NPMAm (g)
1	40	30	1.07	37	0.02
2	30	20	0.48	8	0.06
3	20	30	0.73	21	0.07
4	20	20	0.32	8	0.21
5	20	10	0.29	5	0.21
6	10	30	0.64	18	0.11
7	10	20	0.33	5	0.18

[a] The reaction was carried out in the same manner as described in Table 9. NPMAm, 0.4 g; MMA, 1.88 g;
[b] Based on MMA added

AN may accelerate the bimolecular termination of poly(NPMAm) radicals, since AN is a very good solvent for poly(NPMAm). The reason for the low reactivity of St is obscure at present, although the radical polymerization of St is known generally to proceed much more slowly than those of acrylate monomers (see also Table 6).

Table 10 summarizes some further results on the reaction of poly(NPMAm) radicals with MMA. Longer reaction times led to an increase in the MMA conversion. The temperature effect on the MMA conversion was, however, not substantial in the range 10 °C to 40 °C. The enhanced propagation with increased temperature may be compensated by an accelerated termination reaction between the polymer radicals.

The amount of recovered NPMAm monomer varied considerably for different runs (Tables 9 and 10). This probably originated from the fact that the polymerization of NPMAm in these systems was initiated photochemically and proceeded heterogeneously. However, the amount of NPMAm recovered showed a tendency to decrease with an increase in the MMA conversion. This suggests that a small amount of NPMAm was incorporated in the MMA block in the post-polymerization.

The polymer mixtures formed from the reaction of poly(NPMAm) radicals with MMA and MA were fractionated into three parts by selective precipitation. Part A

Table 11. Fractionation of the polymer mixtures formed in the NPMAm/MMA and NPMAm/MA systems

Run[a]	Part A[b] (%)	Part B[c] (%)	Part C[d] (%)
1	32	34	34
2	5	41	54
3	6	56	38
4	14	28	58
5	3	31	66
6	14	51	35
7	3	37	60
8[e]	10	61	29

[a] Same as described in Table 10;
[b] Part A: insoluble in an AcOEt/MeOH (1:7(v/v)) mixture and soluble in benzene;
[c] Part B: insoluble in the AcOEt/MeOH mixture and insoluble in benzene;
[d] Part C: soluble in the AcOEt/MeOH mixture and insoluble in benzene;
[e] Second monomer: MA (from Table 9)

was insoluble in an AcOEt/MeOH (1:7(v/v)) mixture and soluble in benzene. Part B was insoluble in the AcOEt/MeOH mixture and insoluble in benzene. Part C was soluble in the AcOEt/MeOH mixture and insoluble in benzene. The results of fractionation are summarized in Table 11. IR spectra of parts A and C in the MMA systems were nearly identical with those of poly(MMA) and poly(NPMAm), respectively. As anticipated, part B showed absorption bands due to both, poly(MMA) and poly(NPMAm). Thus, the block copolymer formed in this system was satisfactorily isolated as part B. It constituted 28–56% of the total polymer obtained.

The fractionation of the polymer formed in the poly(NPMAm) radical/MA system was not so satisfactory as in the MMA system. The IR spectrum of part A exhibited

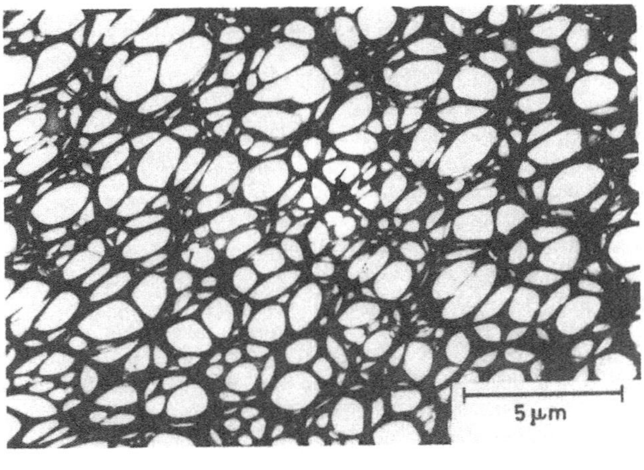

Fig. 30. Transmission electron micrograph of the NPMAm-MMA block copolymer (part B in run 1)

a small absorption peak due to amide groups, while that of part C showed a peak due to ester carbonyl groups. Thus, parts A and C contained small amounts of block copolymer. Hence, the amount of block copolymer in the MA system was more than 60% of the total polymer yield.

GPC data in the MMA system showed that the molecular weight ($\bar{M} = 54.7 \times 10^4$) of part B was much higher than that ($\bar{M} = 5.1 \times 10^4$) of part C (run 1). The latter value was similar to that of original poly(NPMAm).

Figure 30 shows a transmission electron micrograph of a thin film prepared from an AcOEt solution of part B in run 1. A sea-island structure is clearly observed. The occurrence of such a microphase separation supports the conclusion that this polymer is a block copolymer.

7 Conclusions

The radical polymerization of some acrylamide derivatives such as NMAAm and NMMAm yields polymer microspheres involving living propagating radicals of the amide monomers. The efficiency of living radical formation is very high. These living polymer radicals react effectively with other vinyl monomers, giving the stable propagating radicals of the second monomers at room temperature. The ESR spectra of the propagating radicals of α-MeSt and DPE, which cannot be homopolymerized, are also observed easily by using this method. The microsphere radical method will be useful for the clarification of the propagation mechanisms of some interesting polymerization systems, such as alternating copolymerizations via charge-transfer complexes, cyclization polymerizations, ring opening polymerizations, etc. The reactions of the living polymer radicals with acrylate monomers give, in good yields, block copolymers which are amphiphilic, consisting of hydrophilic and lipophilic sequences.

8 References

1. Bamford, C. H., Jenkins, A. D., Symons, M. C. R., Townsend, M. G.: J. Polym. Sci. *34*, 181 (1959)
2. Bamford, C. H., Jenkins, A. D., Ward, J. C.: J. Polym. Sci. *48*, 37 (1960)
3. Minoura, Y., Ogata, Y.: J. Polym. Sci. A-1, *7*, 2547 (1969)
4. Seymour, R. B., Owen, D. R., Stahl, G. A.: Polymer *14*, 324 (1973)
5. Seymour, R. B., Stahl, G. A.: J. Polym. Sci., Polym. Chem. Ed. *14*, 2545 (1976)
6. Seymour, R. B., Stahl, G. A.: J. Macromol. Sci.-Chem. *A11*, 53 (1977)
7. Seymour, R. B.: J. Macromol. Sci.-Chem. *15*, 815 (1981)
8. Alder, G., Ballantine, D., Baysal, B.: J. Polym. Sci. *48*, 195 (1960)
9. Shioji, Y., Ohnishi, S., Nitta, I.: J. Polym. Sci. A, *1*, 3373 (1963)
10. Kücükyavuz, S., Baysal, B.: J. Polym. Sci., Polym. Chem. Ed. *16*, 2901 (1978)
11. Mikulásŏvá, D., Chrástová, V., Citovický, P., Horie, K.: Makromol. Chem. *178*, 429 (1977)
12. Horie, K., Mikulásŏvá, D.: Makromol. Chem. *175*, 2091 (1974)
13. Bigdeli, E., Lenz, R. W., Oster, B., Lundberg, R. D.: J. Polym. Sci., Polym. Chem. Ed. *16*, 469 (1978)
14. Kataoka, S., Ando, T.: Kobunshi Ronbunshu *38*, 821 (1981)
15. Kataoka, S., Ando, T.: Polymer *25*, 507 (1984)
16. Farina, M., Silvestro, G. D.: J. Chem. Soc., Chem. Comm. 842 (1976)

17. Miyata, M., Osaki, Y., Takemoto, K., Kamachi, M., Nozakura, S.: Polymer Preprints Jpn. *33*, 141 (1984)
18. Osada, Y., Iriyama, Y., Takase, M.: Kobunshi Ronbunshu *38*, 629 (1981)
19. Osada, Y., Takase, M., Iriyama, Y.: Polym. J. *15*, 81 (1983)
20. Simionescu, C. I., Simionescu, B., Leanca, M., Ananiescu, C.: Polym. Bull. *5*, 61 (1981)
21. Simionescu, B. C., Popa, M., Ioan, S., Simionescu, C. I.: Polym. Bull. *6*, 415 (1982)
22. Kabanov, V. A.: J. polym. Sci., Polym. Symp.: *50*, 71 (1975)
23. Lee, M., Minoura, Y.: J. Chem. Soc., Faraday Trans. I. *74*, 1726 (1978)
24. Lee, M., Utsumi, K., Minoura, Y.: J. Chem. Soc., Faraday Trans. I, *75*, 1821 (1979)
25. Hungenberg, K.-D., Bandermann, F.: Makromol. Chem. *184*, 1423 (1983)
26. Lee, M., Morigami, T., Minoura, Y.: J. Chem. Soc., Faraday Trans. I, *74*, 1738 (1978)
27. Mun, Y., Sato, T., Otsu, T.: Makromol. Chem. *185*, 1507 (1984)
28. Otsu, T., Yoshida, M.: Makromol. Chem., Rapid Comm. *3*, 127 (1982)
29. Otsu, T., Yoshida, M., Tazaki, T.: Makromol. Chem., Rapid Comm. *3*, 133 (1982)
30. Otsu, T., Yoshida, M., Kuriyama, A.: Polym. Bull. *7*, 45 (1982)
31. Otsu, T., Yoshida, M.: Polym. Bull. *7*, 197 (1982)
32. Otsu, T., Kuriyama, A.: Kobushi Ronbunshu *40*, 583 (1983)
33. Otsu, T., Kuriyama, A.: Polym. Bull. *11*, 135 (1984)
34. Tanaka, H., Sato, T., Otsu, T.: Makromol. Chem. *180*, 267 (1979)
35. Ikegami, T., Hirai, H.: J. Polym. Sci. A-1, *8*, 463 (1970)
36. Kamachi, M., Kohno, M., Liaw, D. J., Katsuki, S.: Polym. J. *10*, 69 (1978)
37. Bartlett, P. B., Bening, E. P., Pincock, R. E.: J. Am. Chem. Soc. *82*, 1762 (1960)
38. Otsu, T., Tanaka, H., Sato, T., Quach, L.: Makromol. Chem. *181*, 1897 (1980)
39. Tanaka, H., Sato, T., Otsu, T.: Makromol. Chem. *181*, 2421 (1980)
40. Chachaty, C., Forchioni, A.: J. Polym. Sci. A-1, *10*, 1905 (1972)
41. Sato, T., Tanaka, H., Otsu, T.: J. Polym. Sci., Polym. Lett. Ed. *18*, 189 (1980)
42. Weigert, F., Shidei, J.: Z. Phys. Chem., Abt. B, *9*, 329 (1930)
43. Sato, T., Miyamoto, J., Otsu, T.: J. Polym. Sci., Polym. Chem. Ed. *22*, 3921 (1984)
44. Tobolsky, A. V., Baysal, B.: J. Polym. Sci. *11*, 471 (1953)
45. Thomas, W. M., Pellon, J. J.: J. Polym. Sci. *8*, 829 (1954)
46. Imoto, M., Takasugi, H.: Makromol. Chem. *23*, 119 (1957)
47. Srinivasan, N. T., Santappa, M.: Makromol. Chem. *26*, 80 (1958)
48. Van Hook, J. P., Tobolsky, A. V.: J. Am. Chem. Soc. *80*, 779 (1958)
49. Kinoshita, M., Miura, Y.: Makromol. Chem. *124*, 211 (1969)
50. Sakagami, M., Sohma, J.: J. Appl. Polym. Sci. *22*, 2915 (1978)
51. Bensasson, R., Marx, R.: J. Polym. Sci. *48*, 53 (1960)
52. Sato, T., Iwaki, T., Otsu, T.: J. Polym. Sci., Polym. Chem. Ed. *21*, 943 (1983)
53. Sato, T., Iwaki, T., Mori, S., Otsu, T.: J. Polym. Sci., Polym. Chem. Ed. *21*, 819 (1983)
54. Chiantore, O., Guaita, M., Trossarelli, L.: Makromol. Chem. *180*, 2019 (1979)
55. Sato, T., Yutani, Y., Otsu, T.: Polymer *24*, 1018 (1983)
56. Matsuda, M., Otsu, T., Imoto, M.: Kobunshi Ronbunshu *16*, 437 (1959)

Received November 5, 1984
Editors: G. Henrici-Olivé, S. Olivé

Polymerization of Butadiene and Butadiyne (Diacetylene) Derivatives in Layer Structures

Bernd Tieke*
Institut für Makromolekulare Chemie der Universität,
Hermann-Staudinger-Haus, Stefan-Meier-Str. 31,
D-7800 Freiburg i. Br., FRG

This article deals with the polymerization of 1,4-disubstituted butadiene and butadiyne (diacetylene) derivatives in layer structures. An outline is given on the polymerization properties in monomolecular layers at the air-water-interface, Langmuir-Blodgett multilayers, bi- and multilamellar aggregates in aqueous dispersion (vesicles, liposomes), perovskite-type layer structures, and further structures with a marked two-dimensional extension.
Recent results concerning photochemistry and -physics of the polymerization reactions, structure and morphology of the layer-type arrays, are compiled. Potential applications of the resulting polymers making use of their unusual physical properties are discussed.

* Present address: Ciba-Geigy AG, Forschungszentrum KA, CH-1701 Fribourg, Switzerland

1 Introduction

Polymerization in organized monomer media has found a widerspread interest in recent years. Reactions were studied in the bulk crystalline state [1-4], as well as in low-dimensional systems [5-21], as for example in multichannel inclusion compounds [5-7], lipid layer structures [8-15], monolayers adsorbed on solid substrates [16, 17], or in molecular crystals with a marked two-dimensional structure [18-20].

Especially suited for the formation of highly regular polymers are topochemical reactions, in which the structure of the product is controlled by an invariable mutual orientation of the monomers in the crystal lattice throughout the entire reaction process [22, 23]. In particular, diolefine [1] and diacetylene derivatives [2-4] are known to polymerize in lattice controlled reactions. In addition, distinct butadiene derivatives have recently been demonstrated to untergo an analoguous reaction [24, 25].

In recent years interest increasingly focussed on polymerization in layer structures. Among the monomers studied for polymerization were in particular butadiene and -diyne derivatives. Polymerization was investigated in lipid layer structures and layer structures of organic inorganic complex salts.

Highly regular polymers as well as ultrathin stable polymer films could be obtained. Research activities concentrated on the study of the photochemistry and -physics of the two polymerization reactions, structure and morphology of the layer-type arrays,

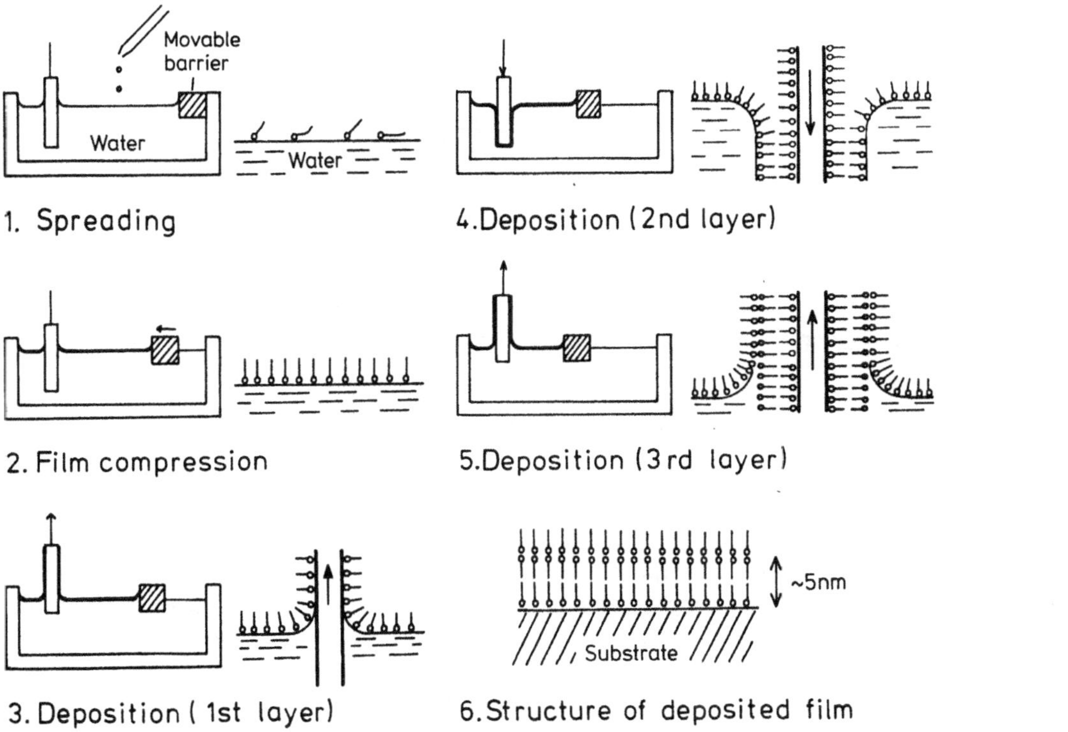

Fig. 1. Successive steps in monolayer and Langmuir-Blodgett multilayer formation [12-15, 34]

and potential applications of the polymerized structures either for technical purposes or as model systems in biochemistry and -physics [26-31].

Purpose of the present article is to summarize the present state of knowledge on the polymerization of butadiene and butadiyne derivatives in layer structures. Literature works are reviewed and combined with the author's own results. The article especially focusses on reactions in Langmuir-Blodgett multilayers and organic-inorganic complex salts, but studies on other layer-type arrays are also discussed.

1.1 Lipid Layer Structures

Polymerization of lipid layer structures was investigated

a) in monomolecular layers at the air-water interface.
   Polymerizable monolayers are obtained by spreading solutions of the amphiphiles at the air-water interface, followed by a film compression in order to orient the molecules at the interface (see also Fig. 1) [32].
b) in Langmuir-Blodgett (LB-)multilayers. LB-multilayers are obtained by successive transfer of oriented monolayers onto hydrophilic or hydrophobic substrates by the LB-technique (see also Fig. 1) [33,34],
c) in bi- and multilamellar layer aggregates of the amphiphiles in aqueous dispersion [35],
d) in spherical aggregates (vesicles, liposomes) obtained upon ultrasonication of amphiphile dispersions at temperatures above the phase transition point of the amphiphile (see also Fig. 2) [27-29],
e) in single crystals of the amphiphiles.

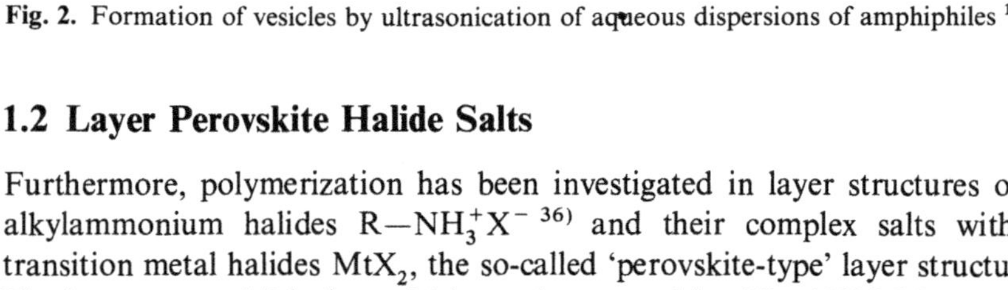

Fig. 2. Formation of vesicles by ultrasonication of aqueous dispersions of amphiphiles [146]

1.2 Layer Perovskite Halide Salts

Furthermore, polymerization has been investigated in layer structures of primary alkylammonium halides $R-NH_3^+X^-$ [36] and their complex salts with divalent transition metal halides MtX_2, the so-called 'perovskite-type' layer structures [37-39]. The latter ones exhibit the stoichiometric composition $[R-NH_3]_2MtX_4$, and consist of an alternating sequence of single layers of corner-sharing X_6 octahedra with the metal ions in their centers, and bilayers of $R-NH_3^+$ ions in a head-head-tail-tail arrangement, where R stands for an organic substituent, typically C_nH_{2n+1}. In case that R represents a long alkyl chain, the structure is best described as a lipid bilayer-type arrangement, sandwiched by the inorganic octahedra layers [38,39]. The structure is schematically represented in Fig. 3.

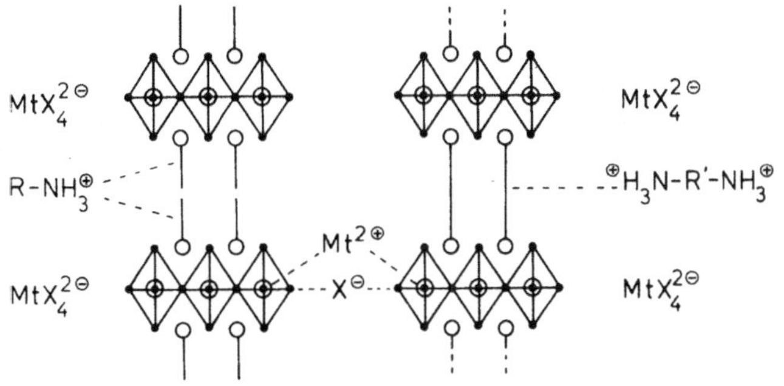

Fig. 3. Schematic representation of the structure of layer perovskites of mono- and diammonium salts [37)]

Figure 3 also indicates that the bilayers of the primary amines can be replaced by single layers of α,ω-diamines.

2 Butadiyne (Diacetylene) Derivatives

Butadiyne derivatives are known to undergo a solid-state 1,4-addition of the unsaturated units according to

$$n\ R-C{\equiv}C-C{\equiv}C-R' \xrightarrow{h\nu} \left[\begin{matrix} R \\ | \\ C \end{matrix}-C{\equiv}C-\begin{matrix} C \\ | \\ R' \end{matrix}\right]_n ,$$

if exposed to UV-light, X- or γ-irradiation, or to certain reactive gases [2–4, 40–42)]. Polymerization also occurs thermally, or upon pressure treatment of the crystals [43–45)]. Macroscopic and in some cases nearly defect-free polymer single crystals are obtained, in which the polyconjugated backbones are all aligned parallel to each other.

Details on mechanism, kinetics and structural aspects of the topochemical reaction have recently been reviewed [3, 46–48)]. Furthermore, the solution properties of the polymer have been investigated [49–54)], and the nature of the intermediate states of the photopolymerization has been characterized [55–58)].

2.1 Amphiphilic Compounds — General Characterization of the Polymerization Behaviour and the Properties of the Polymer

In recent years a variety of diyne surfactants has been synthesized and investigated in lipid layer structures. In Table 1 the compounds are compiled [59–122)]. Among them are single chain amphiphiles with one or two polar headgroups (Structures I and II), and double chain amphiphiles carrying the reactive unit either in the hydrophobic chain (Structure III) or in the headgroup (Structure IV). Moreover, the chain lengths m and n, and the nature of the polar headgroup X of the amphiphiles have widely been varied.

I $CH_3-(CH_2)_{m-1}-C{\equiv}C-C{\equiv}C-(CH_2)_n-X$

II $X-(CH_2)_n-C{\equiv}C-C{\equiv}C-(CH_2)_n-X$

III $CH_3-(CH_2)_{m-1}-C{\equiv}C-C{\equiv}C-(CH_2)_n\urcorner$
$\qquad X$
$CH_3-(CH_2)_{m-1}-C{\equiv}C-C{\equiv}C-(CH_2)_n\lrcorner$

IV $CH_3-(CH_2)_n-X-CH_2-C{\equiv}C-C{\equiv}C-CH_2-X-(CH_2)_n-CH_3$

While single chain amphiphiles of structure I were mainly investigated in the crystalline state, mono- and/or multilayers [59-109], compounds of structure II [110] and III[98, 104-106, 111-121] were preferentially studied in aqueous dispersions of bi- and multilamellar layer aggregates (vesicles, liposomes). Amphiphiles of structure IV have only been studied in the bulk crystalline state [122]. A survey of the physicochemical and physical studies on the individual compounds is given in Table 1.

In the following, some general observations on polymerization behaviour and properties of the polymers are briefly summarized.

2.1.1 Photoreactivity

As a rule, diyne surfactants are highly photoreactive upon exposure to UV-light, X- or γ-irradiation. A thermal polymerization is not observed. Initially the polymerization proceeds rapidly, and slows down at higher conversion to polymer [69, 72, 102, 103, 107]. A complete conversion to polymer is usually not obtained, unless the compounds are exposed to extremely high dosages of X-radiation, as recently reported [100].

Conversion versus dose curves have been determined for a number of diyne surfactants in the bulk crystalline state, after exposing the crystals to penetrating ^{60}Co-γ-irradiation. Conversion data were obtained gravimetrically by weighing the amount of insoluble polymer subsequently to solvent extraction of residual monomer from the crystals.

In Fig. 4a the polymerization properties of several diyne fatty acids and some of their salt derivatives are compared. From the plots it is obvious that the differences in chain lengths m and n and nature of the counterions strongly affect the photoreactivity. Since diyne polymerization is known to strictly proceed under lattice control, the differences in reactivity can be explained by small variations of the packing geometries of the monomers.

In Fig. 4b a strong effect of the polar headgroup on the photoreactivity is demonstrated. Though all the compounds contain the pentacosa-10,12-diynyl group, the alcohol *32* is only little photoreactive, whereas the corresponding ester of 4-pyridine carboxylic (isonicotinic) acid *36* polymerizes rapidly [102].

Furthermore, a salt formation of the pyridine headgroup with hydrochloric acid causes a complete polymerization of the compound [102].

Table 1. Diacetylene amphiphiles — properties and investigations
1.1 Single chain amphiphiles
1.1.1 $CH_3-(CH_2)_{m-1}-C\equiv C-C\equiv C-(CH_2)_n-COOH$

No.	C_{total}	m	n	m.p. (°C)	Ref.
1	29	16	8	73–75	59–62)
2	27	14	8		59, 60, 63–65)
3	26	13	8	68–69	66, 67)
4	25	0	20		68)
5		12	8	61	59, 60, 64)
6	23	10	8	57	59, 60, 64, 69)
7		16	2	93	70)
8	23	18	0		68, 71)
9	22	9	8	48	59, 60, 64)
10	21	8	8		59, 60, 64)
11	20	16	0	75	64, 70)
12		12	3	57	64)
13	19	12	2		70)
14		14	0	63	72)
15	17	10	2		73)
16		12	0	58	74, 75)

Investigations on *1–16*:	Ref.
In the crystalline state	
UV/VIS- and resonant Raman spectra of *6*	69)
X-ray diffraction of *6*	69)
Electron diffraction of *15, 16*	73, 74)
Optical and scanning electron microscopy of *6*	69)
In monolayers	
Π-A-isotherms	59, 60, 64, 70)
Surface properties of *3* (contact angle, stream potential, XPS)	66)
UV/VIS-spectra of *1*	61)
In LB-multilayers	
UV/VIS-spectra of *1, 7*	62, 68)
IR-spectra of *15, 16*	73, 74)
Fluorescence of *2*	65)
Electron diffraction of *15, 16*	73, 74)
IR-dichroism and ESR of *8*	76)
Polarizing microscopy of *1*	77)
Dark- and photoconductivity of *1*	78)

1.1.2 salts of fatty acids
1.1.2.1 $(CH_3-(CH_2)_{m-1}-C\equiv C-C\equiv C-(CH_2)_n-CO_2^-)_2Cd^{2+}$

No.	m	n	Ref.
1a	16	8	62)
2a	14	8	64, 79)
3a	13	8	80)

Table 1. (continued)

No.	m	n	Ref.
5a	12	8	64, 70, 79, 81, 82)
6a	10	8	64, 72, 79, 81 – 83)
7a	16	2	70)
8a	18	0	68, 71)
9a	9	8	79, 81, 82)
11a	16	0	64, 70)

Investigations on *1a–11a*	Ref.
In monolayers	
Π-A-isotherms	64, 70, 80 – 82)
In LB-multilayers	
UV/VIS-spectra	64, 70, 80 – 82)
Resonant Raman spectra of *6a*	84)
X-ray and electron diffration	70, 79, 81, 82)
Electron microscopy	62, 64, 80)
Polarizing microscopy	85)
Quantum yield of photopolymerization of *2a*, *5a*, and *6a*	86, 87)
Dark- and photoconductivity of *5a* and *6a*	88, 89)
Photoreactivity and structure of mixed multilayers with saturated amphiphiles	90)
Dye sensitized photopolymerization of *5a*	91 – 93)
Non-linear optical properties of *1a*	71, 94)

1.1.2.2 $CH_3-(CH_2)_{m-1}-C\equiv C-C\equiv C-(CH_2)_n-COO^-Li^+$

No.	m	n	Ref.
1b	16	8	61, 95)
5b	12	8	
6b	10	8	96, 97)

Investigations on *1b–6b*	Ref.
In the crystalline state	
Studies on reversible thermochromism of *5b* and *6b* by reflectance spectra and DSC	96, 97)
Structure studies of "blue" and "red" polymer phase of *6b*	96)
In monolayers	
Characterization of photoproducts of *1b* by UV/VIS Spectra	61)
In LB-multilayers	
Structure and morphology of *1b*	77, 95)
Dark- and photoconductivity of *1b*	78)

Table 1. (continued)

1.1.2.3 Further salts

No.	Fatty acid	Cation	Notes	Ref.
8c	*8*	NH_4^+		68, 71)
8d		Ag^+		
8e		Na^+		
8f		Cu^{2+}	Polymerizes in pyridine vapor after decarboxylation	
8g		Mn^{2+}		
8h		Hg^{2+}	decarboxylates rapidly	

Investigations on *8c–8h*	Ref.
In the crystalline state and in LB-multilayers Characterization of photoproducts by UV/VIS-spectra and IR-spectra	68, 71)

1.1.3 $CH_3-(CH_2)_{m-1}-C{\equiv}C-C{\equiv}C-(CH_2)_n-\overset{O}{\overset{\|}{C}}-X$

No.	m	X	M.p. (°C)	Ref.
17	13	$-NH_2$	101	98)
18	12	$-NH_2$	98	99)
19	13	$-NH(CH_3)$	79–81	98)
20	13	$-N(CH_3)_2$	49–50	98)
		$-NH(CH_2)_nOH$		
21	13	n = 2	91–93	98)
22	13	n = 3	88–89	98)
23	13	n = 4	94.5–95.5	98)
24	13	n = 5	88–89	98)
25	13	n = 6	95–97	98)
26	13	$-N(CH_2CH_2OH)_2$	55	98)
27	12	$-NH(\underset{CH_3}{\underset{\vert}{C}H}-C_6H_5)$		100)
28	13	$-NH-(CH_2)_6-O\overset{O^\ominus}{\overset{\vert}{\underset{\underset{O}{\|}}{P}}}O-(CH_2)_2-\overset{\oplus}{N}(CH_3)_3$	99–100	98)
29	13	$-O(CH_2)_2-\overset{\oplus}{N}(CH_3)_3\ Br^\ominus$	64	98)
30	13	$-OCH_2-CHOH-CH_2OH$	130–133	98)

Table 1. (continued)

Investigations on *17–31*	Ref.
In monolayers:	
Π-A-isotherms	[98]
UV/VIS-absorption spectra of polymer	[98]
In LB-multilayers:	
Application of *27* as X-ray beam resist	[100, 101]

1.1.4 $CH_3-(CH_2)_{m-1}-C{\equiv}C-C{\equiv}C-(CH_2)_n-OH$

No.	m	n	m.p. (°C)	Ref.
31	13	9	59	[67]
32	12	9	56	[85]
33	10	9	48	[62, 63]
34	10	3		[73]
35	12	1	61	[74]

Investigations on *31–35*	Ref.
In the crystalline state:	
UV/VIS-absorption spectra of polymer *35*	[74]
Infrared-spectra of *34* and *35*	[73, 74]
In LB-multilayers:	
UV/VIS-absorption spectra of mixtures with fatty acids	[63, 64]
Morphology of mixtures with fatty acids	[85]
Electron diffraction of *34* and *35*	[73, 74]

1.1.5 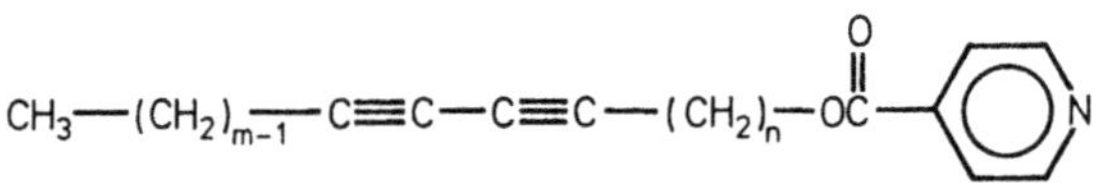

No.	m	n	m.p. (°C)	Ref.
36	12	9	47	[102]
36a	Hydrochloride salt of *36*		71–75	[102]
37	16	3	62	[102]

Table 1. (continued)

Investigations on the compounds:	Ref.
In the crystalline state:	
UV/VIS-absorption spectra of polymer	102)
In monolayers:	
Π-A-isotherms	102)
In LB-multilayers:	
UV/VIS-absorption spectra of polymer, morphology	102)
In solution:	
UV/VIS-absorption spectra and ^{13}C-NMR spectra of polymer	103)

1.1.6 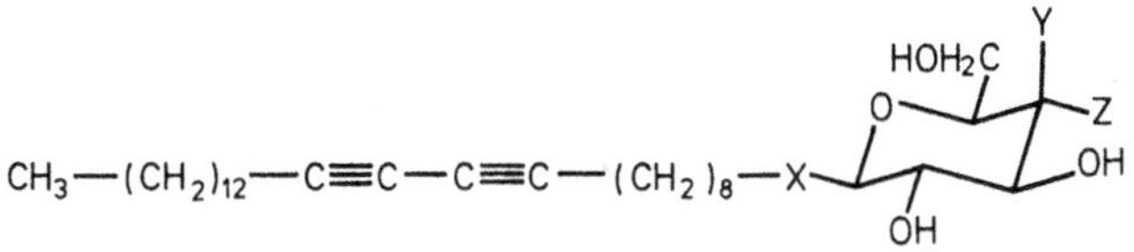

No.	X	Y	Z	m.p. (°C)	Ref.
38	$-CH_2O-$	—OH	—H	73–75	104)
39		—H	—OH	114	104)

Investigations on *38* and *39*:	Ref.
In monolayers:	
Π-A-isotherms of monomer and polymer	104)
In vesicles:	
Interaction of *38* with lectines	104)

1.1.7 $CH_3-(CH_2)_{12}-C{\equiv}C-C{\equiv}C-(CH_2)_9-O\overset{\overset{\displaystyle O}{\|}}{\underset{\underset{\displaystyle R_1}{|}}{P}}O-R_2$

No.	R_1	R_2	m.p. (°C)	Ref.
40	—OH	—H	79–82	67, 105, 106,
41	—OH	$-CH_2-CHOH-CH_2OH$	130–135	98)
	$-\overline{\underline{O}}|^{(-)}$	$-(CH_2)_2-\overset{(+)}{N}X_1X_2X_3$		

Table 1. (continued)

42	$X_1 = X_2 = X_3 = H$	219–225	98, 105, 106)
43	$X_1 = X_2 = H, X_3 = CH_3$	167	98, 106)
44	$X_1 = H, X_2 = X_3 = CH_3$	106	98, 106)
45	$X_1 = X_2 = X_3 = CH_3$	262–272	98, 105, 106)

Investigations on the compounds *40–45*:	Ref.
In monolayers:	
Π-A-isotherms as a function of pH and temperature	98, 105, 106)
In vesicles:	
UV/VIS-absorption spectra of polymer	106)

1.1.8 $CH_3-(CH_2)_{11}-C{\equiv}C-C{\equiv}C-(CH_2)_9-X^+Y^-$

No.	X^+	Y^-	m.p. (°C)	Ref.
46	$-NH_3^{(+)}$	$Cl^{(-)}$	94–95	99)
47	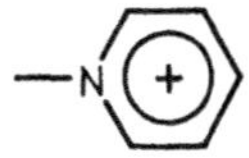	Br$^{\ominus}$	45	107)
48	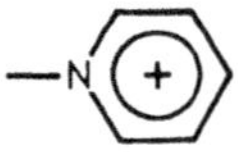	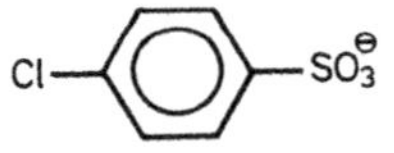	102	107)
49	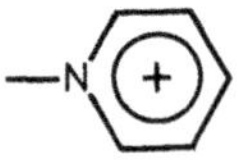	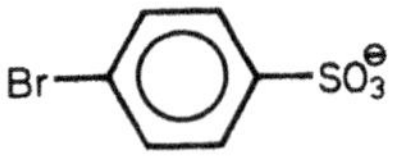	115	107)

[a] If crystallized from chloroform

Investigations on compounds *46–49*:	Ref.
In the crystalline state:	
UV/VIS-absorption spectra of polymer	99, 107)
X-ray diffraction	107)
In monolayers:	
Π-A-isotherms	99, 107)
In LB-multilayers:	
UV/VIS-absorption spectra of polymer	107)
In vesicles:	
Electron microscopy and diffraction of *49*	107)

Table 1. (continued)

1.1.9 $CH_3-(CH_2)_{m-1}-C\equiv C-C\equiv C-(CH_2)_n-\overset{\overset{\displaystyle O}{\|}}{\underset{\underset{\displaystyle OH}{|}}{P}}-OH$

No.	m	n	Ref.
50	11	8	[108]

1.1.10 $R-\overset{\overset{\displaystyle O}{\|}}{C}O-(CH_2)_9-C\equiv C-C\equiv C-(CH_2)_9-OH$

No.	R	m.p. (°C)	Ref.
51	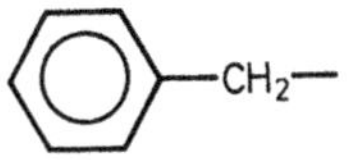	43	[109]
52	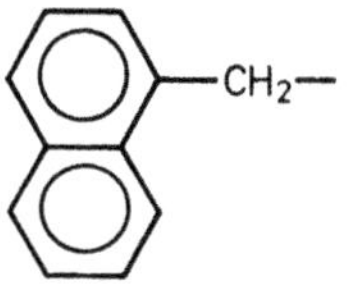	39	[109]
53	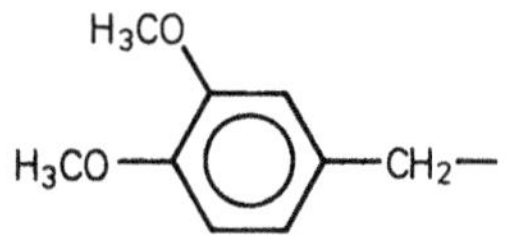	40, 44[a]	[109]

[a] two modifications

1.2 Two-headed single chain amphiphiles
$X-(CH_2)_m-C\equiv C-C\equiv C-(CH_2)_m-X$

No.	m	X	m.p. (°C)	Ref.
54	8	$-COOH$	119	[110]
55	9	$-OH$	82	[110]
56	9	$-OPO_3H_2$	91–93	[110]

1.3 Double chain amphiphiles

1.3.1 $CH_3-(CH_2)_{m-1}-C\equiv C-C\equiv C-(CH_2)_8-\overset{\overset{\displaystyle O}{\|}}{C}-X-(CH_2)_2$

$\qquad\qquad\qquad\qquad\qquad\qquad\qquad\qquad\qquad\qquad\qquad\qquad\quad Y$

$CH_3-(CH_2)_{m-1}-C\equiv C-C\equiv C-(CH_2)_8-\underset{\underset{\displaystyle O}{\|}}{C}-X-(CH_2)_2$

No.	m	X	Y	m.p. (°C)	Ref.
57	13	O	O		[106]
58	13	O	$>N-CH_3$	48	[106]

Table 1. (continued)

No.	m	X	Y	m.p. (°C)	Ref.
59	13	O	$>\overset{(+)}{N}(CH_3)_2Br^{(-)}$	94	67, 106, 111)
60	10	O	$>\overset{(+)}{N}(CH_3)_2X^{(-)}$		112)
61	13	O	$>\overset{(+)}{N}H-(CH_2)_2-SO_3^{(-)}$	109	113)
62	13	O	$-O(CH_2)_2-O(CH_2)_2-O-$	49	67, 111)
63	13	NH	O		106)

Investigations on *57–63*	Ref.
In monolayers:	
Π-A-isotherms	60, 67, 106)
Π-A-isotherms of mixed monolayers of *59* and DSPC, and cholesterol	111)
UV/VIS-absorption spectra of polymer	60, 67, 106)
Enzymatic hydrolysis of mixed monolayers of *59* and DLPC	111)
In vesicles:	
UV/VIS-absorption spectra of polymer	67, 106, 117, 114)
Electron microscopy	67, 114)
Differential scanning calorimetry of pure *59* and *63*, and mixtures of both with DOPC and DSPC	111)
ATP-synthetase activity in polymerized vesicles of *61*	114)

1.3.2 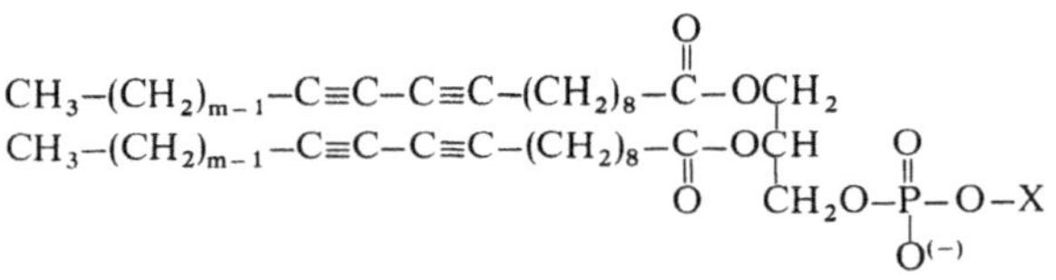

No.	m	X	m.p. (°C)	Ref.
64	13	H	37–38	98, 106)
65	13	$-(CH_2)_2-\overset{(+)}{N}(CH_3)_3$	215	98, 106, 115)
66	12	$-(CH_2)_2-\overset{(+)}{N}(CH_3)_3$	48[a]	116, 117)
67	10	$-(CH_2)_2-\overset{(+)}{N}(CH_3)_3$	38.5[a]	116, 118)
68	13	$-(CH_2)_2-\overset{(+)}{N}H_3$	180	98, 106, 115)

[a] Transition temperature

Investigations on *64–68*	Ref.
In monolayers:	
Π-A-isotherms	98, 115–117)
Π-A-isotherms of mixed monolayers with cholesterol and distearoylcephalin	111)

Table 1. (continued)

Investigations on *64–68*	Ref.
In LB-multilayers:	
UV/VIS-absorption spectra of polymer *66*	117)
In vesicles:	
UV/VIS-absorption spectra of polymer in pure and mixed vesicles	106, 116, 119)
Electron microscopy	112, 118)
Stability and leakage behaviour of *65*	106)
In aqueous dispersions:	
UV/VIS-absorption spectra of monomer and polymer	116)
Circular dichroism of polymer	119)
Differential scanning calorimetry	118)

1.3.3

$$(CH_3)\text{–}(CH_2)_{16}\text{–}\overset{O}{\overset{\|}{C}}\text{–}OCH_2$$
$$CH_3\text{–}(CH_2)_{m-1}\text{–}C{\equiv}C\text{–}C{\equiv}C\text{–}(CH_2)_n\text{–}\underset{O}{\underset{\|}{C}}\text{–}OCH \quad \overset{O}{\|}$$
$$CH_2O\text{–}P(OH)_2$$

No.	m	n	m.p. (°C)	Ref.
69	12	8	37–38	98, 115)

1.3.4

$$\left.\begin{array}{l} CH_3\text{–}(CH_2)_{m-1}\text{–}C{\equiv}C\text{–}C{\equiv}C\text{–}(CH_2)_9 \\ CH_3\text{–}(CH_2)_{m-1}\text{–}C{\equiv}C\text{–}C{\equiv}C\text{–}(CH_2)_9 \end{array}\right] X$$

No.	m	X	m.p. (°C)	Ref.
70	12	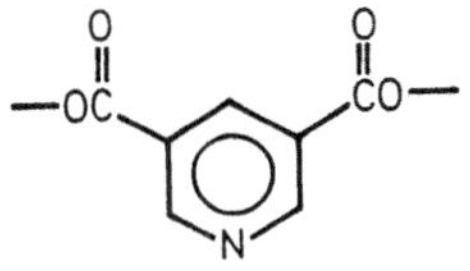	59	103)
71	12	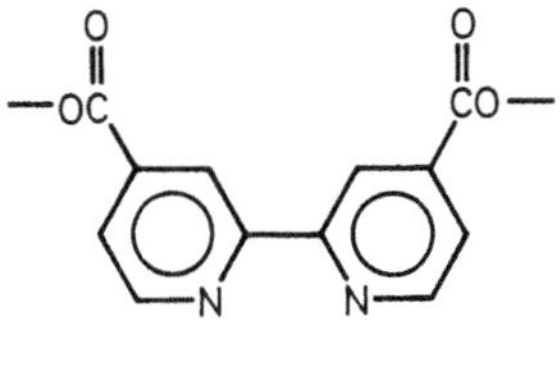	64	103)
72	12	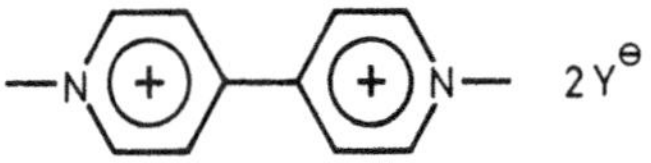 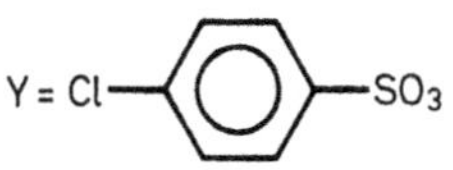	250[a]	107)

Table 1. (continued)

No.	m	X	m.p. (°C)	Ref.
73	12	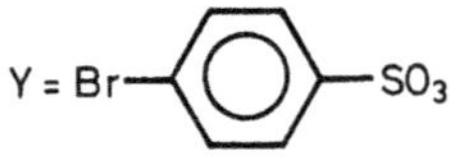	270[a]	107)
74	13	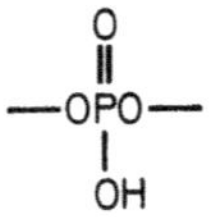		120)
75	10	— ,, —		112, 121)

[a] Melting under decomposition

Investigations on *70–75*	Ref.
In the crystalline state:	
UV/VIS-absorption spectra of polymer *71–73*	103, 107)
X-ray diffraction of *72, 73*	107)
Differential scanning calorimetry of *72, 73*	107)
In monolayers:	
Π-A-isotherms	103, 107)

1.4 $CH_3-(CH_2)_n-\overset{O}{\overset{\|}{C}}OCH_2-C{\equiv}C-C{\equiv}C-CH_2O\overset{O}{\overset{\|}{C}}-(CH_2)_n-CH_3$

No.	n	m.p.	Notes	Ref.
76	4	28–30[a]		122)
77	6	28–32[a]		122)
78	8	46	polymer golden orange, turning red upon solvent treatment of the crystals	122)
79	10	55–56		122)
80	12	59–60		122)
81	14	67–68		122)
82	16	72–74		122)

[a] some sublimation/decomposition apparent

Investigations on compounds *76–82:*	Ref.
Reflectance spectra of polymer *78*	122)
Resonant Raman spectra of polymer *78* and *79*	122)

On the other hand, the double chain bipyridine derivative *71* exhibits a lower photoreactivity than the corresponding single chain analogue *36*. Similar effects were also reported from other double chain amphiphiles and their corresponding single chain analogues [98, 107]. Possibly in the double chain amphiphiles the reaction of one of the diyne units slightly alterates the packing geometry of the other unit and thus renders its polymerization more difficult [102,107].

Especially O'Brien and coworkers [98,121] referred to the observation that polymerization of double chain amphiphiles *57–63* with equivalent alkyl chains proceeds 1000 times as efficient as that of phospholipids *64–68* with non equivalent chains in hydrated bilayers. The authors explained this behaviour by the formation of different types of polymer, as schematically shown in Fig. 5. Because of their different binding to the glycerol group, the two alkyl chains of phospholipids *64–68* do not extend equally into the lipid bilayer. Thus the diyne units are not in the proper stereochemical alignment to allow intramolecular reaction with formation of the αβ-type polymer.

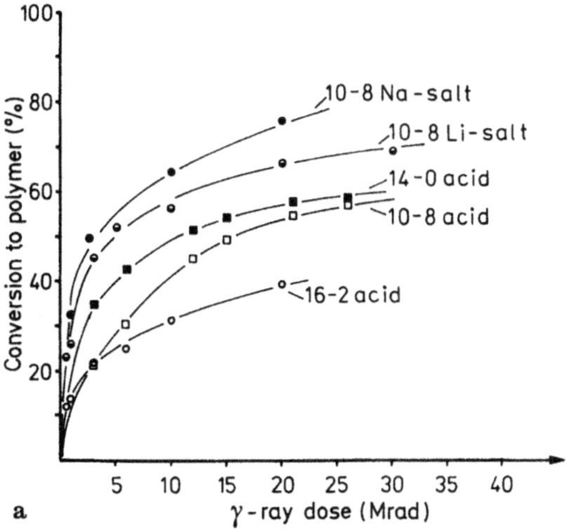

a

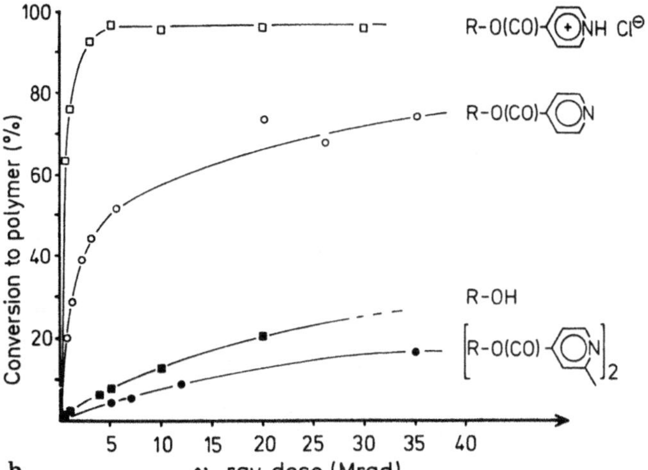

b

Fig. 4a and b. Plots of fractional polymer conversion versus ^{60}Co-γ-ray dose. **a.** C_mC_n-acids and alkali salts; **b.** Compounds with R = $-(CH_2)_9-C{\equiv}C-C{\equiv}C-(CH_2)_{11}-CH_3$ and various polar headgroups

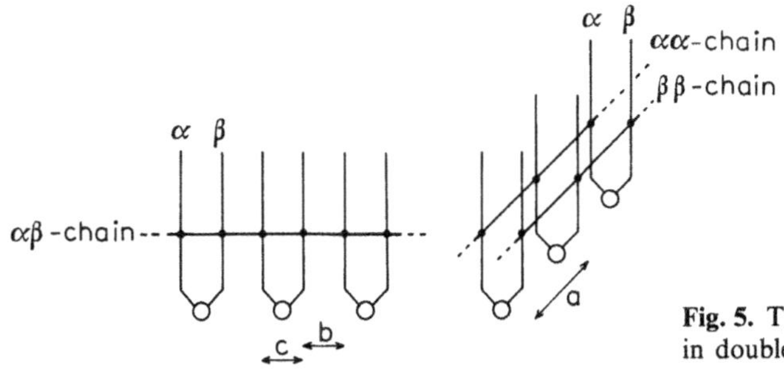

Fig. 5. Types of polymer chain formation in double-chain diacetylene amphiphiles

Instead only the ladder polymer of the αα-/ββ-type can be formed. On the other hand, the authors believe that the higher reactivity of compounds *57–63* arises from the ability to undergo intramolecular polymerization in addition.

2.1.2 Morphology

Diacetylene derivatives are known to polymerize in a lattice controlled reaction. Polymerization proceeds via formation of a solid solution of the polymer in the monomer lattice. For a number of compounds the formation of polymer single crystals has been demonstrated [48].

Structure studies of diyne surfactants are complicated by the fact that single crystals — whenever obtained — are extremely thin and fragile and rapidly polymerize in the X-ray beam. Thus a full structure analysis of any of the compounds either as a monomer or polymer is still lacking up to the present [69, 72]. All informations on structure and structure changes during polymerization are based on morphological studies by polarizing and electron microscopy [69, 73], as well as structure studies by electron diffraction [73, 74, 79, 95] and infrared spectroscopy [75].

These studies indicate that the diyne surfactants only initially polymerize with formation of a solid solution of the polymer in the monomer lattice. At higher conversion to polymer a phase transition takes place. In the new phase the residual monomer is less reactive, which explains the conversion versus dose behaviour shown in Fig. 4.

The phase transition is due to a rearrangement of the paraffin chains. It may be accompanied by a fibrillization and crack formation parallel to the polymer chain direction, as it occurs in crystals of *6* upon prolonged exposure to γ-irradiation (Fig. 6) [69]. However, the extent of morphological changes strongly depends on the concentration of defects and dislocations in the individual crystals [69]. Moreover, morphological changes strongly depend on the chemical structure of the individual amphiphiles [73]. While single crystals and multilayers of alcohol *34* remain unchanged in their appearance, crystals and multilayers of acid *15* essentially show the same features as they were observed for the crystals of acid *6* already. Morphological changes in multilayer samples are described in more detail in chapter 2.3.4.2.

It has been assumed [69, 73] that the phase transitions originate from a mismatch of monomer and polymer structures. Polymerization of *15*, for example, is accompanied by large volume changes of the crystal lattice [73]. Thus, polymerization is believed to induce the formation of strain or stress in the lattice that is relieved in the subsequent phase transition.

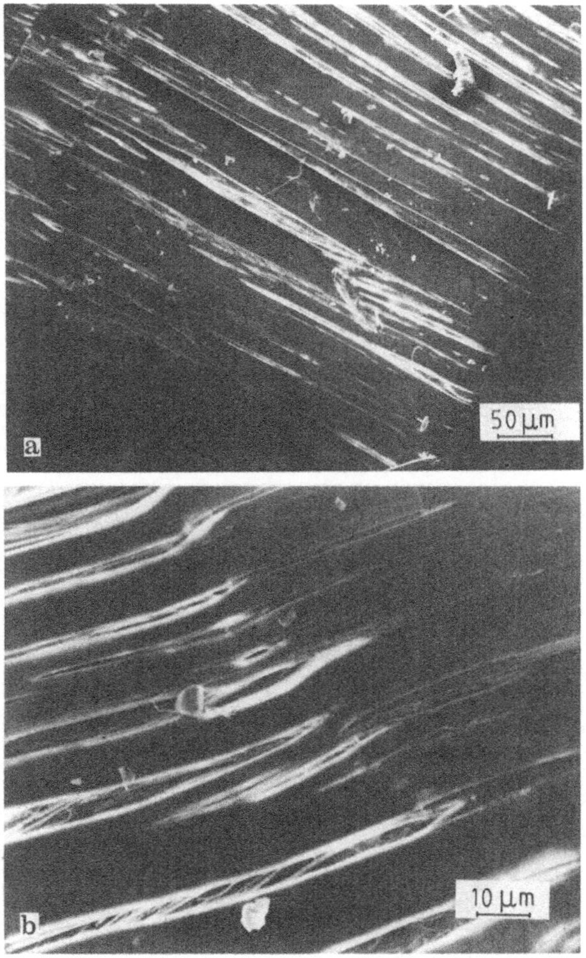

Fig. 6a and b. Scanning electron micrographs of crystals of *6* subjected to a γ-ray dose of 20 Mrad. Deformations and cracks are due to the disruptive phase transition. The major cleavage is along the a-direction, most likely being consistent with the chain direction [69)]

2.1.3 Spectroscopic Characterization

Due to the conjugated structure the polymer chains exhibit a strong absorption in the visible. Polymerization of diyne surfactants is therefore indicated by a rapid greenish-blue [69, 72)], purple [64, 102)], or yellow [107)] coloration of the crystals. At higher conversion to polymer, crystals usually turn deep purple or black and appear golden-metallic in reflection.

The polymerization of diyne amphiphiles can therefore easily be characterized by UV/VIS-absorption spectroscopy. In Fig. 7 absorption spectra of mono- and multilayers are shown which were monitored after different sample treatments.

In Figs. 7a and b the effect of UV-exposure is demonstrated. Already after short exposure times a strong absorption builds up in the visible indicating increasing conversion to polymer. Upon prolonged exposure times the bandshapes change and the absorption maxima shift towards shorter wavelengths. These shifts amount to about 100 nm for compounds shown in Fig. 7a and b.

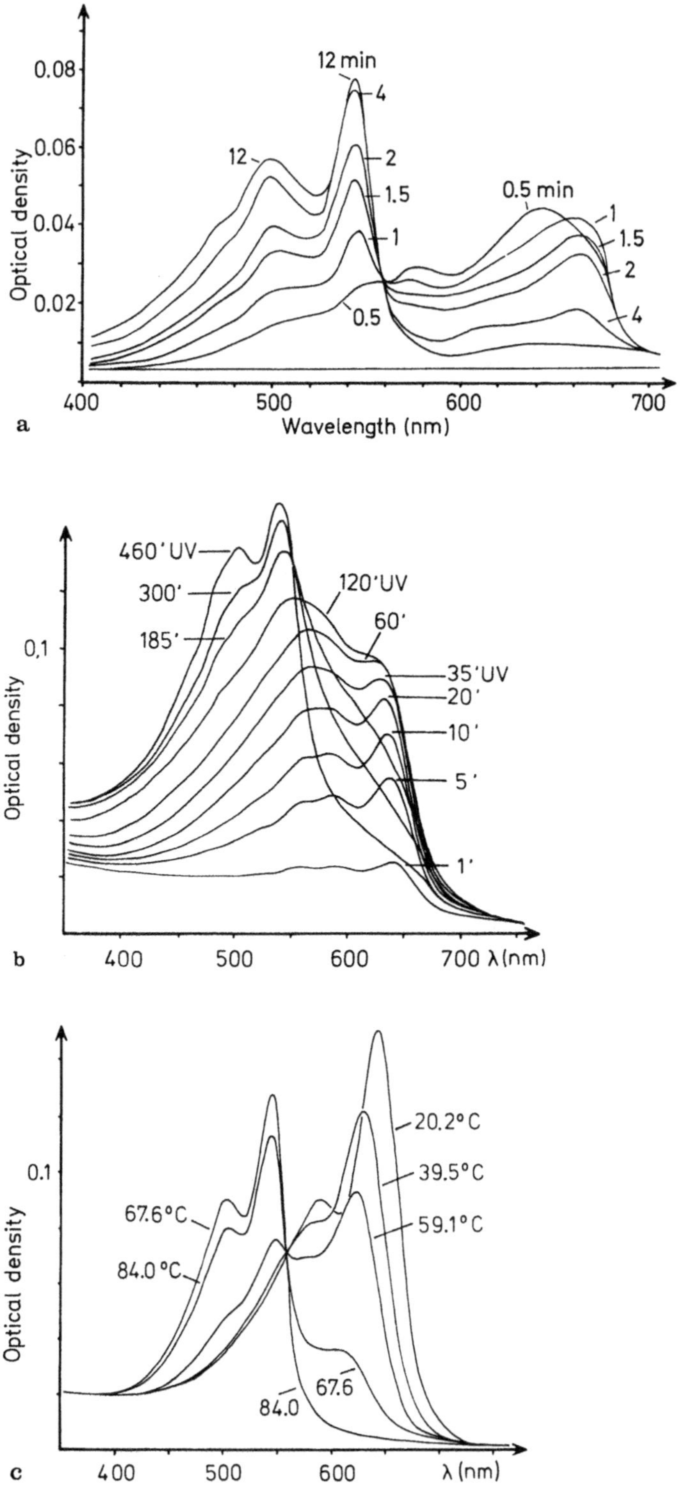
0.08
0.06
0.04
0.02
Optical density
12 min
4
12
2
1.5
1
0.5
0.5 min
1
1.5
2
4
400
500
600
700
Wavelength (nm)
a
0,1
Optical density
460' UV
300'
185'
120'UV
60'
35'UV
20'
10'
5'
1'
400
500
600
700 λ(nm)
b
0.1
Optical density
20.2°C
39.5°C
59.1°C
67.6°C
84.0 °C
67.6
84.0
400
500
600
λ (nm)
c

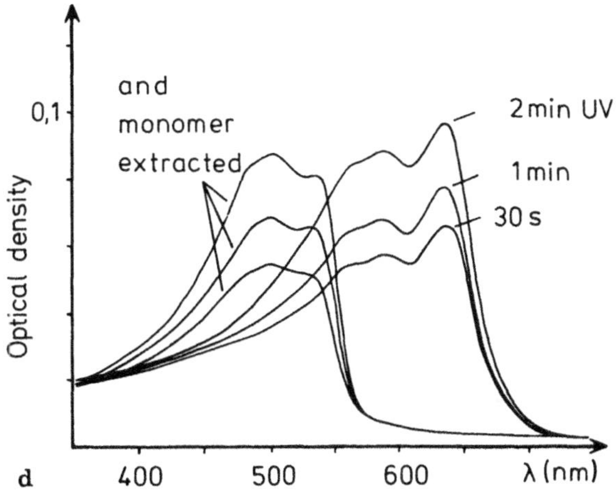

d

Fig. 7a–d. VIS-absorption spectra of mono- and multilayers of diacetylene futty acids and their cadmium salts. Change of the polymer absorption

a, b. As a function of the irradiation time. Samples: (**a**) Monolayer of *59* at the air-water interface [60], and (**b**) multilayer of *2a* on a quartz support, thickness 32 layers;

c. As a function of annealing at various temperatures. Sample: Multilayer of *5b*, partially polymerized;

d. Upon solvent treatment with ethanol. Samples: Multilayers of *5a*, partially polymerized by UV-irradiation for different time periods, thickness 28 layers [64]

Similar colour changes occur, if partially polymerized samples are heated (Fig. 7c) or treated with a solvent able to dissolve the monomer and unable to dissolve the polymer (Fig. 7d) [64]. Colour changes upon heating of the samples are reversible, unless the annealing temperature exceeds the melting point of the residual monomer. Reversible thermochromism is especially observed for high melting salts as for example the lithium salt *6b* [96, 97].

Some of the colour changes are accompanied by isosbestic points in the optical spectra (Figs. 7a and c). Isosbestic points indicate a single-step transition process. In case of the diyne amphiphiles they imply a transition of the electronic structure of the backbone. Each polymer chain in its "blue" form is converted into another one in the "red" form without occurance of any intermediate state [79].

Resonance Raman spectroscopy represents another method which is excellently suited to characterize the polymerization of diyne surfactants. Raman spectra were recorded as a function of the UV-irradiation time at constant laser frequency (Fig. 8a) and with changing laser frequency for a partially polymerized sample (Fig. 8b) [84].

The spectrum of a multilayer irradiated for 30 minutes shows two main bands ν_1 and ν_2 at 1448 and 2099 cm^{-1}, which can be ascribed to the C=C- and C≡C-stretching modes of the polymer backbone (Fig. 8a). With increasing UV-irradiation time these bands change continuously to 1455 and 2092 cm^{-1}. Simultaneously a new set of backbone vibrations ν_1' and ν_2' at 1486 and 2110 cm^{-1} appears, which can be ascribed to the formation of a new polymer species.

Changing the laser frequency to wavelengths above 17100 cm^{-1} even allows to detect a third set of backbone vibrations ν_1'' and ν_2'' at 1514 and 2120 cm^{-1}. In Fig. 8c

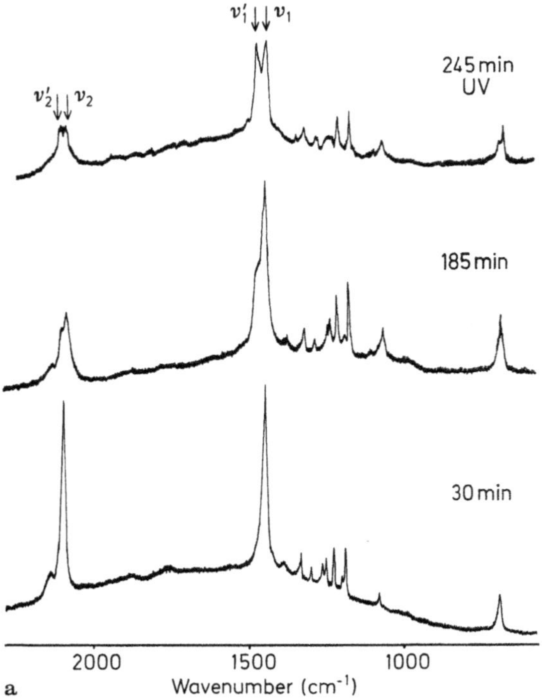

a

Fig. 8a. Raman spectra of a multilayer of *6a* (16 layers). Spectra recorded with a fixed laser frequency of 16480 cm^{-1} after different UV-irradiation times (6W low pressure mercury lamp). Increasing UV-irradiation creates different polymer species with distinct C=C- and C≡C-stretching modes [84]

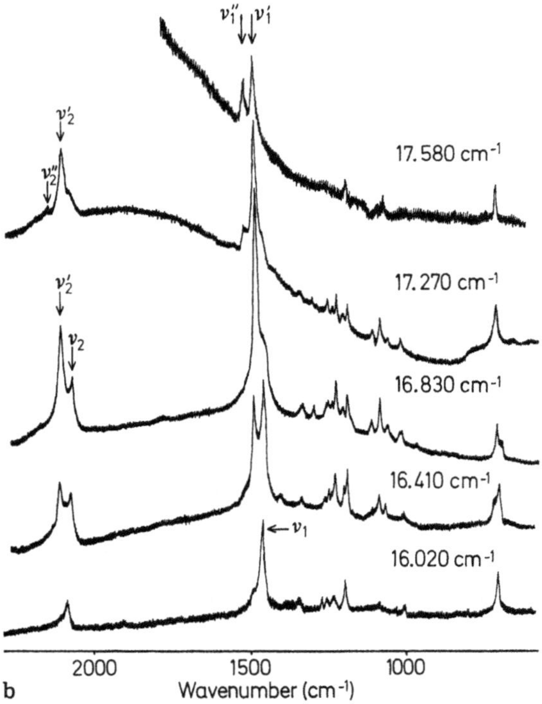

b

Fig. 8b. Raman spectra of a multilayer of *6a* (16 layers). Spectra recorded with different laser frequencies after 245 min. UV-irradiation. Three sets of C=C- and C≡C-stretching modes occur, depending in their intensity on the excitation wavelength of the laser line [84]

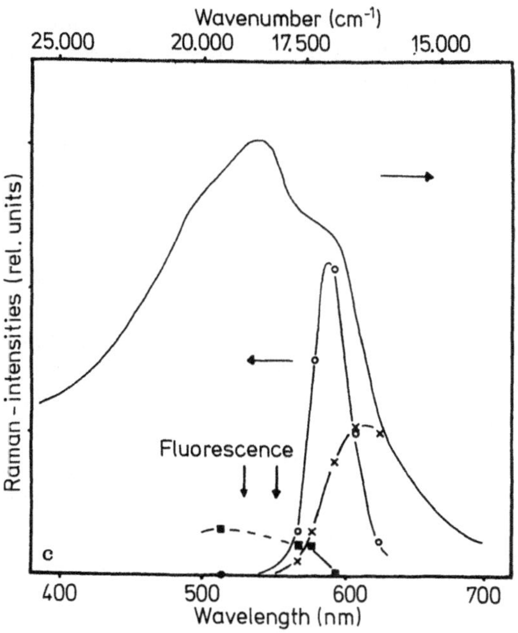

Fig. 8c. Raman scattering excitation profiles, deduced from the intensities of the C=C-stretching modes ν_1 (×), ν_1' (○) ν_1'' (■), for different laser frequencies, and the absorption spectrum of a sixteen layer *6a* multilayer after 245 min. UV-irradiation [84]

the intensities of the C=C stretching modes derived from the spectra of Fig. 8b are plotted as a function of the excitation wavelength. In this way, excitation profiles of the three different polymer species are obtained, which can be correlated with the optical spectrum of the sample also shown in the same figure.

The plot of Fig. 8c indicates that ν_1 corresponds to a "blue" polymer species giving rise to the long wavelength tail of the optical spectrum. ν_1' correlates with a "purple" polymer causing the shoulder at 590 nm, and ν_1'' can be acribed to a "red" polymer with an absorption maximum at 540 nm.

Table 2. Raman frequencies and absorption maxima of amphiphilic polydiacetylenes

No.	$\nu_{C=C}$ (cm^{-1})	$\nu_{C\equiv C}$(cm^{-1})	ν (cm^{-1}) of Laser excitation	λ_{max} (nm)	Notes	Ref.
6	1452	2100	16103	680	γ-ray dose: <0.1 Mrad	[69]
	1515	2120	16103	540	γ-ray dose: 26 Mrad	
6a	1448	2099	16480	640	t_{UV}: 30 min[a]	[84]
	1486	2110	16830	580[b]	t_{UV}: 245 min	
	1514	2120	18520	540	t_{UV}: 245 min	
6b	1447	2093	16500	640	24 °C, at low conversion	[96]
	1480	2084	19420	580	90 °C, at low conversion	

[a] 6 W low pressure mercury lamp; [b] shoulder

Due to a strong fluorescence in that region the excitation profile of ν_1'' cannot be determined accurately. The fluorescence is most likely due to a disordered structure of the "red" polymer. Disorder formation may originate from the previous transitions of the "blue" and "purple" modifications into the "red" one.

C=C- and C≡C-stretching frequencies determined for some diyne amphiphiles as well as the corresponding wavelengths of the absorption maxima are compiled in Table 2.

The actual origin of the backbone colour changes is still unclear. Chain end effects were considered [63,64], as well as bond isomerization [123], conformational transitions [124,125], and aggregation phenomena [119] without reaching a definite conclusion. Studies on surfactant diacetylenes at least indicated that the colour changes are due to changes in the electronic structure of the backbone, which occur simultaneously with the rearrangement of the paraffinic side groups. Thus, the backbone colour change actually is part of a phase transition of the whole crystal lattice [70].

Further speculating on its origin, it has been assumed [69] that the polymer chains become strained or stressed during the phase transition. Such a deformation could affect the electronic structure of the backbone and cause the colour changes. Though not proven yet, these considerations are at least consistent with the increase in exciton energy observed for PTS¹-fibers in extension [126].

Despite of the unsolved problem of the colour changes the absorption behaviour of the polymer can be used to deduce information on the state of order in lipid layer structures and on phase transitions in those structures. This has recently been demonstrated studying the morphology of mixed multilayers [90].

2.1.4 Solubility and Molecular Weight

Determination of the molecular weight of polymers is complicated by the fact that most of the amphiphilic polymers are either completely insoluble, or only partially soluble in common organic solvents. The following amphiphiles have been reported to form polymers that are partly soluble: Acid *6* in hot 1,4-dioxane or cyclohexanone [96], acid *16* in dimethylformamide [74], alcohol *35* in decaline [74], the pyridinium salts *48* and *49* in a 1:1 mixture of chloroform and methanol [107], and the phospholipids *66* and *67* in chloroform [119].

Complete solubility of the polymer independently of the γ-ray dose received by the crystals is only observed for compound *36* [102,103]. ^{13}C—NMR spectra of the polymer correspond to those of other soluble polydiacetylenes [127,128]. Signals of the unsaturated carbon atoms of the polymer backbone appear at $\delta = 99.5$ and 129.3 ppm, supporting an en-ine structure, and those of the monomeric butadiyne unit at $\delta = 65.3$ and 77.3 ppm. Viscosity measurements indicate an η-value of 570 ml g^{-1}, which is equivalent to a weight average molecular weight of about 200,000 [103], in good agreement to values reported for other polydiacetylenes in solution [50-53].

¹ poly-(1,6-di-p-toluene sulfonoyloxy-2,4-hexadiyne)

2.2 Monomolecular Layers at the Air-Water Interface

2.2.1 Monolayer Stability

The lattice-controlled diyne polymerization can occur in lipid bilayers only, if the amphiphiles are present in the solid-condensed state. Several studies dealt with the characterization of the monolayer properties of diyne surfactants. Fatty acids, for example, were investigated concerning influences of a chain length variation and headgroup ionization on the film stability (Fig. 9a) [59, 60, 62–64, 70]. For a number of compounds with the hexacosa-10,12-diynyl group the effect of a headgroup variation on the Π-A-isotherms was studied (Fig. 9b) [98, 106].

In Fig. 9c the collapse pressures of monolayers of diyne fatty acids in the solid condensed state are plotted versus their chain lengths. A prolongation of the chain length m of the hydrophobic substituent — $(CH_2)_{m-1}$—CH_3 causes a linear increase of the film stability. For acids with a short length n of the "hydrophilic" chain — $(CH_2)_n$—COOH this increase is steeper than for those with long C_n chains. Chain prolongation by an increase of n at a constant m, however, changes the film stability only slightly [70].

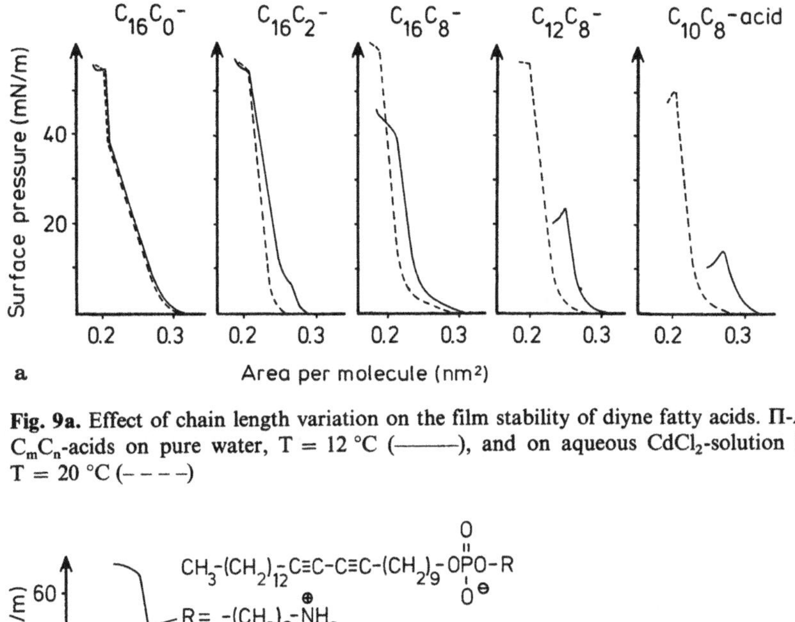

Fig. 9a. Effect of chain length variation on the film stability of diyne fatty acids. Π-A-isotherms of C_mC_n-acids on pure water, T = 12 °C (———), and on aqueous $CdCl_2$-solution (10^{-3} moles/l), T = 20 °C (- - - -)

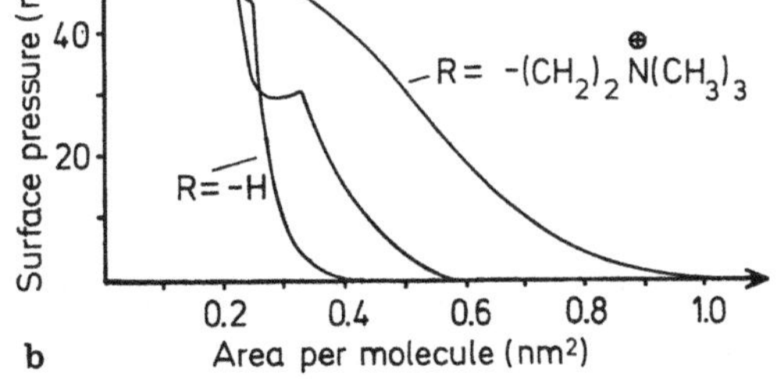

Fig. 9b. Effect of headgroup size on the film stability of diyne amphiphiles. Π-A-isotherms of *40*, *42* and *45* onpure water, T = 20 °C. Increasing size of the headgroup decreases the film stability (from Ref. [105]).

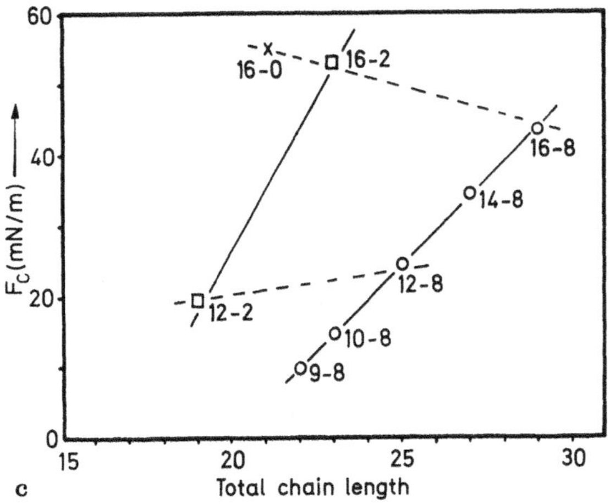

Fig. 9c. Plot of collapse pressure Π of condensed phases of C_mC_n acids as a function of their chain lengths, T = 12 °C [70]

Though the diagram shown in Fig. 9c is far from being complete one can derive that acids with a position of the diyne unit amidst the aliphatic chain exhibit a minimum in monolayer stability, whereas the stability is strongly increased, if the reactive unit is shifted towards the polar headgroup [70]. In the latter case the polar headgroup becomes activated by the diyne unit.

Presence of divalent metal ions (e.g. Cd^{2+}) in the subphase increases the film stabilities of fatty acids considerably (Table 3) [62-64, 70, 79, 80, 96]. This is due to a salt formation with the metal ions. The extent of salt formation is controlled by the pH-value of the subphase. Only at pH-values ≧6.5 the salt formation of the monolayer is complete [79].

Bulky headgroups occupying a large area on the water surface reduce the packing density of the paraffin chains [98, 102, 105-107]. As a consequence, the formation of liquid-expanded phases is favoured. For example, the single chain phosphatidic acid analogue *40* with the headgroup being $-OPO_3H_2$ shows a single condensed phase on water at temperatures below 35 °C, whereas the corresponding lecithin analogue *45* with $-O\overset{O}{\overset{\|}{P}}O(CH_2)_2\overset{+}{N}(CH_3)_3$ (with O^- on P) as the polar headgroup exhibits a liquid expanded phase in the whole temperature region between 2 and 50 °C (Fig. 9b) [98]. It may be added that only diyne phospholipids with a total chain length ≧26 carbon atoms exhibit film stabilities comparable to dipalmitoyl and distearoyl lecithin [98].

2.2.2 Photopolymerization

Photoreactions in monolayers are usually indicated by changes in the Π-A-isotherms of the amphiphiles. Monolayers of most of the diyne surfactants exhibit a film contraction upon exposure to UV-light. While the amphiphiles are highly reactive in the solid-condensed state, no reaction occurs in the liquid-expanded state [59, 60, 67, 104, 105, 113].

To overcome the problem of product identification Day and Ringsdorf [61] constructed a monolayer absorbance device combined with a UV-spectrometer. Absorption spectra monitored after different UV irradiation times under constant film pressure are shown in Fig. 7a. The spectra indicate that the "blue" polymer formed initially at the air-water interface is converted into a red one at long exposure times.

Simultaneously with the colour change a decrease of film area occurs indicating an increase of packing density of the monolayer. Increase of the red polymer absorption and decrease of film area could be directly correlated to one another. The isosbestic point involved in the colour change indicates a two-phase process without formation of intermediate states.

The same authors also detected that polymerization in monolayers is strongly affected by oxygen. They assumed that the conjugated polymer chains are affected by ozon, which is simultaneously formed by the intense UV-irradiation of the sample.

2.3 Langmuir-Blodgett Multilayers

2.3.1 Formation and Photoreactivity

Because of their easy handle Langmuir-Blodgett multilayers are best suited for a characterization of the diyne polymerization in lipid layer structures. The built-up layers consist of monolayers successively deposited on substrates, the number of layers being determined by the number of dipping cycles of the substrate [34]. The method however is restricted to amphiphiles that are able to form highly stable, solid-condensed films at the air-water interface.

Especially suited for the formation of multilayers are diyne fatty acids and their cadmium salts [62-64, 70, 79, 80, 96], phospholipids [117], and amphiphilic esters of isonicotinic acid [102]. In Table 3 multilayers of the cadmium salts of fatty acids are briefly characterized.

Exposing the multilayers to UV-light causes a rapid polymerization, indicated by the successively blue, purple, and red coloration of the samples. Absorption spectra of a multilayer of *5a* monitored after various UV-irradiation times are shown in Fig. 7b.

Table 3. Characteristic data of monolayers and LB-multilayers of cadmium salts of diyne C_mC_n-acids

No.	Acid		Collapse pressure of monolayer	photoreactive in multilayer?	Layer spacing of multilayer (nm)		Ref.
	m	n	$(mNm^{-1})^a$		Monomer	Polymer	
1a	16	8	55	+			[62]
2a	14	8	52	+	6.03		[62-64, 79]
3a	13	8	45^b	+			[80]
5a	12	8	50	+	5.53	5.89	[62-64, 79]
6a	10	8	35	+	5.15	5.54	[62-64, 70, 79]
7a	16	2	42	$(+)^c$	5.56	—	[62-64, 79]
9a	9	8	19	+	4.96		[62]
11a	16	0	50	+	4.25	5.51	[62, 70]

[a] subphase of all compounds except *3a*: aqueous solution of $CdCl_2$ (10^{-3} mol/l), T = 20 °C, pH = 7.0;
[b] subphase of *3a*: aqueous solution of $CdCl_2$ (5×10^{-3} mol/l), T = 20 °C, pH = 5.5;
[c] turns yellow after UV-exposure for a long time period

It may be pointed out that in contrast to the observations in monolayers the colour changes in multilayers do not involve an isosbestic point [62-64, 70, 87]. Possibly this difference is due to a higher mobility of the molecules in monolayers allowing rearrangements to occur more easily.

The polymerization in multilayers is only little sensitive to oxygen. In general, polymerized multilayers are stable for several weeks without showing a bleaching of the backbone chromophor. If occurring at all, radiation damage is likely restricted to the topmost layer [64].

2.3.2 Photochemical Studies

Because of their high transparency and exactly defined thickness multilayers have been proven to be excellently suited for quantitative studies of the polymerization process. The action spectrum and quantum yield of the photopolymerization were determined. Self-sensitization as well as sensitization of the photopolymerization by surfactant dye molecules were studied in great detail.

2.3.2.1 Quantum Yield and Action Spectrum of Photopolymerization

Multilayers of diyne fatty acids and their cadmium salts readily polymerize if they are exposed to UV-light of wavelengths $\leqq 300$ nm, The action spectrum of the photopolymerization [63, 64] indicates two maxima at $\lambda = 256$ and 242 nm, which are in good agreement with the first two maxima of the monomer absorption. Hence, photoreactivity is primarily due to photon absorption by the monomer diyne unit.

In additional studies [86, 87] the quantum yield Φ of the UV-photopolymerization was determined. Φ was found to be $\geqq 10$, which was evidence that the absorption of a single photon initiates the formation of a whole polymer chain.

Separate studies of X-ray induced polymerization of amphiphilic diacetylenes [100] have demonstrated quantum efficiencies as high as 10^{12}. In contrast to UV photons, absorbed X-ray photons provide a natural gain mechanism by creating large numbers of secondary electrons and other photons that actually cause polymerization. The acid *5*, for example, can be polymerized completely at an X-ray dose of 1.5×10^{-3} mJ cm^{-2} [101].

2.3.2.2 Self-Sensitization of the Photopolymerization

Monomeric crystals of diyne surfactants not only polymerize in UV-light, but also exhibit a rapid blue coloration upon exposure to daylight through a glass window. This effect has recently been investigated in multilayers [129] by measuring the action spectrum of photopolymerization either for freshly prepared, completely monomeric samples, or samples preirradiated with UV-light for a short time period. It was found that in the preirradiated samples the photoreactivity was extended to wavelengths $\leqq 500$ nm. Since photopolymerization in the visible cannot proceed via direct excitation of the monomer diyne unit, it was concluded that oligomers formed upon the previous UV-irradiation are able to cause a self-sensitization of the polymerization.

Interestingly, the action spectrum of self-sensitization follows the photoelectric gain curve of the multilayers [129]. It was therefore concluded that the same transition that causes excitation of an electron from the valence to conduction band of the polymer backbone is also responsible for the 'self-sensitization'. As a possible mechanism it was suggested that the optical excitation of the oligomers causes a reactivation of 'dead' chain ends either by formation of radicals and/or carbenes [129].

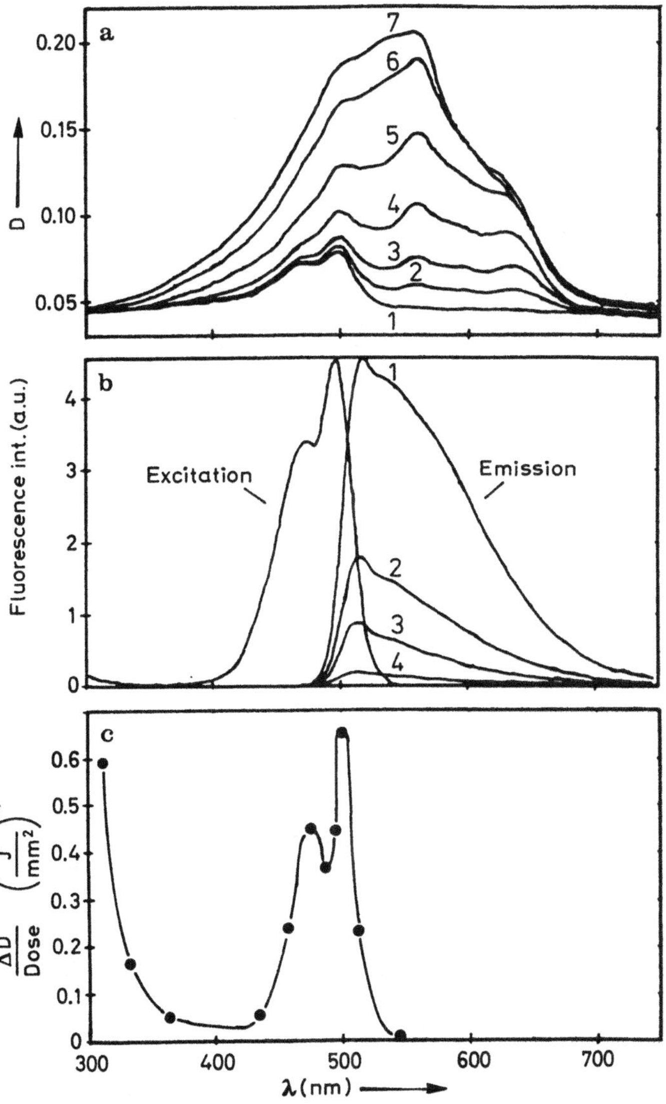

Fig. 10a. Absorption spectra of 40 layers of *5a* containing 1 mol-% dye *III* after different irradiation times (514.5 nm-laser): 1: 0 s, 2: 5 s, 3: 20 s, 4: 1.5 min, 5: 4.5 min, 6: 12 min, 7: 24 min. **b.** Excitation and emission spectra of the dye fluorescence. The numbers of the emission spectra refer to the irradiation times of part a. The emission spectra were observed by excitation of the dye at 450 nm The excitation spectrum was obtained by monitoring the emission at 560 nm as a function of the excitation wavelength. **c.** Action spectrum of the diacetylene polymerization [91]

Table 4. List of amphiphilic dye molecules and their characteristic properties in multilayers of *5a*

No.	Dye ($R = -(CH_2)_{17}-CH_3$)	Concentration in multilayer (mole-%)	Absorption λ_{max} (nm)	Sensitizes polymn. ?	Quantum yield Φ	Ref.
I	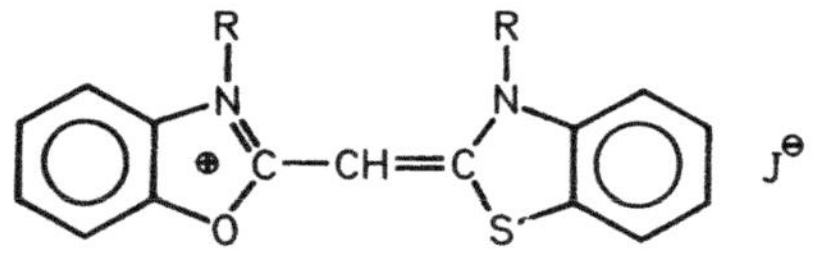	0.16	408	+	17×10^{-3}	[92]
II	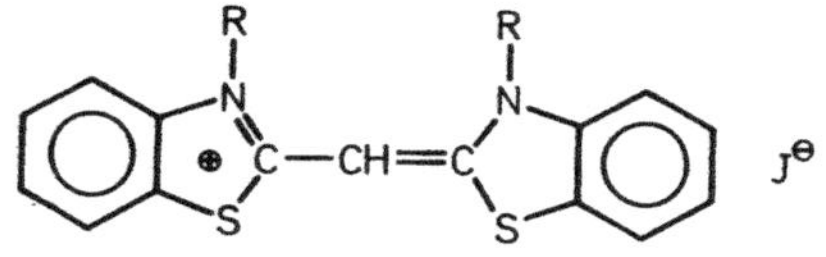	0.16	430	+	14×10^{-3}	[92]
III	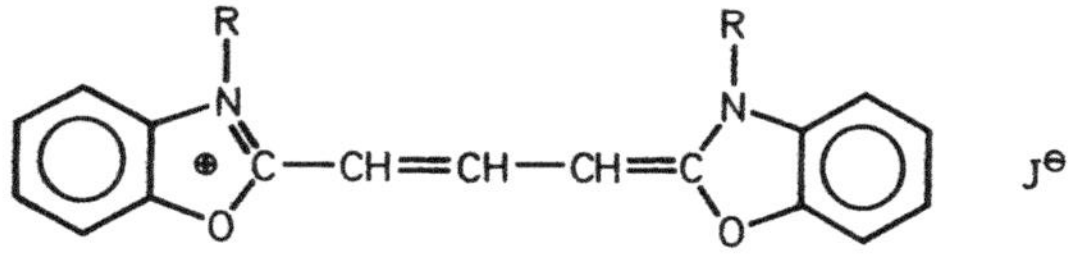	0.02–3.0	498	+	0.7×10^{-3}	[92]
IV	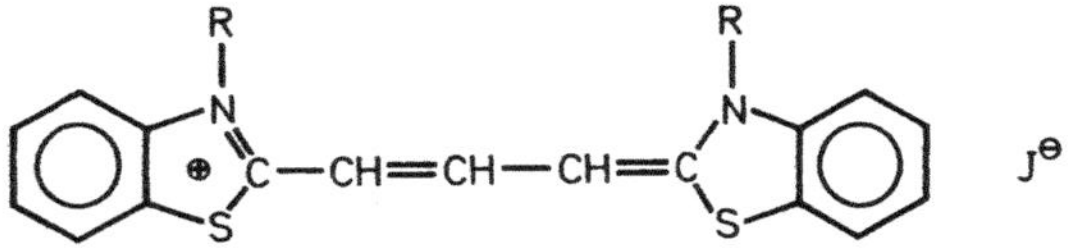	0.15	572	+	1.2×10^{-3}	[92]
V	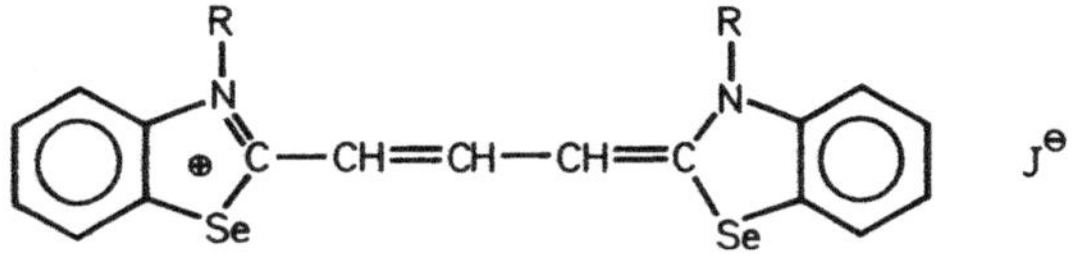	0.14	586	+	1.4×10^{-3}	[92]

VI	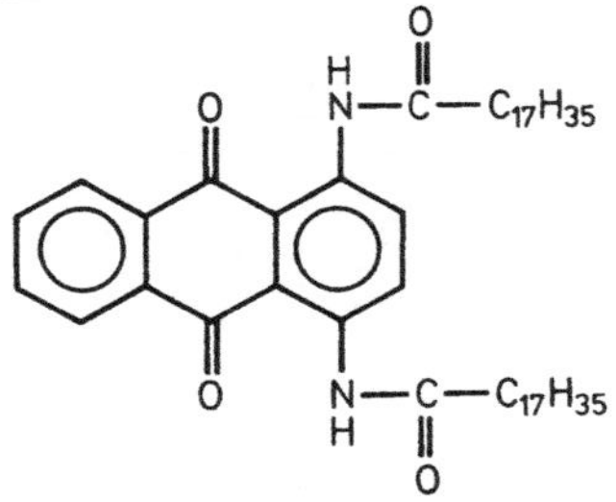	3.7	575	+		92)
VII	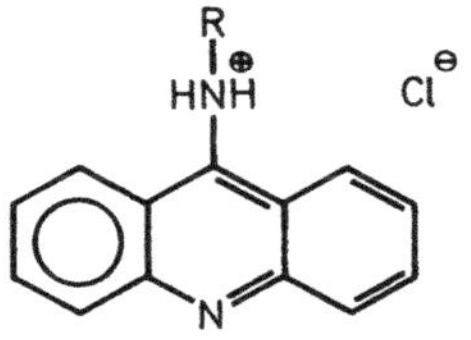	1.6	439	+	2×10^{-3}	92)
	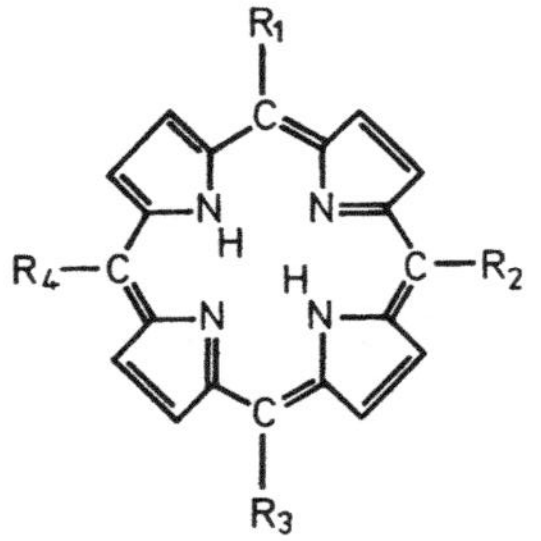					
VIII	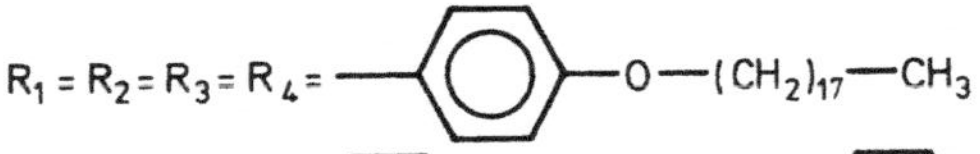	1.46	430[a]	+		130)
		1.15	440[a]	(+)[b]		130)
IX	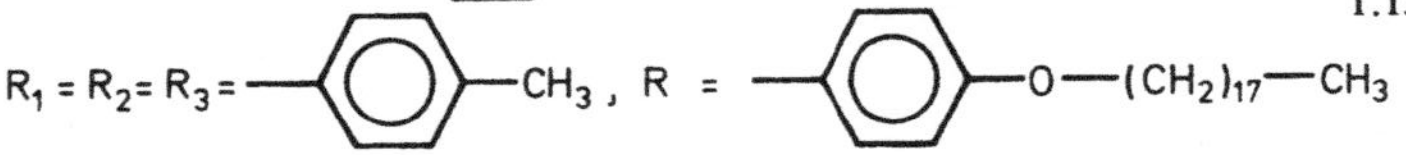					

[a] Soreth-band; [b] low sensitization capability

2.3.2.3 Sensitization by surfactant dye molecules

Fouassier et al. [87] observed for the first time that multilayers of diyne surfactants "doped" with amphiphilic cyanine dyes are photactive in the visible in the region of the dye absorption. In subsequent investigations Bubeck et al. [91,92] determined the sensitization capability of a series of donor and acceptor dyes and tried to elucidate the mechanism of the sensitizing effect.

In Fig. 10 the sensitized polymerization is demonstrated for a multilayer sample consisting of compound *5a* doped with one mole percent of dye III. Spectrum 1 of Fig. 1a was measured prior to irradiation of the sample and only shows the dye absorption at about 500 nm. Spectra 2–7 were monitored after exposure of the multilayer to the 514 nm line of an argon laser for different time periods. Formation of polymer is indicated by the occurrance of an additional broad absorption in the visible.

Figure 10b shows the dye excitation and emission spectra recorded at various stages of conversion. Apparently the fluorescence is quenched more rapidly than the dye decomposes, which indicates an energy transfer from the dye molecules to the conjugated polymer.

The photoresponse curve of the dye-sensitized photopolymerization is shown in Fig. 10c. Clearly the curve shapes are identical with the absorption spectrum of the dye and prove its sensitization capability.

Quantum yields Φ of the sensitized photopolymerization are low (Table 4). Individual dye molecules must absorb about 10^3 photons to allow the reaction of one monomer molecule [91].

In order to explain the sensitization effect several mechanisms have been discussed considering either energy transfer by a Förster mechanism or vibrational energy, or a photoinduced electron transfer, in analogy to the photographic process [131].

Bubeck et al. favour an electron transfer for the following reasons:

a) A Förster mechanism seems unlikely, because the electronically excited singlet and triplet states of the diyne unit cannot be populated directly with the low quantum energy sufficient only to excite the singlet transition of the dyes [91].
b) Transfer of vibronic energy from the dye to the diyne unit seems unlikely, because no correlation could be found between the Stokes shift of the dye and the quantum yield Φ of the sensitized polymerization. In addition, it was observed that non-fluorescent azo dyes, which deactivate radiationlessly, do not sensitize the polymerization [132].

2.3.2.4 Fluorescence

Fluorescence of polydiacetylenes was first reported qualitatively by Baughman and Chance[133], and has also been abserved in multilayers of polymerized *5a* [84]. While the blue form of polymerized amphiphiles is non-fluorescent, the red form exhibits a strong fluorescence with emission maxima at 570 and 640 nm. The luminescence is likely due to a polymer degradation product formed in disordered areas of the sample [127]. In multilayers of *2* the fluorescence quantum yield has been determined to be $(2.0 \pm 0.5) \times 10^{-2}$ [65].

2.3.3 Dark- and Photoconductivity

Studies on the dark conductivity of polydiacetylene bilayers were carried out after depositing the layers onto lock and key devices [78]. Samples exhibited an initial conductivity in the order of $10^{-2}\ \Omega^{-1}\ cm^{-1}$, which decayed at varying rates depending on the sample composition and its water content to $10^{-11}\ \Omega^{-1}\ cm^{-1}$. Doping with iodine caused a current increase of over three orders of magnitude [78, 89]. After removal of the iodine the current flow decreased back to the original value.

Polydiacetylene multilayers were also found to be photoconducting [88]. Due to their low thickness d the condition $\alpha d < 1$ (α being the absorption coefficient) can be established and the photoelectric gain ought to reflect the absorption spectrum of the ionizing transition directly.

However, absorption and photoconduction spectra do not reveal any correspondence [88], which suggests that the dominant optical transition is photoelectrically inactive and cannot be a valence-to-conduction band transition. Further analysis of the data in terms of a model considering a valence-to-conduction band transition, which is buried under the vibronic side bands of the dominant excitonic transition, yielded a gap energy $E_g = 2.5 \pm 0.1$ eV for the "blue" layer (absorption peak at 1.94 eV) and $E_g = 2.6 \pm 0.1$ eV for the "red" layer (absorption peak at 2.33 eV), in good agreement to other polydiacetylenes [71].

2.3.4 Structure and Morphology

The occurance of the lattice controlled diyne polymerization in lipid layers provoked the question for the state of order of these structures. Several studies therefore dealt with the investigation of structure and morphology mainly of multilayer samples using X-ray [62, 64, 70, 79, 90] and electron diffraction [64, 70, 73, 74, 79, 90, 95], electron [62, 64, 73, 80] and polarizing microscopy [62, 77, 85], infrared spectroscopy [73–75] and infrared dichroism measurements [76].

2.3.4.1 General Appearance

From X-ray studies of the multilayers the layer (d_{001}) spacings could be determined [62, 70, 79]. Corresponding data obtained for samples of the cadmium salts of fatty acids are listed in Table 3. Since the layer spacings are considerably shorter than the double length of the molecules derived from molecular models, it was concluded that consecutive layers are Y-structured, the paraffin chain axes of the molecules being tilted with regard to the layer plane. Tilting angles vary between 50 and 70° depending on the molecular structure of the amphiphile [62]. A structure model for a multilayer of *6a* is shown in Fig. 11.

Further informations on texture and mutual orientation of the molecules in multilayer were obtained from electron diffraction studies. These studies indicate a unique orientation of the molecules in crystallites (domains) with a lateral extension of the order of 10^{-1} to 10 micrometers in diameter. In adjacent domains the orientation of the molecules is different. From the diffraction patterns regular subcells of the paraffin chain packing have been determined [79, 82].

In multilayers of the cadmium salts of fatty acids the cations are not located at fixed lattice sites, and the crystallinity is therefore restricted to the paraffinic portions of the acid anions. Probably a stacking is hindered by the presence of water in the interlayer regions, which partially solvates the cations. As a consequence, an epitaxial deposition with formation of three-dimensionally ordered crystallites is not observed [79]. A structure model derived from the studies is shown in Fig. 12.

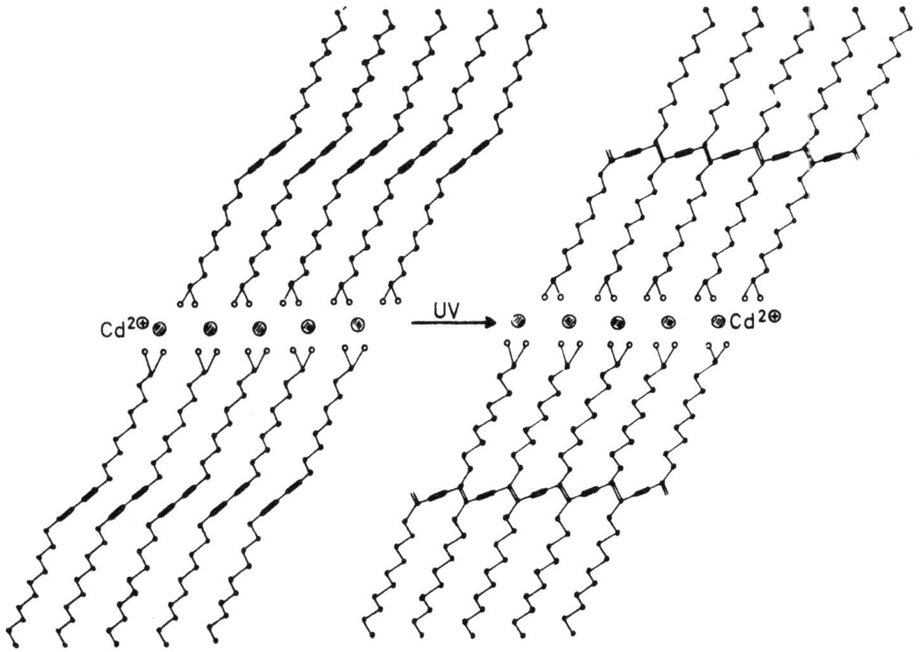

Fig. 11. Structure model of monomeric and polymerized multilayers of cadmium salts of diacetylene fatty acids. Polymer chains grow within the layer plane, the aliphatic chains are tilted with regard to the layer plane [87,96]

The polymer chains formed in the multilayers are stretched out in the plane of the substrate (Fig. 11). Because of the crystalline domain structure the individual chains are not extended over macroscopic dimensions, but are restricted in their length by the size of the domains. Within a domain all chains are aligned parallel to each other, but the mutual orientation in adjacent domains is different [62].

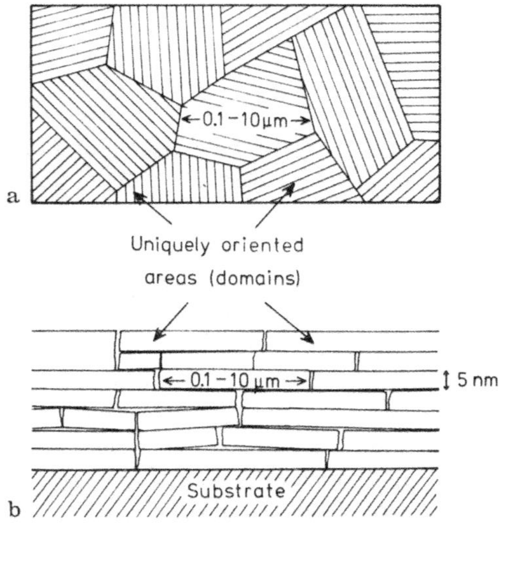

Fig. 12a and b. Model of the multilayer morphology. View (**a**) perpendicular and (**b**) parallel to the substrate [79]

Since the polymer ist strongly birefringent, the crystalline morphology of individual monolayers can easily be visualized between crossed polarizers in an optical microscope [62, 77, 85, 102, 132]. Monolayers polymerized at the air-water interface and subsequently transferred onto copper grids, for example, exhibit a domain structure consisting of either mosaic blocks or spherulites, depending on the chemical structure of the amphiphile [77]. On the other hand, monolayers built up on quartz substrates by the LB-technique, polymerized subsequently, and viewed on the substrate in an optical microscope, exhibit numerous dislocations, twin boundaries and a strong orientation of the domains in the dipping direction (Fig. 13a). It is assumed that these features are a consequence of a bending and shearing of the film during the transfer process [62, 85, 132].

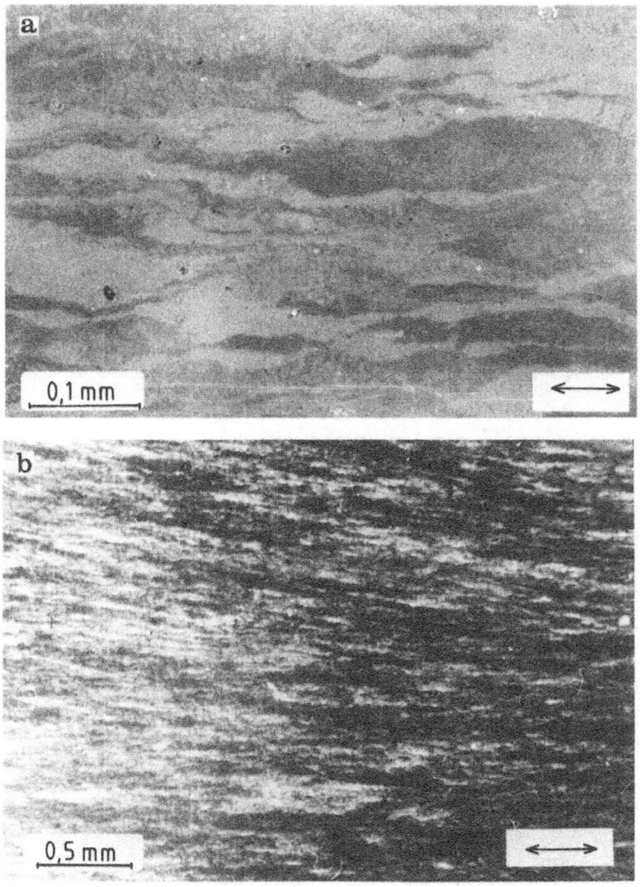

Fig. 13a and b. Polarizing micrographs of polymerized multilayers of *5a*, deposited on a quartz substrate precoated by three layers of cadmium arachidate. Deposition direction is indicated by ↔. **a.** Single layer of *5a*; **b.** 5 double layers of *5a*. In thick samples the lack of epitaxial deposition obstructs the morphological details seen in the LB-monolayer sample [85]

Only in monolayers of the diyne surfactants morphological details can be observed. In multilayers they are lacking because of the superposition of differently oriented domains in consecutive layers (Fig. 13b) [85].

2.3.4.2 Structure of Polymerized Multilayers

Polymerization of diyne surfactants is accompanied by structure changes, which in their extent are dependent on the molecular structure of the amphiphiles [62, 70]. In most of the cadmium salts of fatty acids, for example, the layer spacing increases (see also Table 3) [70, 79]. This increase has been explained by a slip of adjacent paraffin chains by one or more zig-zag periods increasing the tilting angle of the longest molecular axis with regard to the layer plane [79].

In compound *11a* the layer spacing increases by approximately 30%, whereas the structure changes of *5a* are relatively small [62, 96]. As a consequence, polymerized multilayers of *11a* show features such as holes, cracks, and phase boundaries, whereas samples of *5a* exhibit a smooth and fairly perfect surface, as indicated by the electron micrograph of Fig. 14. The few bulges seen in the micrograph are probably caused by the preparation of the sample for electron microscopy. The diffraction patterns of areas of comparable sizes indicate different sample textures as well [62, 96].

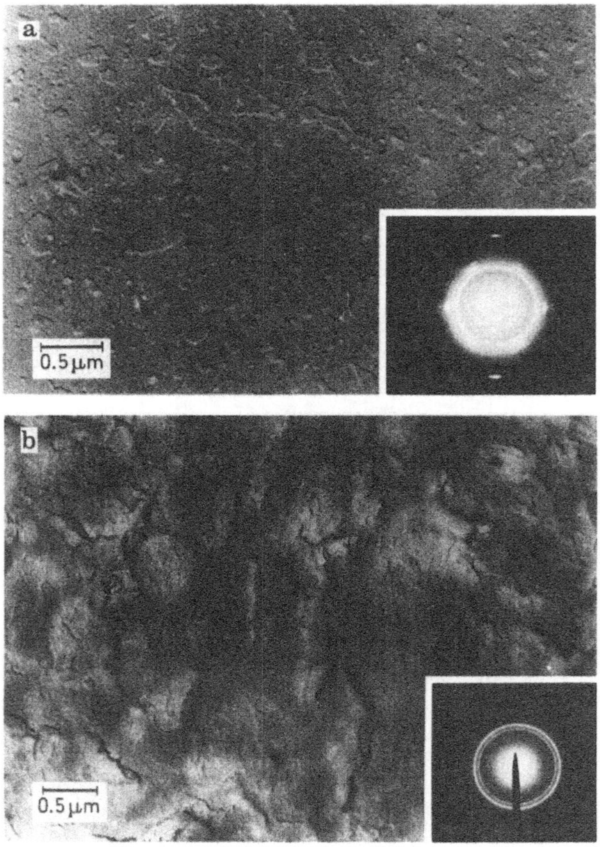

Fig. 14a and b. Electron micrographs of polymerized multilayers of (**a**) *5a*, and (**b**) *11a* (thickness 20 layers), showing different morphologies as also indicated by the electron diffraction patterns [62, 96]

2.3.4.3 Spontaneous Crystallization

Multilayers of saturated fatty acids are known to rearrange spontaneously into three-dimensionally ordered crystallites [134]. The same is true for multilayers of monomeric diyne surfactants. It was found that the tendency to rearrange into microcrystals increased with a decrease of the total chain length of the amphiphile. In a series of multilayers of fatty acid cadmium salts the highest tendency to rearrange was found for the compound with the shortest chain *9a* [82,96]. Similar effects may occur in multilayers of the short chain compounds *15* and *34*, and explain that their packing geometry in multilayers is equal to that in the bulk crystalline state [73].

2.3.5 Specific Variation of the Morphology

A technical application of multilayers requires a possibility to control the morphology. In several publications efforts to specifically affect the domain structure have been reported, either by a variation of the preparation conditions of monolayers [62,77,85,135,136], by annealing of multilayers [85], or by formation of mixed multilayers [85,90].

2.3.5.1 Preparation Conditions

Any manipulations of the monolayers starting from the spreading at the air-water interface till the deposition on substrates influences their morphology. Fig. 15 schematically summarizes the typical changes in morphology during the formation of mono- and multilayers.

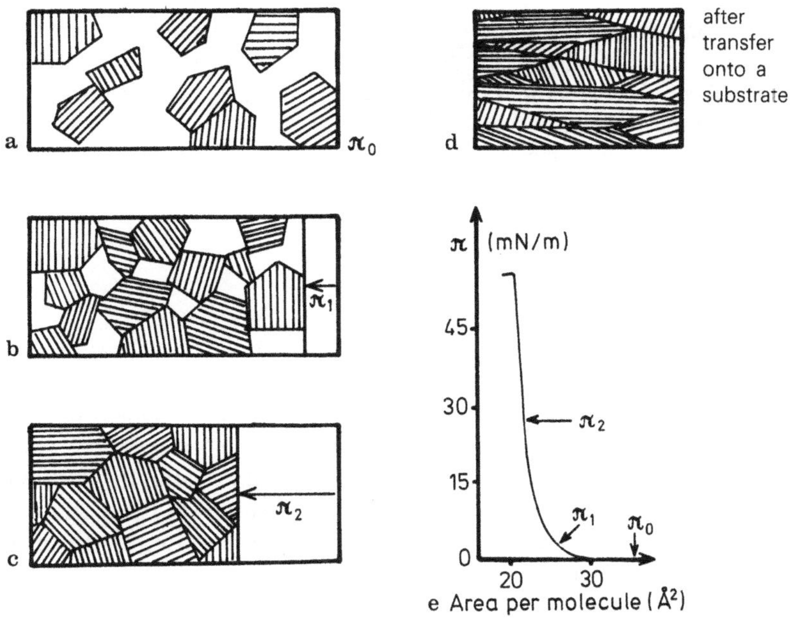

Fig. 15a–e. Schematic representation of the morphology of a single layer as dependent on various manipulations: **a.** In the "expanded" state directly after spreading at the air-water interface; **b, c.** After compression to surface pressures Π_1 and Π_2; **d.** After transfer onto a substrate by the LB-technique; **e.** In addition: Π-A-isotherm of *5a* (T = 13 °C), subphase 3×10^{-4} moles/l $CdBr_2$ [85]

Domain formation already occurs at the air-water interface directly after spreading. Film compression therefore only reduces the free area present between the domains. Hole free films are only obtained, if the monolayers are compressed to Π-values Π_2, which correspond to the steep rise of the Π-A-isotherm shown in Fig. 15. Finally, upon transfer of the film onto substrates the domains become oriented in the dipping direction.

Similar to normal, three-dimensional crystallization processes the monolayer morphology is dependent on the conditions of crystallization [77,85]. For example, the nature of the spreading solvent and the presence of impurities have a strong influence on the size of the domains. A slow evaporation of the solvent in a solvent-saturated atmosphere favours the formation of large domains [77]. On the other hand, simple correlations between vapor pressure of the spreading agent and domain sizes obtained do not exist [85]. In general, cyclic spreading agents such as cyclohexane or toluene or benzene seem to favour the formation of large domains, whereas n-hexane or n-octane, for example, are partially incorporated in the films and disturb the structural perfection [85].

Attempts to increase the domain sizes by cospreading diacetylene amphiphiles and hexadecane [136] could not be reproduced [85]. As already found for n-hexane or n-octane, the paraffin partly remains in the film and the domains of the diyne compound are considerably reduced in size [85].

In general, materials involved in the monolayer formation, such as spreading solvent, subphase and amphiphiles ought to be extremely purified. Use of zone refined amphiphiles has been reported to yield monolayers with considerably increased domain sizes [135].

Annealing of monolayers at the air-water interface, or monomeric Langmuir-Blodgett multilayers, did not increase the domain sizes, as it was desired [85]. Instead, the opposite was observed. A considerable amount of molecules became disordered and was no longer photoreactive.

Application of electric fields during spreading of the amphiphiles has also been reported to increase the domain sizes. Uniquely ordered film regions approaching 1 mm in diameter were obtained [135].

2.3.5.2 Binary Mixtures

The strongest variation of the film morphology, however, could be achieved by cospreading diyne surfactants with other amphiphiles. Already small amounts of additives as low as 0.1 to 1 mole percent reduced the average domain sizes considerably [85].

Binary mixtures of diyne surfactants and non-polymerizable, saturated or unsaturated amphiphiles exhibit phase separations at all compositions of the mixture [85; 90]. Obviously the packing of the two compounds is incompatible due to the diyne unit representing a rod of a length of 0.67 nm.

Polarizing micrographs shown in Fig. 16 indicate that the morphologies of binary mixtures are considerably different from those of the pure compounds. For example, in micrographs of a mixed monolayer of *6a* and cadmium stearate (Fig. 16a) only large stripes are seen extended over the whole specimen and oriented in the transfer direction. The stripes indicate a unique alignment of the polymer chains over large

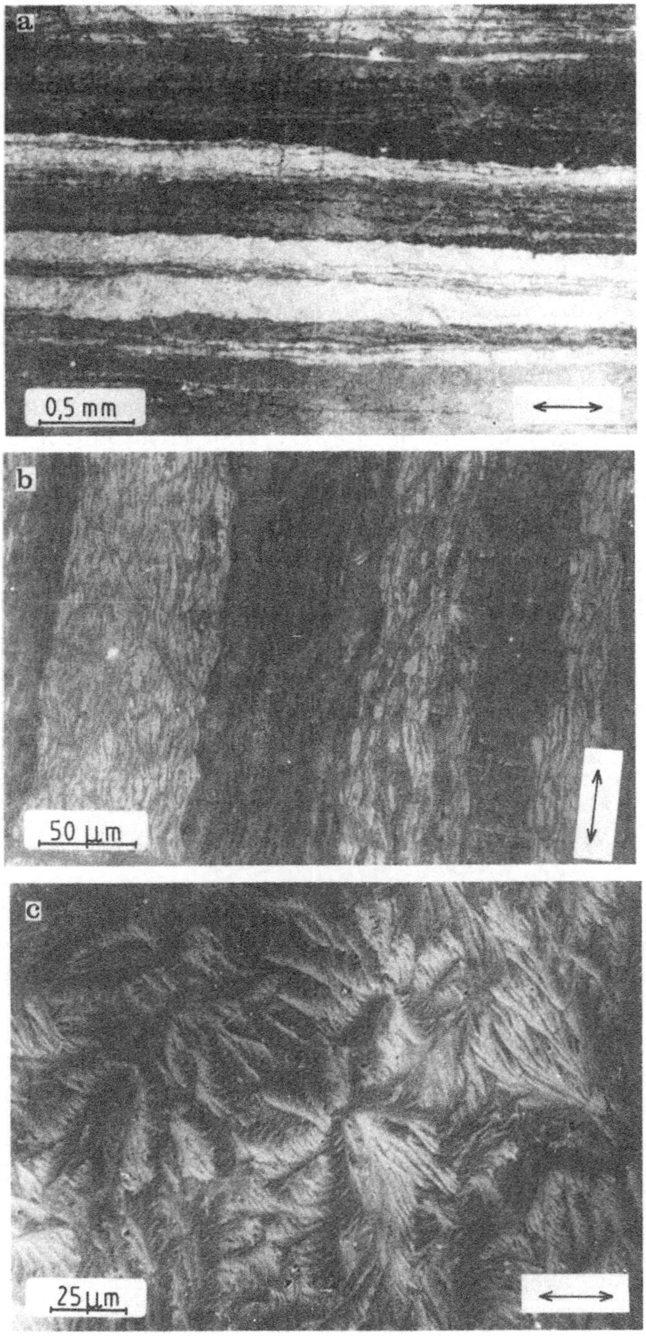

Fig. 16a–c. Polarizing micrographs of polymerized LB-monolayers of binary mixtures of amphiphiles. **a.** *6a* and cadmium stearate in a molar ratio of 1.27:1 at low magnification; **b.** the same mixture at higher magnification; **c.** *6a* and *7a* in a molar ratio of 2.3:1. Deposition direction: ↔ [85]

distances. On the other hand, at higher magnifications a phase separation of the two compounds can be recognized (Fig. 16b). The grey regions not showing any birefringence can be attributed to cadmium stearate, acting as a matrix for the dark and bright regions of the diyne surfactant.

Phase separation on one hand, and formation of a superstructure of the diyne compound on the other hand, have been explained [85] by assuming strong interactions between the paraffin chains of the otherwise immiscible compounds. Such interactions could originate from similarities in the packing geometry of the paraffinic portions of the hydrophobic chains, as it was indicated by electron diffraction studies of the mixed multilayers [90].

Quite a different morphology has been observed for binary mixtures of the isomeric cadmium salts *6a* and *7a*. Mixed monolayers of these compounds form a texture consisting of microdomains of only a few micrometers in diameter or even below (Fig. 16c). Orientation phenomena due to the deposition process do not occur [85].

The morphology most likely originates from the ability of the two isomers to form a 1:1 complex. Complex formation was evidenced by UV/VIS-spectroscopy, X-ray and electron diffraction studies of the multilayers [90]. It is most likely caused by the similar chemical structures of the two isomers.

2.3.6 Potential Applications

Various applications of diacetylene mono- and LB-multilayers have been discussed, making use of their insulating properties, their ability to polymerize, the exceptionally large second and third order non-linear susceptibilities of the polymer [137,138], and their permeability.

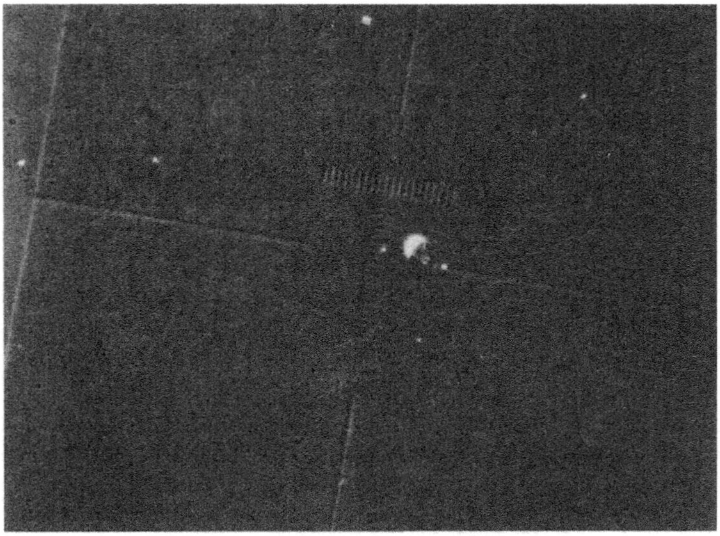

Fig. 17. Photomicrograph of a UV-lithographic T feature showing 1 μm-wide spaces and 2 μm wide bars in the center of a 300 μm wide square (sample: 10 monolayers of polymerized *5a*, deposited on an oxidized silicon wafer by the LB-technique). (Reproduced with permission from Ref. [101], Copyright 1982, Laser Focus)

G. G. Roberts and coworkers [139] deposited monolayers of diyne fatty acids on the surface of the narrow band gap semiconductor $Hg_xCd_{1-x}Te$, utilized in the fabrication of infrared detection devices. Admittance data determined after polymerization of the amphiphiles compared favourably with data obtained for equivalent devices with inorganic insulators. Hence, an application for passivating semiconductor surfaces seems feasible.

A. F. Garito and coworkers [100, 101] and a Japanese group [140] demonstrated the suitability of diyne LB-multilayers as high resolution UV-, X- and electron beam resist materials. An example of a micron size test pattern of a developed multilayer of *5a* deposited on an oxydized silicon wafer by the LB-technique is shown in Fig. 17. Such patterns are developed by liquid reagents, but can also be processed by dry plasma etching.

Several authors [101, 141–143] suggest the use of polydiacetylene multilayers as active guided wave structures. G. M. Carter et al. [142] determined the degenerate third order nonlinear susceptibility of the polymer of *1*. For this purpose, multilayers of a thickness of 500 nm were built up on a silver overcoated grating etched on a silicon wafer. The grating allowed coupling of a freely propagating laser beam into the planar waveguide structure formed by the polydiacetylene on the metal. By changing the intensity of the laser beam ($\lambda = 670$ nm) a change in coupling angle $\Delta\Theta$ to the waveguide mode was observed. From $\Delta\Theta$ the nonlinear index of refraction of the polymer near the absorption edge could be estimated to be of the order of 10^{-5} $(MW/cm^2)^{-1}$. Further studies to elucidate $\chi^{(3)}$ as a function of λ of the incident light are currently under progress [143].

Waveguiding combined with $\chi^{(2)}$ and $\chi^{(3)}$ processes in polydiacetylene multilayers may open a wide range of applications to optical switches and frequency multipliers in integrated optics, data storage, and memory technologies. Current research activities are focussed on amphiphilic compounds exhibiting a phase matched second harmonic generation [144], as for example

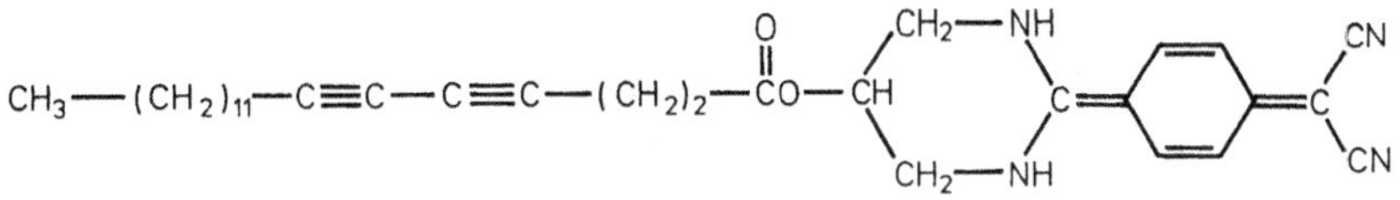

K. Heckmann et al. [145] suggested the use of monolayers of *3* as separating layers in a hyperfiltration membrane. The polymeric monolayers are crosslinked with the molecules of the supporting layer, e.g. an asymmetric Sartorius cellulose membrane, via hydrophilic groups.

2.4 Bilayer Type Aggregates (Vesicles, Liposomes)

Stable vesicles are formed either by slow injection of surfactant solutions in alcohol into thermostated water, or by the ultrasonic dispersal of the surfactants in water kept above the phase transition temperature of the vesicle (see also Fig. 2) [27–29]. Besides naturally occurring phospholipids also synthetic surfactants were demonstrated to form globular aggregates. In general, double chain amphiphiles are better suited for vesicle formation than single chain ones.

In recent years polymerization of diyne surfactants in aqueous dispersion has been investigated by several research groups [67,105,112,114,116,118–121,146]. Compounds which have been studied are listed in Table 1.3.

2.4.1 Aggregate Formation and Photoreactivity

Formation of vesicles was demonstrated for compounds *59*, *61*, and *62* by electron microscopy [67,112,114,118]. For compound *67* contradictive results were obtained. Recent studies by optical microscopy [147] indicated the formation of cylinder-shaped aggregates ('tubules'), whereas in earlier studies [116,118] a formation of vesicles was reported. Yager and Schoen [147] believe that *67* forms vesicles only at temperatures above the phase transition point, which on subsequent reduction of the temperature rearrange into tubules of about one micrometer in diameter and up to hundreds of micrometers in length. On heating, these structures are readily converted into liposomes again [147].

Structural rearrangements similar to those described above were also reported from aqueous dispersions of the single chain alkadiynyl pyridinium salt *49* [107]. While *49* forms a micellar solution at elevated temperature, the molecules aggregate upon cooling and form vesicles, which rearrange into disc-shaped lamellae of several micrometers in diameter within a few hours. Electron micrographs of vesicle dispersions and multilamellar layer aggregates are shown in Fig. 18.

Exposure of the aqueous vesicle dispersions to UV-light causes a rapid coloration indicating the polymerization of the surfactants. Polymerization was reported from vesicles, cylinder- and disc-shaped aggregates. However, it only occurred at temperatures below the phase transition point of the surfactants. Photoreactivity of vesicles was found to be restricted to larger particles exhibiting diameters $\geqq 100$ nm [105]. It is believed that in smaller aggregates the topochemical reaction is hindered by the high surface curvature [105,112].

Polymerization properties of amphiphiles *57*–*63* resemble those of the fatty acids and their derivatives already discussed above, whereas the phospholipids *65*–*67* polymerize only slowly and turn orange in aqueous dispersion (see also Chapt. 2.1.1).

Polymerized aggregates are much more rigid than the monomeric ones. DSC-traces indicate that in polymerized samples the phase transitions are completely lacking [106]. Tubules of *67*, however, represent an exception. Prolonged UV-exposure did not cause a strong reduction in peak intensity though the sample was deeply coloured. It therefore was concluded that in those aggregates the conversion to polymer remains incomplete [147].

Vesicles of single chain α,ω-dipolar diyne surfactants *54*–*56* are completely photoinactive except they are cosonicated with cholesterol [110]. The authors believe the photoinactivity to originate from the high surface curvature of the vesicles. Cholesterol, however, becomes incorporated in the outer half of the membrane, decreases the surface curvature and renders the vesicles photoreactive [110].

2.4.2 Utility as Membrane Model

Various studies on polymerized vesicles are focussed on their suitability as synthetic, stable models for biomembranes, which may allow to mimick cell membrane functions and cell-cell interactions [26–28,120,146].

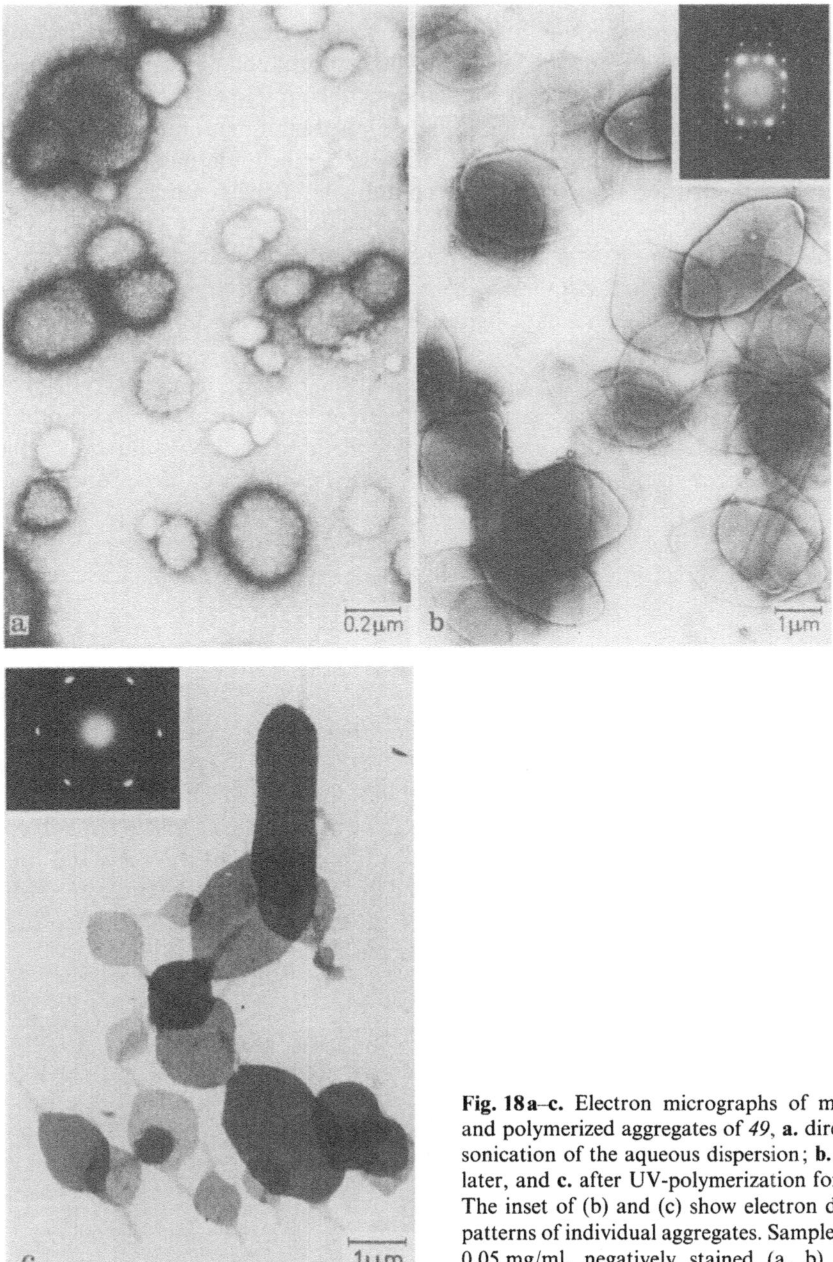

Fig. 18a–c. Electron micrographs of monomeric and polymerized aggregates of *49*, **a.** directly after sonication of the aqueous dispersion; **b.** six hours later, and **c.** after UV-polymerization for 5 hours. The inset of (b) and (c) show electron diffraction patterns of individual aggregates. Sample solutions 0.05 mg/ml, negatively stained (a, b) with 1% uranyl acetate (by weight) [107]

The specific surface recognition of polymerized liposomes has been demonstrated by a simple model reaction [104]: A dispersion of polymerized liposomes of glycolipid *38* was mixed with the lectin Concanavalin A, which caused a rapid agglutination and precipitation of the liposomes. This effect was not observed for polymerized liposomes without sugar headgroups.

Studies on the biological activity of polydiacetylene vesicles were reported by Wagner et al. [114]. F_0–F_1-ATPase from Rhodospirilium rubrum could easily be incorporated into polymerized liposomes under retention of the protein activity. While monomeric liposomes showed a lower activity than corresponding natural phospholipids, it increased upon polymerization to the value determined for natural liposomes. Different from the natural ones the ATPase containing polymers are stable over long time periods.

Moreover, the chromoprotein bacteriorhodopsin from Halobacterium halobium has been incorporated into liposomes of the polymerizable sulfolipid *61*. Bacteriorhodopsin was found to be active as a light-driven proton pump in the polymerized liposomes [148].

An additional method to model biomembranes was recently described by Leaver et al. [149]. Polymerizable diacetylene fatty acids were biosynthetically incorporated into Acholeplasma laidlawii cells and their polymerization via UV-irradiation could be realized. In order to determine the effect of polymerization on the properties of the membrane, the activity of intrinsic and extrinsic membrane-bound enzymes, NADH oxidase and ribonuclease were studied. The NADH oxidase activity decreased rapidly upon polymerization of the lipid environment whereas the ribonuclease activity was unaffected.

2.5 Black Lipid Membranes

Planar bimolecular lipid membranes, also called "black lipid membranes" (BLM's) represent a further membrane model system, which is stable only for a few minutes up to several hours. All attempts to further stabilize these BLM's by diyne polymerization failed [150]. Diacetylene compounds which have been studied are listed in Table 5.

The photoinactivity of BLM's is likely due to the presence of the molecules in a liquid-expanded state, which is too disordered to allow a topochemical reaction. Similar attempts to polymerize diacetylene derivatives in the liquid crystalline state or in micellar dispersions failed for the same reason [107].

Table 5. Lifetime and photoreactivity of diacetylene amphiphiles in BLM's [150]

No.	Compound	Photo-reactivity[a]	Lifetime of BLM	T (°C)
38	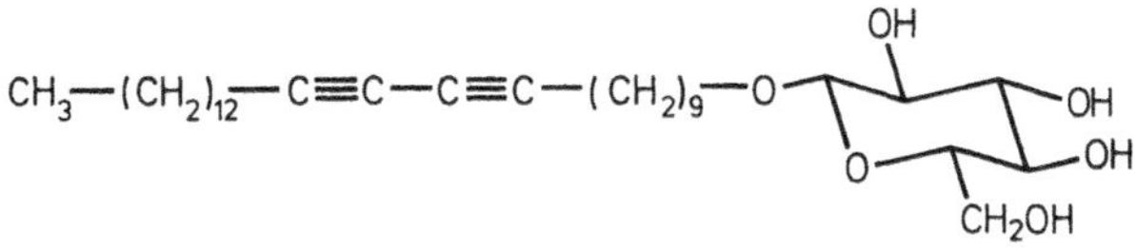	—	1–2 h	40
83	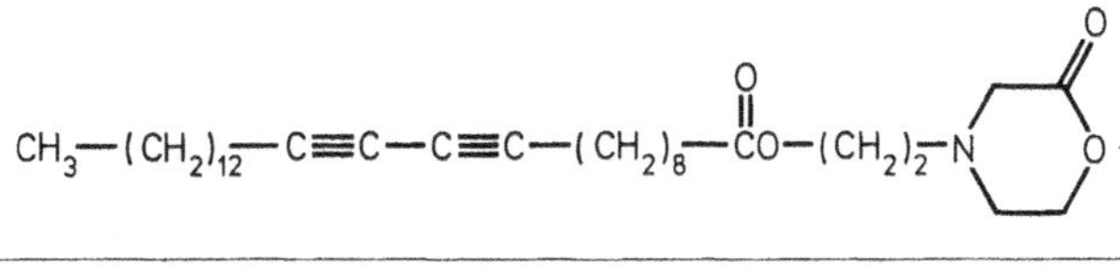	—	1–3 h	25

[a] upon UV-irradiation ($\lambda = 254$ nm)

2.6 Crystalline Complexes with Phenazine

Long chain 2,4-diynoic acids such as compounds *11*, *14*, and *16* are able to form crystalline complexes with phenazine in 2:1 stoichiometry, as listed in Table 6 [72, 151]. The following factors contribute to the complex formation [72]:

a) The 2,4-diynoic acids are activated by the diyne unit, which favours the formation of hydrogen bonds of the carboxylic acid unit with the nitrogen atoms of the phenazine moiety.

Table 6. Characteristic data of phenazine-complexes of 2,4-diynoic acids [72]

No.	Acid	M.p. °C	Photoreactivity[a]	Unit cell dimensions					
				a	b	c (nm)	α	β	γ (°)
84	*11*	87	+						
				Monomer:					
85	*14*	85	+	0.474	0.715	3.64	84.1	92.1	106.1
				Polymer:					
				0.4866	0.7066	3.434	89.3	94.8	103.4
				Monomer:					
86	*16*	83	+	0.477	0.709	3.32	91.5	91.6	104.5

[a] Upon UV-(λ = 254 nm) and γ-irradiation

b) The diyne acids exhibit lattice parameters in the layer plane which are very similar to those of α-phenazine in the b,c-plane [72]. The crystal structure of the complex can therefore be explained as a compromise between the structures of the pure compounds. This also explains why acridine or pyrazine, which have very different packings, do not form complexes.

The complex crystals grow up to centimeter lengths, but are extremely thin and fragile, which prevented a full structure determination as yet [72]. However, from the unit cell parameters (Table 6) a structure model was derived, considering an alternating stack of bilayers of the fatty acids in a head-head-tail-tail-fashion and single layers of phenazine intercalated in the hydrophilic interlayers, the phenazine molecules being hydrogen bonded to the carboxylic acid units (Fig. 19) [72].

Upon γ-irradiation the complex crystals polymerized under retention of the single crystalline character. Maximum conversion to polymer of 65% could be obtained [72]. The photoresponse curve of the UV/VIS initiated polymerization, shown in Fig. 20, indicates high reactivities in the region of the monomer absorption at wavelengths ≦300 nm, and in the region between 380 and 450 nm, due to a sensitizing effect of the phenazine molecules [151].

Braunschweig and Bässler [152] studied the photopolymerization of the complex crystals. They determined the quantum yield of the UV-initiated reaction at 280 nm and the phenazine-sensitized reaction at 450 nm to be Φ(280) = 0.5 and Φ(450) = 0.1, respectively. The activation energies of either reactions are 0.17 and 0.21 ± 0.02 eV.

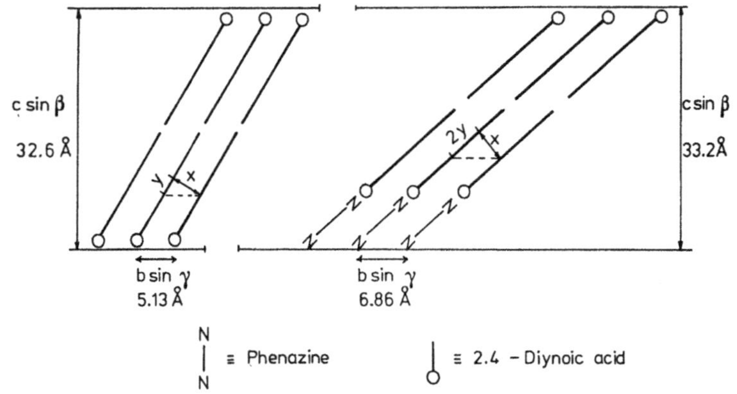

Fig. 19. Structure models of 2.4-diynoic acids and their 2: 1 complex crystals with phenazine in 100-projection, i.e. perpendicular to the polymer chain direction (a-axis) [72]

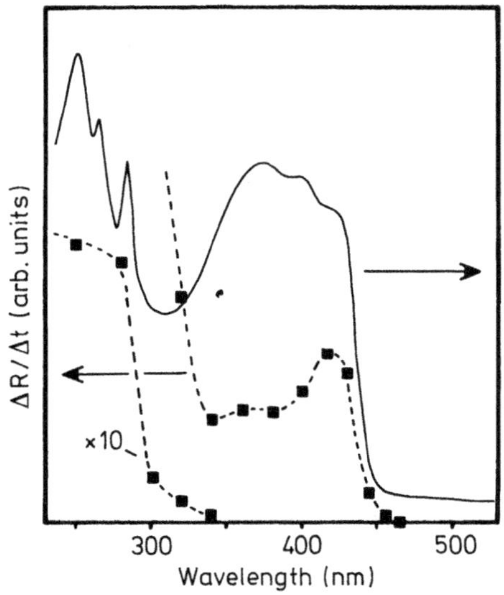

Fig. 20. Diffuse reflectance spectrum of monomeric complex *86* and its action spectrum of UV/VIS-photopolymerization.

While the former value attributes to the chain growth, the difference value between the two processes of 0.04 eV was associated with the formation of a chain initiation species as a result of the phenazine reaction.

Mondong and Bässler [153] irradiated partially polymerized crystals of *84* through a two-dimensional grid by UV-light. After dissolution of unreacted monomer surface structures could be detected by scanning electron microscopy, which indicated that polymer chains grow also into unirradiated crystal areas. From the etch profile an average chain length of ≤ 3.5 μm could be estimated, being equivalent to a degree of polymerization $n \simeq 10^4$. This allowed to calculate the primary quantum yield $\Phi^{(i)}$ of chain initiation. For photopolymerization via excitation of phenazine $\Phi^{(i)420}$

$\simeq 10^{-5}$ was obtained, and for polymerization via excitation of the diyne moiety $\Phi^{(i)280} \simeq 5 \times 10^{-5}$, respectively.

The nature of the chain initiation species was studied in detail by optical and ESR measurements at low temperatures [154]. UV irradiation at 4 K causes the formation of stable triplet diradicals. Most likely one of the two unpaired electrons is located at the phenazine molecule, and the other one at the carboxylic acid group or the diyne unit of the fatty acid molecule. Moreover, it was observed that the photopolymerization via phenazine-excitation proceeds under decarboxylation of the acid molecule.

The authors [154] conclude that the sensitized polymerization is initiated by the phenazine (Ph) acting as an electron and proton acceptor in its excited state. $PhH^{\cdot}$ or $PhH_2^{+\cdot}$ radicals are created, as well as acid radicals and anions, which could act as precursor states of the decarboxylation ($R = CH_3—(CH_2)_m—$ with m = 11, 13, 15):

$$\begin{aligned}&R{-}C{\equiv}C{-}C{\equiv}C{-}COOH + Ph^* \rightarrow\\ &\rightarrow PhH^{\cdot} + R{-}C{\equiv}C{-}C{\equiv}C{-}CO_2^{\cdot} \qquad \text{or}\end{aligned}$$

$$\begin{aligned}&2R{-}C{\equiv}C{-}C{\equiv}C{-}COOH + Ph^* \rightarrow\\ &\rightarrow PhH_2^{+\cdot} + R{-}C{\equiv}C{-}C{\equiv}C{-}CO_2^{\cdot} + R{-}C{\equiv}C{-}C{\equiv}C{-}CO_2^{-}\end{aligned} \qquad (1)$$

and

$$\begin{aligned}&R{-}C{\equiv}C{-}C{\equiv}C{-}CO_2^{\cdot} \rightarrow [R{-}C{\equiv}C{-}C{\equiv}C^{\cdot} \leftrightarrow R{-}\dot{C}{=}C{=}C{=}Cl]\\ &+ CO_2\uparrow\end{aligned}$$

$$R{-}C{\equiv}C{-}C{\equiv}C{-}CO_2^{-} \rightarrow [R{-}C{\equiv}C{-}C{\equiv}\overset{(-)}{C}l \leftrightarrow R{-}\overset{(-)}{\underline{C}}{=}C{=}C{=}C] + CO_2\uparrow \qquad (2)$$

It is further concluded that the diacetylene radicals and carbenes formed upon the decarboxylation serve as active centers able to start the solid state polymerization. Radicals and carbenes are known from other diacetylene derivatives to be intermediates active in the polymerization process [55,56].

2.7 Layer Perovskite Halide Salts

In the past layer perovskite halide salts $(R—NH_3)_2MtX_4$ and $(NH_3—R'—NH_3)MtX_4$ ($Mt = Cd^{2+}, Cu^{2+}, Mn^{2+}, Fe^{2+}$, $X = Br^-, Cl^-$, R and R' being organic moieties, e.g. aliphatic chains) were used as model systems for the study of magnetic properties of two-dimensionally extended structures [155,156], as well as structural phase transitions of lipid bilayer membranes embedded in a crystalline, inorganic matrix [39]. Only recently the utility of the organic-inorganic complex salts as a template for solid state reactions has been demonstrated [24,157].

The structure of the layer perovskites has been introduced already in Chapt. 1. According to Fig. 3 the $R—NH_3^+$ cations are arranged in bilayers, sandwiched by single layers of corner sharing Cl_6-octahedra with the Mt^{2+} ions in the centers. Hence, the inorganic layers can be looked at as a matrix forcing the $R—NH_3^+$ ions into specific packing patterns.

Table 7. List of complex salts of diacetylene amphiphiles

7.1 Layer perovskite halide salts $[R_1-C{\equiv}C-C{\equiv}C-R_2]_2MtX_4$

No.	Compound				Photoreactivity	Notes	Ref.
	R_1	R_2	Mt	X			
87	$C_{12}H_{25}-$	$-(CH_2)_9-NH_3$	Cd	Cl	+	Purple after 1 h UV	24, 99)
88	$C_{12}H_{25}-$	$-(CH_2)_9-NH_3$	Mn	Cl	+	Thermochromic (blue at T < 20 °C, red at T > 40 °C)	99)
89	$C_{12}H_{25}-$	$-(CH_2)_9-NH_3$	Cu	Cl	+	Purple after 10 min UV	24, 99)
90	C_4H_9-	$-(CH_2)_3-NH_3$	Mn	Cl	+		24)
91	$ClCH_2-$	$-CH_2-NH_3$	Cd	Cl	+	Red after 1 h UV, black after 8 h x-irradiation	157)
92	$H_3NCH_2C(=O)O-CH_2-$	$-CH_2OC(=O)-CH_2NH_3$	Cd	Cl	–	Pale brown after 1 h UV	157)

7.2 Complex salts of unknown structure

No.	Compound	M.p. (°C)	Photore-activity	Notes	Ref.
93	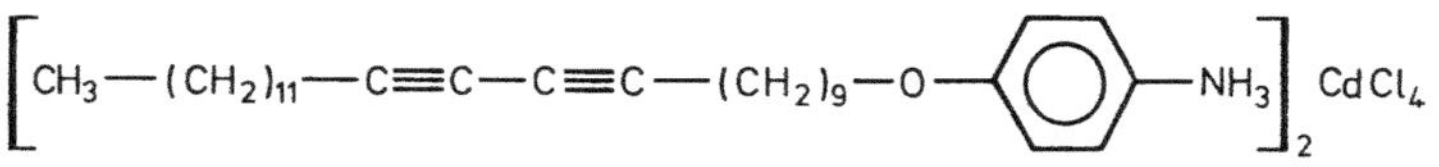	217[a]	+	Pink after 30 min. UV	24)
94	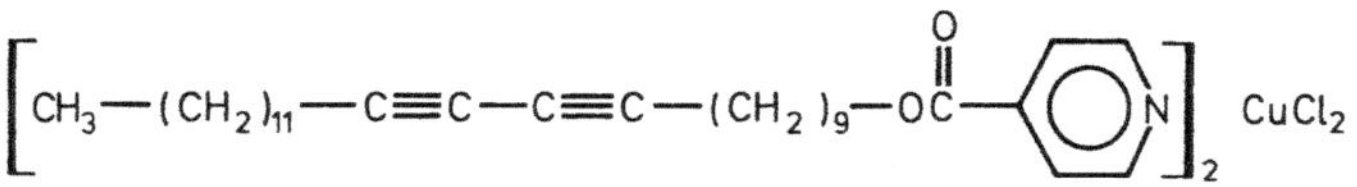	135[a]	+	Orange at a γ-ray dose of 30 Mrad (10% conv.)	103)

95	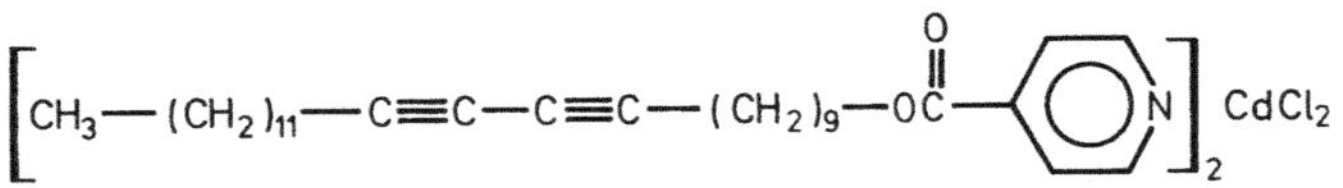	160[a]	+	Red at a γ-ray dose of 30 Mrad (40% convers.)	153)
96	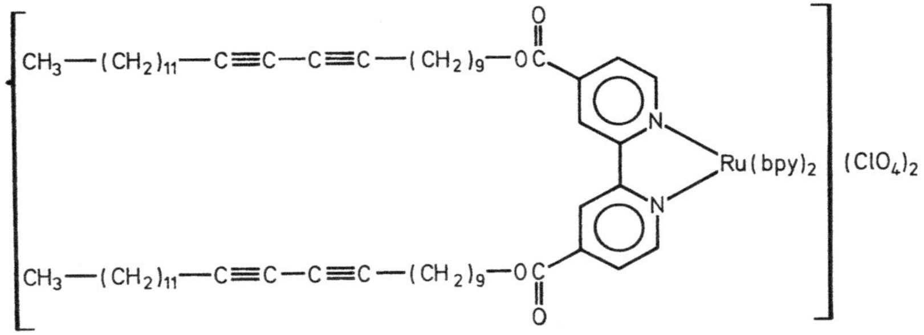	232	—	—	103)

[a] Melting under decomposition

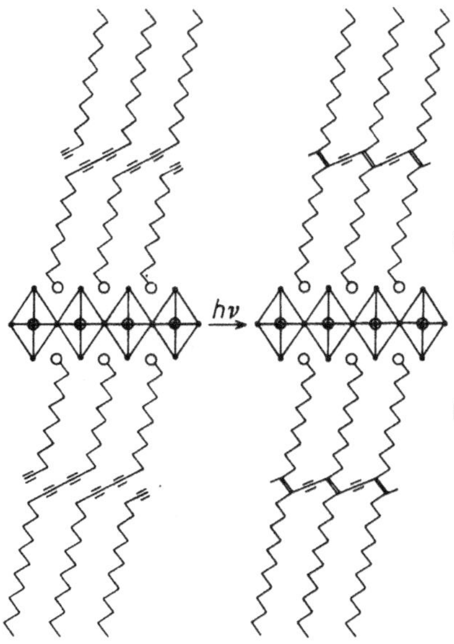

Fig. 21. Schematic representation of the diacetylene polymerization in layer perovskites. Model of the 010 projection ($M = Mn^{2+}, Cd^{2+}, Cu^{2+}, X = Cl^-$, for R see Tab. 7a)

Several primary amines carrying diyne units in their alkyl chains have been studied concerning their photoreactivity in the complex salts (Table 7.1). Compounds *87–89* are photoreactive upon UV- or γ-irradiation, but reaction rate and colour of the polymer strongly vary with the nature of the inorganic counterions.

This indicates a topochemical control by the inorganic part of the complex salt. The reaction in layer perovskites is schematically represented in Fig. 21.

The α,ω-diamino-alkadiyne *92* is photoinactive, apparently due to a packing geometry of the monomer unfavourable for the photopolymerization. Moreover, sterical reasons may play a role, since the molecules to be polymerized are held at both ends so that any configurational change is hindered and likely would break up the lattice[157].

2.8 Further Complex Salts

Besides the layer perovskite halide salts further complex salts of unknown structure have been reported. Since they all are formed by amphiphilic diacetylenes, it may be inferred that they exhibit a bilayer type arrangement as well. The complex salts are listed in Table 7.2.

The inorganic part of the complexes *94* and *95* is assumed to exhibit a band structure in analogy to the corresponding complex of unsubstituted pyridine, $Cu(py)_2Cl_2$ [102]. This complex consists of edge sharing, distorted octahedra, formed by four chlorine and two nitrogen atoms of the pyridine units, with the metal atoms in the centers [158, 159].

Both the complexes *94* and *95* are only little photoreactive, which probably is a consequence of an unfavourable packing geometry of the diyne moieties. Exposure to a γ-ray dose of 30 Mrad converts about 40 percent of *94*, and 10 percent of *95* into insoluble polymer [102]

It may be added that metal complexes of the polymer of *36* can also be obtained from solution. The polymer of *36* is readily soluble in chloroform, and rapidly forms metal complexes, if ethanolic solutions of metal halides such as $CdCl_2$, $CuCl_2$, or $MnCl_2$ are added to the polymer solution[103].

Ruthenium tris-2,2′-bipyridine complexes are known to play an active role as photochemical redox agents in the solar energy conversion [160]. The corresponding amphiphilic complex *96* does not polymerize, presumably due to sterical reasons. The Ru^{2+} ions are octahedrally coordinated by the three 2,2′-bipyridine ligands, which causes a very bulky size of the head group. As a consequence, the distances between adjacent diyne units likely become too large for a solid-state 1,4-addition [102].

3 Butadiene Derivatives

3.1 General Remarks on the Photoreactivity

While the solid state photochemistry of diacetylene derivatives has been investigated in great detail, this applies much less to butadiene derivatives. Studies on photoreactions of butadiene derivatives indeed showed the formation of dimers and polymers upon exposure to UV-light, but neither systematic studies on the structure and properties of the polymers nor on the reaction mechanism were reported yet.

Comprehensive studies on the dimer formation were carried out by G. M. J. Schmidt and coworkers [161–163]. They observed for a variety of compounds, such as mucodinitrile, muconic acid monomethyl ester, and sorbic acid amide, a [2 + 2]-cyclodimerization upon exposure to UV-light. Besides dimers small amounts of oligomers were formed, in addition. Careful analysis of the photoproducts as well as crystal structure determinations revealed that the photoreactions proceed under lattice control.

However, photoreactivity of butadienes is not restricted to the bulk crystalline state, but also occurs in less ordered systems, as for example in the liquid-expanded state of monolayers [164], in bilayer aggregates above the phase transition point [165], and in host guest complex systems, in which the guest monomers frequently are not located at fixed lattice sites [5–7].

In addition to that, spontaneous thermal polymerization in the solid state was reported from 2,4-pentadienoic acid [166], leading to atactic, partially crosslinked polymers.

Only recently, examples of a lattice controlled polymerization of butadiene derivatieves have been reported [24, 25, 167–171]. UV- or γ-irradiation of butadienes crystallized in perovskitetype layer structures yielded erythro-diisotactic 1,4-trans-polymers [25]. Furthermore, crystalline 1,4-trans-polymers could be obtained upon UV- or γ-irradiation of native halide salts of unsaturated primary amines [167], and long chain butadienes in lipid layer structures [170, 171].

Polymerization proceeds as a 1,4-addition of the conjugated double bonds according to

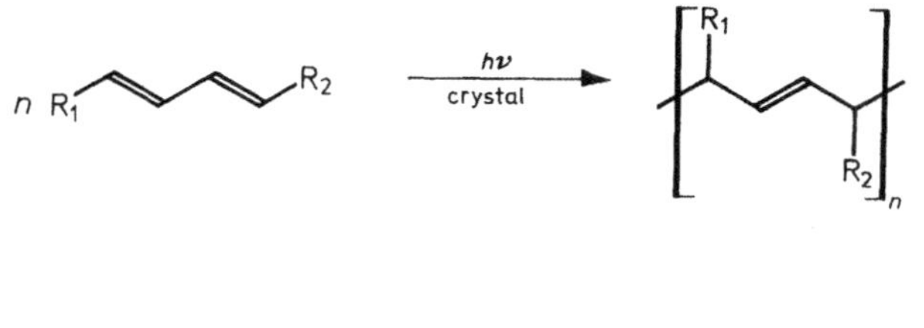

in analogy to the solid state polymerization of butadiyne derivatives [2-4]. However, lattice-controlled polymerization only occurs if certain requirements are fulfilled:

Firstly, monomers must be arranged in layer structures restricting the photoreactivity to only two dimensions.

Secondly, monomers should exhibit a specific packing within the layer plane allowing the 1,4-addition, but being unfavourable for the competing [2 + 2]-cycloaddition.

Thirdly, monomers ought to be hydrogen-bonded mutually, either in the layer plane or to molecules or ions in adjacent layers. Strong intermolecular interactions prevent molecular diffusion and conformational rearrangements during polymerization and favour the formation of stereoregular polymers.

In the following an outline is given on the polymerization of butadiene derivatives in various layer structures.

3.2 Layer Perovskite Halide Salts

3.2.1 Various Derivatives

Butadiene derivatives containing aminomethyl groups as pendant substituents are able to form layer perovskite halide salts [162-169], as similarly reported from aminodiacetylenes (chapter 2.7) or other primary amines R-NH_2 with unsaturated units in R [24,157]. The structure of layer perovskites has been described in chapt. 1.3. already.

In Table 8 compounds are listed which have been studied concerning their reactivity upon exposure to UV-light or ^{60}Co-γ-irradiation. All compounds were found to be the butadiene chromophor was extended by a C=O unit, shifting the absorption maximum to about 260 nm. A thermal polymerization was not observed for any of the compounds.

The photoreactivity was detected by spectroscopic methods. In Fig. 22 infrared spectra of *100a* are shown, which were monitored after various γ-ray doses received

Table 8. Butadiene derivatives and their reactivities in layer perovskites [R—CH=CH—CH=CH—$CH_2NH_3]_2MtX_4$ [24,169]

No.	R of organic compound	Inorganic salt MtX_2	Reactivity upon irradiation with	
			UV (λ = 254 nm)	^{60}Co-γ-rays
97a	H	$CdCl_2$	—	+
98a	CH_3	$CdCl_2$	—	+
99a	COOH	$CdCl_2$	+	+
99b	COOH	$CuCl_2$	—	—
99c	COOH	$FeCl_2$	+	+
99d	COOH	$MnCl_2$	+	+
99e	COOH	$CdBr_2$	—	—
100a	$COOCH_3$	$CdCl_2$	+	+
100b	$COOCH_3$	$CuCl_2$	—	—
101a	$COOCH(CH_3)_2$	$CdCl_2$	+	+
102a	$COOCH_2CH(CH_3)_2$	$CdCl_2$	+	+
103a	$COO(CH_2)_3CH_3$	$CdCl_2$	+	+

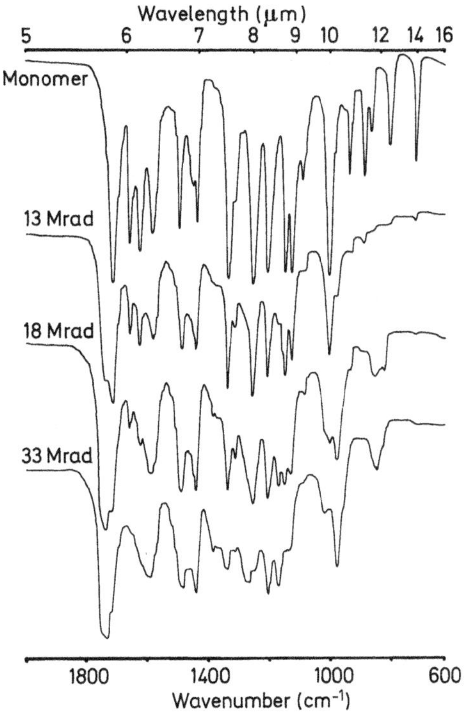

Fig. 22. Section of IR-spectra of the layer perovskite compound *100a* as monomer and after various γ-ray doses [169]

by the crystals. A reaction of the butadiene groups is indicated by the decrease of the C=C stretching and wagging modes, typically occurring at about 1620 and 1000 cm^{-1}, respectively [24, 167–169]. Simultaneously, the wagging mode of an isolated C=C trans unit appears at about 960 cm^{-1}. ^{13}C-NMR spectra [169] of the reaction product of *99a* show two signals in the region of unsaturated carbon atoms (δ = 135.4 and 134.2 ppm), and two signals in the region of saturated carbon atoms (δ = 49.8 and 49.2 ppm), which have been attributed to a 1,4-trans-polybutadiene backbone.

Conversion versus dose curves of compounds $[R'-CH=CH-CH=CH-CH_2NH_3]_2\ MtX_4$ are shown in Figs. 23a and b. They demonstrate effects on the photoreactivity due to a variation of the substituent R′, and the inorganic salt MtX_2 [24]. From Fig. 23b a topochemical control of the reaction caused by the nature of the Mt^{2+} ions is evident. Similar effects are also reported for corresponding diacetylene derivatives (Chapt. 2.7). The butadiene polymerization in layer perovskite halide salts is schematically represented in Fig. 24 [169].

3.2.2 Structure and Morphology

Crystals of the layer perovskite halide salts are typically obtained as small platelets which are frequently twinned. Preliminary X-ray investigations [24] showed that the three-dimensional order of the compounds *97–103* is retained during the polymerization process. As Fig. 25 indicates, crystals of compounds *101a* actually remain unchanged in their appearance. This may originate from the relatively small changes of the unit cell parameters during polymerization [24, 167].

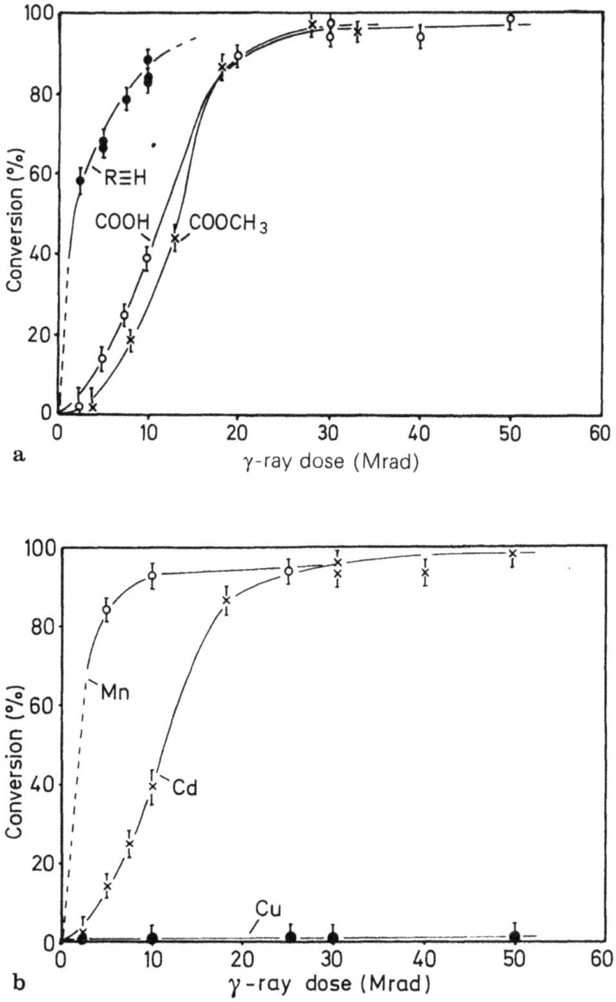

Fig. 23a and b. Conversion vs. dose curves of layer perovskites [R—CH=CH—CH=CH—$CH_2NH_3]_2$ MtX_4. **a.** Effect of variation of R (Mt = Cd^{2+}, X = Cl^-), and (**b**) effect of variation of Mt (R = COOH) on the reaction kinetics [24]

The crystal structure of the polymerized compound *99a* has been determined by X-ray diffraction [25]. The crystals with space group P $\bar{1}$ (a = 0.7214, b = 0.7247, c = 1.8594 nm, α = 104.49°, β = 96.63°, γ = 95.71°, Z = 2) consist of two types of layers which alternate in the third dimension. One type of layers consists of corner sharing $CdCl_6$ octahedra, whereas the other type contains the organic polymer (Fig. 26a). Consecutive layers are H-bonded between the ammonium and Cl-atoms (Fig. 26b).

The Cl-octahedra are distorted, and the octahedra layer deviates from the regular chessboard like arrangement. Consecutive octahedra are rotated against each other about the layer normal by approximately 30°.

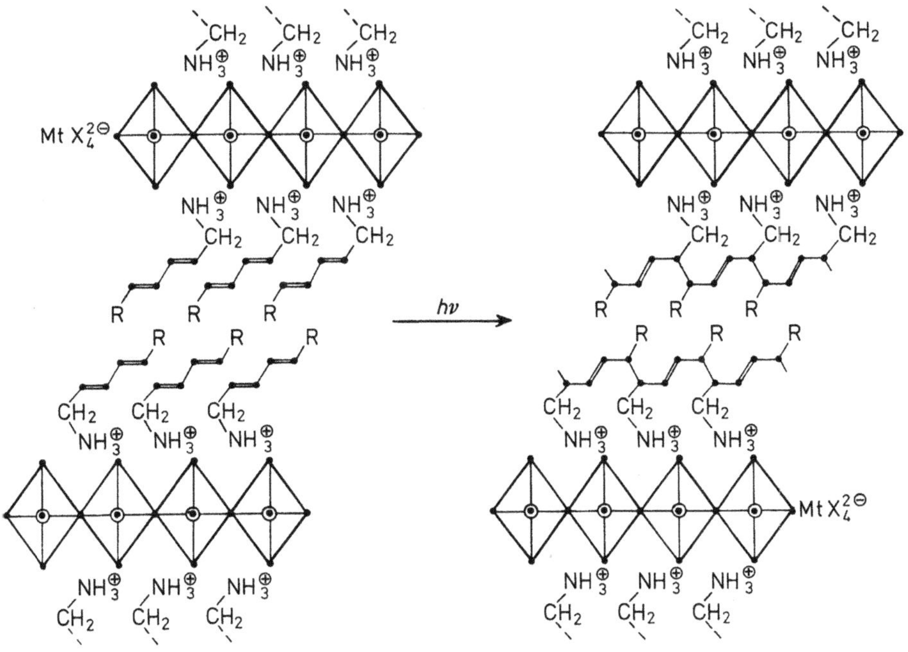

Fig. 24. Schematic representation of the butadiene polymerization in layer perovskites. Model of the 010 projection (Mt = Mn^{2+}, Cd^{2+}, X = Cl^-, R see Tab. 8) [169]

From comparative X-ray studies of monomeric crystals a contraction of the lattice constant parallel to the polymer chain direction by approximately 8% is evident. The contraction induces the rotation of the octahedra against each other, which creates the two types of different nitrogen cavities, and thus explains the existence of the two crystallographically independent parts [25].

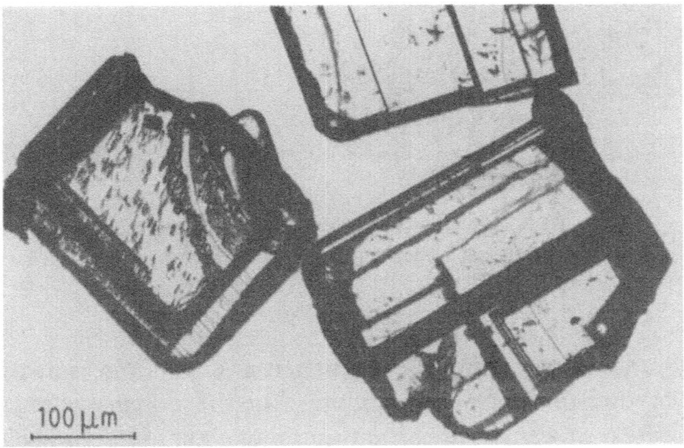

Fig. 25. Polymerized layer perovskites. Crystals of the layer perovskite compound *99a* after exposure to a γ-ray dose of 30 Mrad [24]

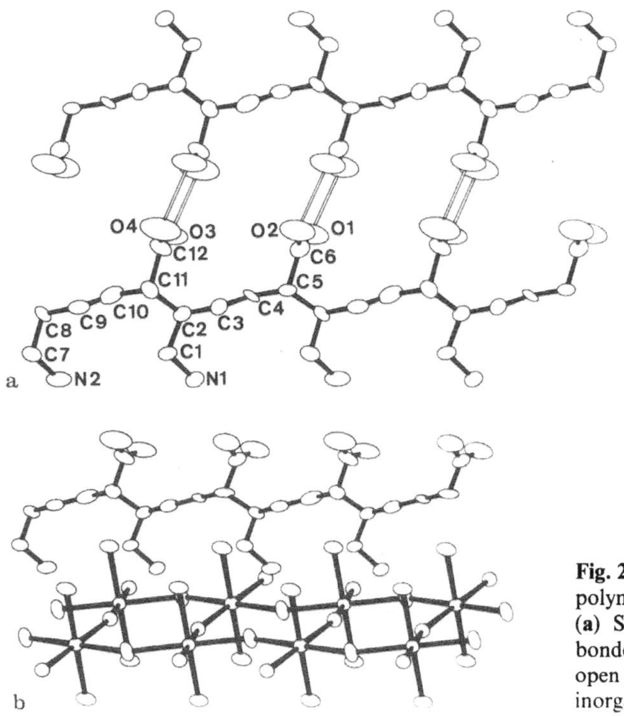

Fig. 26 a and b. Crystal structure of the polymerized layer perovskite *99a* [25]. (**a**) Section of the polymer with H-bonded oxygen atoms represented by open bonds; (**b**) Perspective view of the inorganic and polymer layers

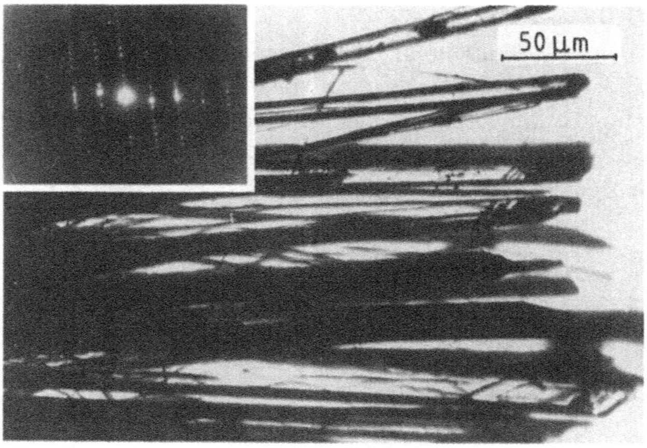

Fig. 27. Evidence for one-dimensional chain growth in layer perovskites: Fibrous structure of polymerized compound *99a*, obtained upon mechanical cleavage of crystals shown in Fig. 25. Insert: Electron diffraction pattern of a polymeric fiber [24]

Both the carboxylic acid and aminomethyl substituent groups are in an isotactic arrangement, causing an erythro-diisotactic configuration of the polymer chains. Indeed, the 1,4-addition induces two centers of chirality per monomer unit. However, due to the center of symmetry of the space group the unit cell contains a racemic mixture of the polymer.

As a consequence of the one-dimensional chain growth, the polymeric crystals can easily be cleaved mechanically along the crystal axis parallel to the chain direction (a-axis). This is shown in Fig. 27. The electron diffraction pattern shown in the same figure demonstrates the crystalline perfection of the polymer [24, 167].

3.2.3 Thermal Properties

and polymeric layer perovskites are listed. As indicated by the Table, the heat stability of the complex crystals considerably increases upon polymerization [24, 167]. DSC-studies indicate an increase of the decomposition temperature by about 60° for compound *99a* and 100° for compound *100a*. The two polymeric compounds decompose at a temperature of about 620 K.

Table 9. Thermal properties of monomeric and polymeric halide and double halide salts of butadiene derivatives [24, 167]

No.	Phase transition of monomer [K]	M.p. [K][a]	
		Monomer	Polymer
99a	—	562	620
99b	420	499	—
99e	381, 463	497	—
100a	—	502	614
100b	—	420	—
103a	—	420	
106a	375, 455	505	607
106b	424	504	—

[a] Melting always under decomposition

3.2.4 Solution Properties of the Polymers

Only the acid derivative *99a* is reported to form a soluble polymer [25]. After separation from the cadmium ions by a precipitation as insoluble CdS the polymer behaves as an ampholyte, being soluble in acidified and alkaline aqueous media and insoluble in weak acids or bases, neutral water, and organic solvents. Depending on the pH of the solvent three differently ionized structures I–III are formed:

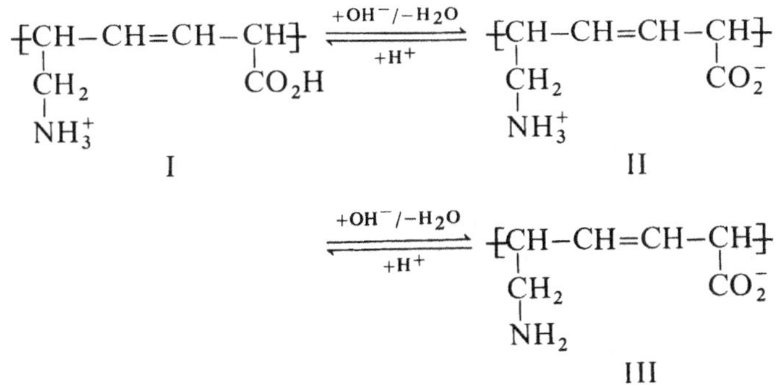

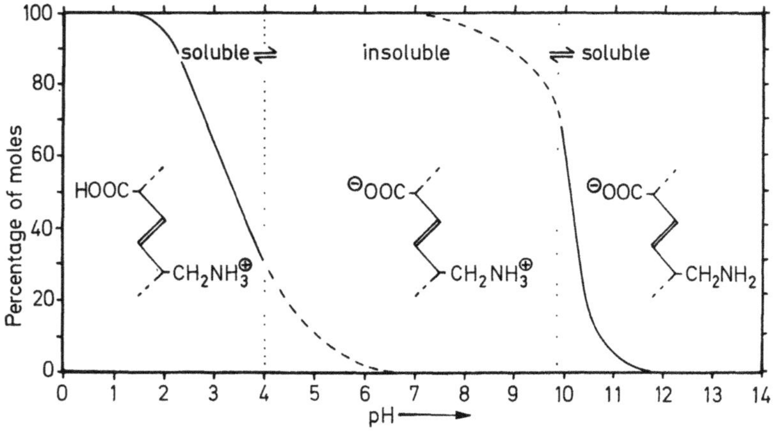

Fig. 28. Ionization of poly(6-amino-2,4-hexadienoic acid) as a function of the pH-value of the aqueous solution. In the pH-range 4.0–9.9 the polymer precipitates from solution [25]

As the titration curve (Fig. 28) indicates the polymer exhibits a broad isoelectric region ranging from pH 6.0 to pH 7.5. Apparently the interactions with the solvent molecules are only weak in this region, very likely due to the ability of the zwitterionic structure II to form an intramolecular salt IV [25]:

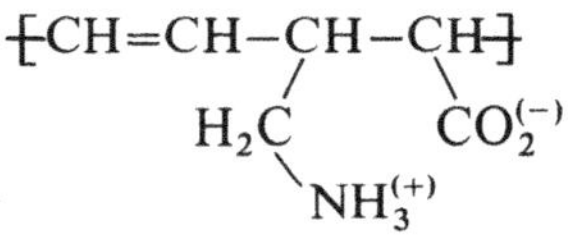

IV

The pK_a- and pK_b-values of the COOH- and the NH_2-group have been determined to be 3.3 and 3.9, respectively.

Further characterizations of the polymer by viscosity measurements [25] and gel permeation chromatography [169] revealed an average molar mass of about $4 \cdot 10^4$ g/mol, in good agreement.

3.2.5 Chelate Complexes with Transition Metals

The regular structure of the polymer not only causes the formation of intramolecular salts IV, but also favours the formation of chelate complexes with transition metal ions. As recently reported, divalent metal ions dissolved in an aqueous solution are readily taken up by the neutral polymer suspended in the same solution [25].

The structure of the polymer-metal complexes could not be determined yet. However, from infrared spectra an ionization of the carboxylic acid units is evident [25]. Since the aminomethyl groups are in positions adjacent to the carboxylate units, they likely act as ligands and contribute to the formation of chelate complexes with the metal ions, as schematically represented in Fig. 29.

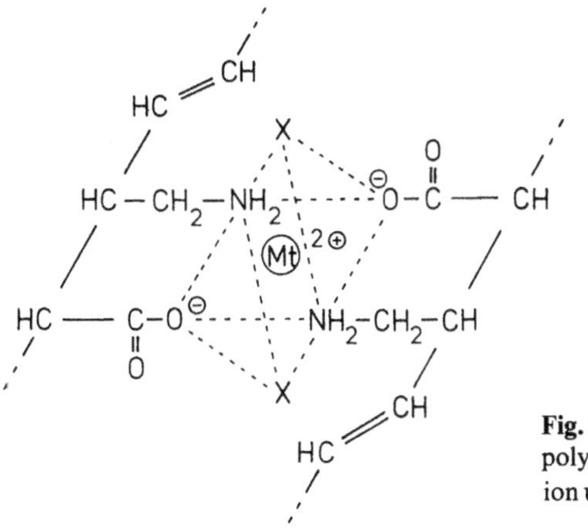

Fig. 29. Possible structure of the Mt^{2+}-complex of poly(6-amino-2,4-hexydienoic acid) formed by metal ion uptake from solution. X can be OH^-, Cl^-, H_2O [25)]

3.3 Native Halide Salts

Besides the complex salts also native salts of the butadiene compounds were studied concerning their reactivity upon UV- and ^{60}Co-γ-irradiation [167, 170)]. As listed in Table 10, only the hydrochlorides *104* and *106a* are reactive upon exposure to γ-rays. In addition, compound *106a* is photoreactive in UV-light. As in the layer perovskites 1,4-trans-polybutadienes are obtained [167)].

Table 10. Butadiene derivatives and their reactivities as halide salts [R—CH=CH—CH=CH—CH$_2$—NH$_3$]X [167, 170)]

No.	R	X	Reactivity upon irradiation with	
			UV (λ = 254 nm)	^{60}Co-γ-rays
104	H	Cl	–	+
105	CH_3	Cl	–	–
106a	COOH	Cl	+	+
106b	COOH	Br	–	–
107	$COOCH_3$	Cl	–	–
108	$COO(CH_2)_3CH_3$	Cl	–	–

The polymerization of compound *106a* was studied more in detail. A γ-ray dose of 30 Mrad is sufficient to nearly cause a complete conversion to polymer. However, the reactivity strongly depends on the morphology of the crystals [167)], and a considerable amount of polymer is formed due to a post-irradiative polymerization [172)]. This is in contrast to the photoreaction in layer perovskites [167)].

From X-ray powder patterns of *106a* a layer structure with a head-head-tail-tail orientation of the monomers is evident [167)]. This packing is different from the structure of saturated alkyl ammonium halides [36, 173)].

During polymerization the layer spacing discontinuously decreases from 1.44 to 1.20 nm. The onset of the phase transition correlates with an increase of the reaction rate, indicating a packing geometry of the new phase being more favourable for the 1,4-addition [167]. However, polymer single crystals are not obtained. The structural changes of the layer spacings induce cracks in the crystals running parallel to each other, apparently in a direction parallel to the polymer chains (Fig. 30).

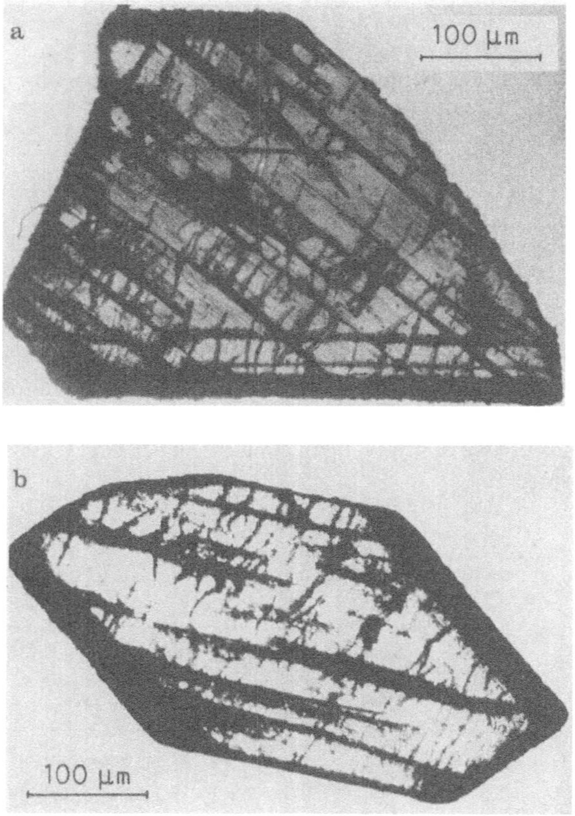

Fig. 30a and b. Polymerized crystals of *106a*, after exposure to a γ-ray dose of 30 Mrad [167]

Polymerization increases the heat stability of the halide salt *106a* strongly. As Tab. 9 indicates, the decomposition temperature is raised by 100°.

3.4 Amphiphilic Compounds

Several amphiphilic butadienes have been synthesized and studied concerning their polymerization properties in monomolecular layers, LB-multilayers, bilayer type aggregates (vesicles), and in the crystalline state.

3.4.1 Photoreactivity in the Crystalline State

Compounds which were studied in the crystalline state are listed in Table 11 [164, 171].

Table 11. List of amphiphilic butadienes $R_1-CH=CH-CH=CH-R_2$ and their photoreactivities in the crystalline state

No.	R_1	R_2	M.p. (°C)	Reactivity upon: UV-light[a]	Reactivity upon: ^{60}Co-γ-rays	Layer spacings in nm: Monomer	Layer spacings in nm: Polymer	Ref.
109	$CH_3-(CH_2)_{12}-$	$-CH_2OH$	58	—	—			164)
110	$CH_3-(CH_2)_{12}-$	$-CHO$	33	—	—			164)
111	$CH_3-(CH_2)_{12}-$	$-COOH$	71	—	—			164)
112	$CH_3-(CH_2)_{16}-C(=O)-NH-CH_2-$	$-COOH$	145	+	+	5.42	5.27	171)
113	$CH_3-(CH_2)_{17}-OC(=O)-$	$-CH_3$	45	+	+	4.92	4.67	171)
114	$CH_3-(CH_2)_{17}-NH-C(=O)-$	$-CH_3$	99	+	+	4.74	5.31	171)
115	$CH_3-(CH_2)_{17}-OC(=O)-$	$-C(=O)O-(CH_2)_{17}-CH_3$	88–89	—	—	3.95	—	171)
116	$CH_3-(CH_2)_{17}-NH-C(=O)-$	$-C(=O)-NH-(CH_2)_{17}-CH_3$	185–188	—	—			171)

[a] λ = 254 nm

Some of the amphiphiles are photoreactive, if exposed to UV- or γ-irradiation, but the doses sufficient for a complete conversion to polymer vary considerably.

Figure 31 shows infrared spectra of *113* recorded after exposure of the crystals to various γ-ray doses [171]. The reaction of the butadiene unit is indicated by the decrease of the C=C-stretching and wagging mode intensities at 1610–1640 and 1000 cm^{-1},

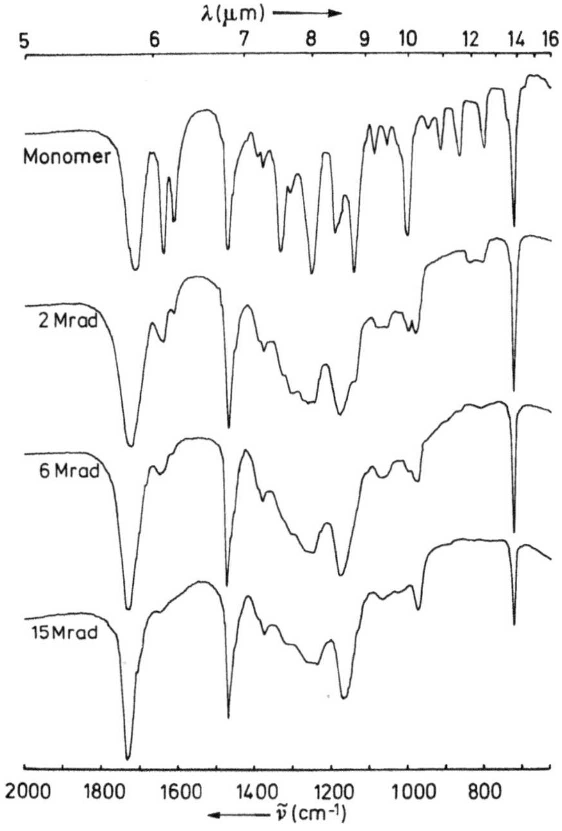

Fig. 31. Section of IR-spectra of *113* as monomer and after various γ-ray doses [171]

respectively, and by the simultaneous appearance of a new mode at 970 cm^{-1} which can be ascribed to an isolated 1,2-trans-C=C-unit. The ^{13}C-NMR-spectrum of the irradiation product (Fig. 32) [171] indicates the formation of a 1,4-trans-adduct, showing two signals (δ = 126.8 and 137.1 ppm) of the unsaturated and two signals (δ = 39.4 and 56.0 ppm) of the saturated carbon atoms of the polymer backbone. Ether or peroxide linkages are not observed in any of the polymers. Effects on the chain growth due to oxygen can thus be ruled out [171].

Preliminary X-ray structure studies of monomer and polymer crystals indicate that the three-dimensional order is essentially retained during polymerization. The layer spacings of individual compounds are listed in Table 11. However, the few informations on the structure available at present do not allow to draw conclusions on the stereoregularity of the polymer [171].

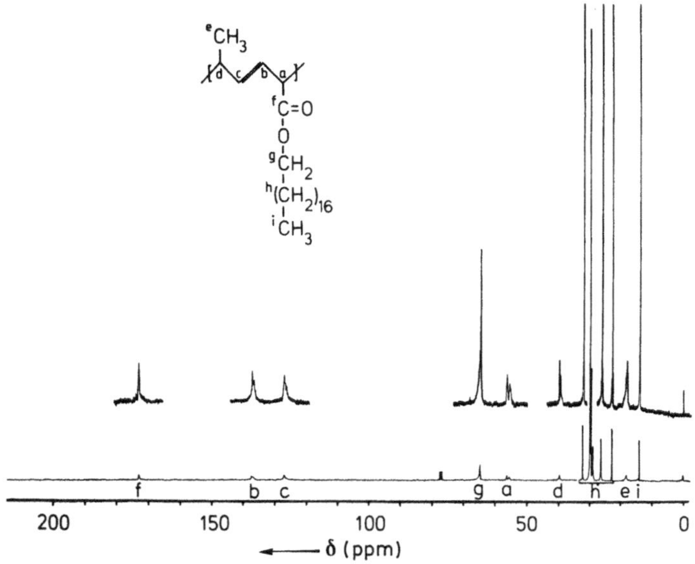

Fig. 32. ^{13}C-NMR-spectrum of irradiation products of *113* after dissolution in $CDCl_3$ (γ-ray dose received by the crystals: 15 Mrad) [171)]

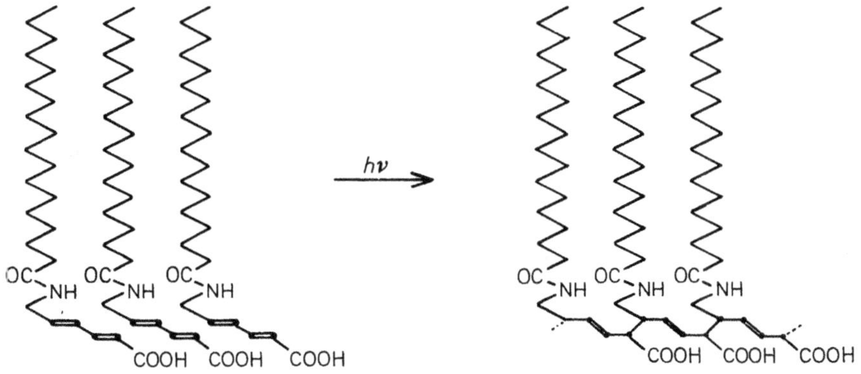

Fig. 33. Schematic representation of the solid state polymerization of a single layer of *112* [170)]

Only for compound *112* the formation of a completely crystalline polymer is reported [171)]. Apparently the strong hydrogen bonding of either substituents renders structural rearrangements during polymerization difficult and causes only small changes in the layer spacing [171)]. Structure and polymerization of a single layer of *112* is schematically represented in Fig. 33.

The polymer of *112* is extremely stable against strong acids and bases and decomposes only at temperatures higher than 325 °C [171)].

3.4.2 Mono- and Multilayers

Amphiphilic butadienes, which have been studied concerning their polymerization in mono- and multilayers are listed in Table 12. The photoreactivity at the air-water

Table 12. List of amphiphilic butadiene derivatives and their photoreactivities in monolayers at the air-water interface and LB-multilayers

No.	R_1	R_2	Photoreactivity[a] in		Ref.
			Monolayers	Multilayers	
a) single chain amphiphiles $R_1-CH=CH-CH=CH-R_2$					
109	$CH_3-(CH_2)_{12}-$	$-CH_2OH$	+	–[b]	[164]
110	$CH_3-(CH_2)_{12}-$	$-CHO$	+	–[b]	[164]
111	$CH_3-(CH_2)_{12}-$	$-COOH$	+	–[b]	[164]
112	$CH_3-(CH_2)_{16}-C(=O)-NH-CH_2-$	$-COOH$	+	+	[171]
117	$CH_3-(CH_2)_{17}-OC(=O)-$	$-H$	not reported	+	[174]
118	$CH_2=CH-(CH_2)_{20}-OC(=O)-$	$-H$	not reported	+	[174]
b) double chain amphiphiles $R_1-CH=CH-CH=CH-C(=O)O-R_2-OC(=O)-CH=CH-CH=CH-R_1$					
119	$CH_3-(CH_2)_{12}-$	$-CH_2-CH(-)-CH_2OH$	+	–[b]	[165]
120	$CH_3-(CH_2)_{12}-$	$-CH_2-CH(-)-CH_2-OP(=O)(O^{(-)})O-(CH_2)_2-\overset{(+)}{N}(CH_3)_3$	+	–[b]	[165]

[a] Upon UV-irradiation ($\lambda = 254$ nm); [b] No formation of multilayers

interface was studied by measuring the Π-A-isotherms prior and subsequent to UV-irradiation.

Monolayers of the amphiphiles *109–111* exhibit relatively large A-values at their collapse points, which is an indication that the butadiene units are in a flat position on the water surface [164]. Only compound *112* forms a condensed phase with a dense packing of the monomers and a collapse pressure near 55 mN/m.

The amphiphiles *109–111* rapidly polymerize in the solid condensed and in the liquid expanded state. Polymerization is accompanied by a large contraction of the film area. Ringsdorf and Schupp [164] believe the change in the film area to be due to a transition of the polar headgroups from a flat into an erected position during polymerization.

The packing density of *112* changes only very little during polymerization at the air-water interface. Reorientation effects of the headgroups can thus be ruled out, and the corresponding polymer film is likely packet more regularly [171].

The polymers formed in the liquid expanded as well as in the condensed state are assumed to exhibit a 1,4-trans-structure [164]. However, a detailed analysis of the products is still lacking, probably because the amount of material formed in the monolayers is extremely small.

In Langmuir-Blodgett-multilayers the photoproducts can be determined more easily. While compounds *109–111* could not be deposited by the LB-technique [164], presumably due to the short lengths of the hydrophobic tails, compound *112* could easily be transferred onto various hydrophobic substrates [171]. The polymerization in multilayers was followed by UV- and IR-spectroscopy. During exposure to UV-

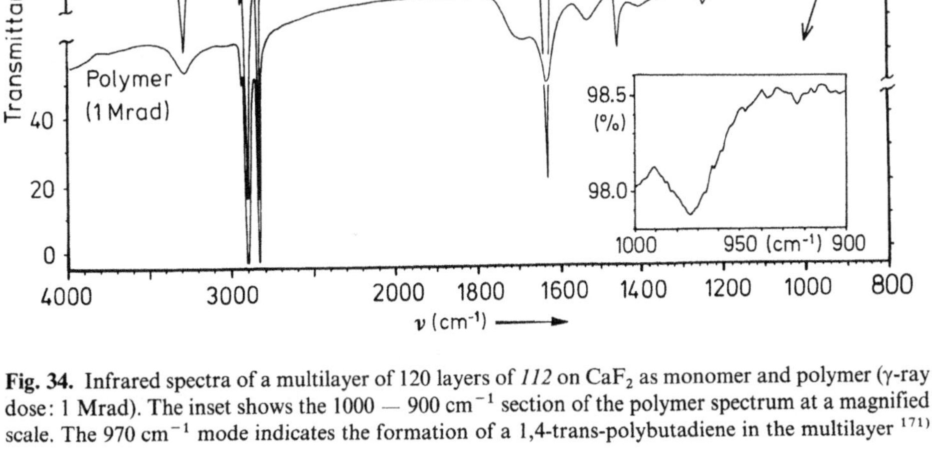

Fig. 34. Infrared spectra of a multilayer of 120 layers of *112* on CaF_2 as monomer and polymer (γ-ray dose: 1 Mrad). The inset shows the 1000 — 900 cm^{-1} section of the polymer spectrum at a magnified scale. The 970 cm^{-1} mode indicates the formation of a 1,4-trans-polybutadiene in the multilayer [171]

light the broad absorption of the sorbate chromophor with shoulders at 260 and 220 nm disappeared. Infrared spectra monitored prior and subsequent to γ-irradiation (Fig. 34) indicate the same spectral changes as they occur in the crystalline state. This allowed to identify the irradiation product as a 1,4-trans-adduct (compare also Fig. 31).

The docosyl- and ω-tricosenyl esters of penta-2,4-dienoic acid *117* and *118*, respectively, are also reported to rapidly polymerize in multilayers [174]. Irradiation with UV-light at $\lambda = 258$ nm $\pm$ 10 nm causes a [2 + 2] dimerization of the butadiene units. The dimers of structure I or II (Fig. 35) with two remaining acrylic double bonds absorb at 215 nm and behave as bifunctional four center monomers. Irradiation with the full UV lamp spectrum initiates a second reaction, in which slowly about 80% of the C=C stretching mode intensity at 1652 cm^{-1} vanishes. A polymer of hypothetic structure III or IV is formed (Fig. 35) [174].

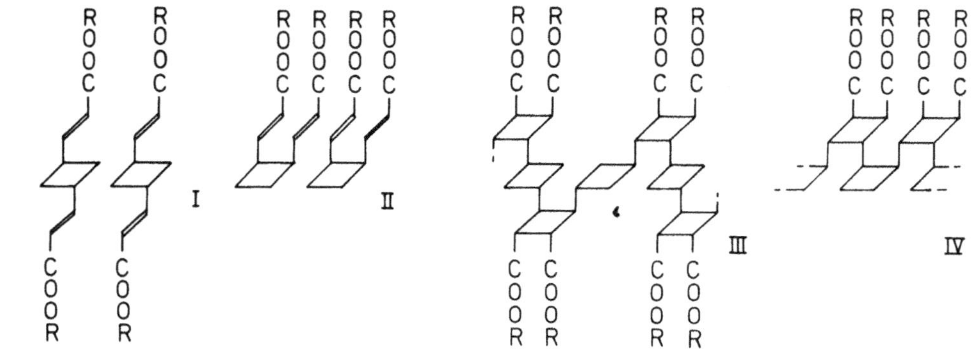

Fig. 35. Possible structures of 4-center-type dimers and polymers of *117* in multilayers ($R = C_{18}H_{37}$) [174]

In compound *118* even a third reaction takes place, in which the ω-double bonds are crosslinked via electron beam radiation with doses of 30 to 60 Mrads [174].

The [2 + 2] cycloaddition is supposed to proceed under lattice control [174], though detailed structural studies are lacking. Further studies seem to be necessary to fully establish the structure of the polymer, especially since a completely different reaction behaviour has been observed for the corresponding compound *113* (see Chapt. 3.4.1., Figs. 31 and 32).

3.4.3 Bilayer-Type Aggregates (Vesicles, Liposomes)

Polymerizable surfactant aggregates such as vesicles and liposomes have been prepared from amphiphiles with diyne (Chapt. 2.4.), methacryloyl [175,176], vinyl [177,178], styrene [179] and butadiene units. The butadiene amphiphiles are listed in Table 13. All compounds readily polymerize upon exposure to UV-light as the decrease of the absorption of the sorbate unit ($\lambda_{max} = 257$ nm) to less than 5 percent of the initial value indicates [180]. During polymerization the spherical shape of the vesicles is preserved [124].

Table 13. List of vesicle forming butadiene derivatives

$$\begin{array}{l} CH_3-(CH_2)_m-CH=CH-CH=CH-\overset{O}{\overset{\|}{C}}O- \\ \qquad\qquad\qquad\qquad\qquad\qquad\qquad\qquad X \\ CH_3-(CH_2)_m-CH=CH-CH=CH-\underset{O}{\underset{\|}{C}}O- \end{array}$$

No.	m	X	M.p. (°C)	Photoreactivity[a]	Ref.
119	12	$-CH_2$ $-CH$ CH_2OH	59	+	106,165)
120	12	$-CH_2$ $-CH$ $CH_2-O\overset{O}{\overset{\|}{P}}O-(CH_2)_2-\overset{(+)}{N}(CH_3)_3$ $O^{(-)}$	220	+	106,165)
121	10	$-CH_2$ $-CH$ $CH_2-O\overset{O}{\overset{\|}{P}}O-(CH_2)_2-\overset{(+)}{N}(CH_3)_3$ $O^{(-)}$		+	180)
122	12	$-(CH_2)_2-\underset{CH_3}{N}-(CH_2)_2-$	94	+	165)
123	12	$-(CH_2)_2-\overset{CH_3}{\underset{CH_3}{N^{(+)}}}-(CH_2)_2-Br^{(-)}$	36	+	106,165)
124	10	$-(CH_2)_2-\overset{(+)}{N}H-(CH_2)_2-$ $(CH_2)_2-SO_3^{(-)}$		+	180)
125	9	$-(CH_2)_2-\overset{(+)}{N}H-(CH_2)_2-$ $(CH_2)_2-SO_3^{(-)}$		+	146)

[a] upon UV-irradiation ($\lambda = 254$ nm)

The polymer is expected to be at least occasionally cross-linked [180]. Since the polymer chains are very flexible, the amphiphilic polybutadienes represent a better approach to biological membranes than the corresponding polydiacetylenes [146].

In particular of interest are partially polymerized membranes, which consist of polymerizable butadiene lecithines and saturated lecithines (e.g. dimyristoyl-lecithin, DMPC). While vesicles of DMPC exhibit a typical "ripple" structure, the surface of the polymerized vesicles appears homogeneous. Electron micrographs of mixed vesicles show a phase separation, indicated by the simultaneous presence of regions with a smooth surface due to the polymer, and "rippled" regions due to DMPC [146]. Instead of DMPC also a cystine derivative with fluorocarbon chains has been used to form mixed vesicles with the butadiene lecithine. This compound can be converted into a water-soluble lysolipid upon treatment with reductive agents, e.g. dithionite [146]. After polymerization of the butadiene compound the mixed vesicles were treated with dithionite. As a consequence, the domains of the non-polymerizing cysteine derivative were pulled out of the membrane and left holes in the vesicles. This experiment clearly demonstrates the high stability of polymerized vesicles.

The increased stability, e.g. towards detergents, could also be demonstrated by 6-carboxy-fluorescein (6-CF) entrapped in the vesicles as a marker for the membrane permeability [146]. While vesicles of the monomer *119* are even more sensitive toward treatment with sodium dodecylsulfate (SDS) than those of the saturated analogue dipalmitoyl lecithin (DPPC), the polymerized vesicles show only a small increase in 6-CF permeability up to very high concentrations of the detergent [146].

Moreover, polymerized vesicles of *120* effectively entrap ^{3}H-glucose for at least a week and are not disrupted by treatment with surfactants [180].

4 Concluding Remarks and Outlook

In a sense, polymerization of monomers preoriented in layer structures belongs to the research area of biomimetic chemistry, which increasingly has become of interest in recent years [35]. The polymerization of butadiene and -diyne derivatives in lipid layers has found a considerable attraction, mainly because of potential technical applications of the polymerized structures, and of their suitability as stable models for biomembranes.

Polymerized Langmuir-Blodgett multilayers represent ultrathin, highly ordered polymers of exactly defined thickness. Technical device applications as high resolution X-ray and electron beam resists, guided wave structures, optical switches, and as passivating layers for semiconductor devices are presently under investigation. Moreover, multilayers have been proven to be excellently suited for quantitative photochemical and -physical studies of the polymerization processes, as well as studies of interactions between the conjugated polydiacetylene backbones and the environment.

Moreover, the diyne surfactants represent excellent probes for the study of the physical state of order of surfactant assemblies, due to the anisotropic properties of the polymer, and the topochemical control of the polymerization process, which easily allows to distinguish between ordered and disordered regions.

Polymerized surfactant assemblies exhibit enhanced stabilities, and decreased permeabilities as compared with the corresponding monomeric ones. Potential applications as drug carriers and devices for solar energy conversions have been discussed [32, 120]. Moreover, previous studies indicated their suitability as synthetic models for biomembranes, able to mimick simple cell membrane functions and cell-cell interactions.

The reactivity of butadiene and -diyne derivatives in layer perovskites demonstrates the suitability of these complex salts to act as templates for solid state reactions. Besides the polyaddition also other chemical reactions should be able to occur with a high stereoselectivity, as for example isomerizations, or addition reactions, such as hydrations, halogenations, oxidations, or polycondensations. In any case it will be worth while to seak for further template reactions, which could either be photolyses of thermolyses.

The polymerization in the layer perovskites results in the formation of materials with novel, interesting properties. Extremely tough, high temperature resistent, one-dimensionally extended polymers with highly anisotropic properties are formed. After separation from the inorganic matrix polyelectrolytes with an erythrodiisotactic configuration of the substituents are obtained. Polymerization in layer perovskites therefore represents a novel method to directly synthesize well-defined, highly ordered polyelectrolytes.

Poly(6-amino-2,4-hexadienoic acid), for example, represents an amphoteric polyelectrolyte with unusual properties in solution. It can be used as a complexing agent for transition metal ions with a selectivity for divalent ions, such as Cu^{2+}, Co^{2+}, or Ni^{2+}.

The reactions in halide and double halide salts, as well as in lipid layer structures, are taylor-made reactions. Lattice-controlled polymerization previously only known from molecular crystals has been applied to layer-type arrays. Other molecular assemblies and organic inorganic complexes are imaginable, which also can serve as templates for solid state reactions. TaS_2 is known to accomodate amines and amino acids [181]. Alkylammonium titanoniobates (nC_nH_{2n+1}—NH_3) ($TiNbO_5$) form crystalline layer structures [182], as well as zirconium phosphates, e.g. $Zr(HPO_4)$ (n-$C_nH_{2n+1}PO_4$) [183]. Several other inorganic layer compounds are known to accommodate alkyl bimolecular layers between the inorganic layers such as mica-type layer silicates [184], niobates [185], vanadates [186], transition metal disulfides [187], and divalent metal phosphosulfides [188]. The alkyl chains between these layers are in the form of alkanols, alkylamines, and alkylammonium ions. Cu^{2+} and Ag^{+} form layered and chained complexes with organic dinitriles, e.g. $Cu(NC(CH_2)_nCN)_2NO_3$, which may also serve as a matrix for solid state reactions [189].

Moreover, lipid layer structures can also represent a matrix for solid state reactions, as recently demonstrated by Regen et al. [190], and Aliev et al. [191]. These authors utilized the outer and inner polar surface of vesicles as a template for polymerization of either acrylates or vinylpyridines, bound to the surface by ionic forces.

Acknowledgement: Many helpful discussions with Prof. G. Wegner and Dr. G. Lieser at the University of Freiburg, West-Germany, and Prof. D. Bloor at Queen Mary College, University of London, are gratefully acknowledged. Furthermore, the author is greatly indebted to Miss Karin Weiss, who essentially contributed to the preparation of this article, including helpful suggestions, a critical reading, and the typing of the manuscript.

5 References

1. Hasegawa, M.: Chem. Rev. *83* (1983) 507
2. Wegner, G.: in "Molecular Metals", W. E. Hatfield ed., Plenum Press, New York 1979, p. 209
3. Bloor, D.: in "Developments in Crystalline Polymers", D. C. Bassett ed., Appl. Science Publ. London 1982
4. for recent reviews see Cantow, H. J. ed.: "Polydiacetylenes", Adv. Polym. Sci. Ser. No. 63, Springer Verlag Berlin—Heidelberg 1984, and references cited herein

5. Farina, M., DiSilvestro, G., Sozzani, P.: Mol. Cryst. Liq. Cryst. *93* (1983) 169
6. Takemoto, K., Miyata, M.: J. Macromol. Sci. Rev. Macromol. Chem. *C18* (1980) 83
7. Chatani, Y.: Progr. Polym. Sci. Jpn. *7* (1974) 149
8. Gee, G.: Proc. Roy. Soc. A *153* (1935) 129
9. Scheibe, G., Schuller, H.: Z. Elektrochem. *59* (1955) 861
10. Cemel, A., Fort jr., T., Lando, J. B.: J. Polym. Sci. *A1*, *10* (1972) 2061
11. Hatada, M., Nishii, M., Hirota, K.: J. Coll. Interf. Sci. *45* (1973) 502
12. Ackermann, R., Naegele, D., Ringsdorf, H.: Makromol. Chem. *175* (1974) 699
13. Puterman, M., Fort, T., Lando, J. B.: J. Coll. Interf. Sci. *47* (1974) 705
14. Barraud, A., Rosilio, C., Ruaudel-Teixier, A.: J. Coll. Interf. Sci. *62* (1977) 509
15. Naegele, D., Lando, J. B., Ringsdorf, H.: Macromolecules *10* (1977) 1339
16. Blumstein, A.: Bull. Soc. Chim. France *1961*, 899
17. Friedlander, H. Z., Frink, C. R.: J. Polym. Sci. B*2* (1964) 475
18. Jakabhazy, S. Z., Morawetz, H., Morosoff, N.: J. Polym. Sci. *C-4* (1964) 805
19. Morosoff, N., Morawetz, H., Post, B.: J. Amer. Chem. Soc. *87* (1965) 3035
20. Shibasaki, Y., Fukuda, K.: J. Polym. Sci. Polym. Chem. Ed. *18* (1980) 2437
21. Blumstein, A.: "Advances in Macromol. Chem.", Vol. 2, W. M. Pasika ed., Acad. Press. London 1970, p. 123
22. Hirshfeld, F. L., Schmidt, G. M. J.: J. Polym. Sci. *A-2* (1964) 2181
23. Morawetz, H.: J. Polym. Sci. *C-12* (1966) 79
24. Tieke, B.: Mol. Cryst. Liq. Cryst. *93* (1983) 119
25. Tieke, B., Chapius, G.: J. Polym. Sci. Polym. Chem. Ed. *22* (1984) 2895
26. Fuhrhop, J., Mathieu, J.: Angew. Chem. *96* (1984) 124
27. Fendler, J. H.: Acc. Chem. Res. *13* (1980) 7
28. Fendler, J. H.: "Membrane Mimetic Chemistry", Wiley-Interscience, New York 1982
29. Fendler, J. H., Tundo, P.: Acc. Chem. Res. *17* (1984) 3
30. Breton, M.: J. Macromol. Sci. Rev. Macromol. Chem. *C 21* (1981) 61
31. Vincett, P. S., Roberts, G. G.: Thin Solid Films *68* (1980) 135
32. Langmuir, I.: J. Am Chem. Soc. *39* (1917) 1848
33. Blodgett, K. B.: J. Am. Chem. Soc. *57* (1935) 1007
34. Kuhn, H., Möbius, D., Bücher, H.: in "Physical Methods of Organic Chemistry", A. Weissberger, P. Rossiter eds., Vol. 1, part 3B, Wiley Interscience 1972, p. 577
35. Kunitake, T., Okahata, Y., Shinomura, M., Yasunomi, S., Takarabe, K.: J. Am. Chem. Soc. *103* (1981) 5401
36. Kind, R., Blinc, R., Arend, H., Muralt, P., Slak, J., Chapuis, G., Schenk, K. J. Zeks, B.: Phys. Rev. A, *26* (1982) 1816
37. Arend, H., Huber, W., Mischgofsky, F. H. Richter-van Leeuwen, G. K.: J. Cryst. Growth *43* (1978) 213
38. Kozelj, M., Rutar, V., Zupancic, I., Blinc, R., Arend, H., Kind, R., Chapuis, G.: J. Chem. Phys. *74* (1981) 4123
39. Needham, G. F., Willett, R. D., Franzen, H. F.: J. Phys. Chem. *88* (1984) 674
40. Prock, A., Shand, M. L., Chance, R. R.: Macromolecules *15* (1982) 238
41. Eckhardt, H., Prusik, T., Chance, R. R.: Macromolecules *16* (1983) 732
42. Patel, G. N., Bhattacharjee, H. R., Preziosi, A. F.: J. Polym. Sci. Polym. Lett. Ed. *19* (1981) 11
43. Patel, G. N., Chance, R. R., Turi, E. A., Khanna, Y.-P.: J. Am. Chem. Soc. *100* (1978) 6644
44. Lochner, K., Hinrichsen, T., Hofberger, W., Bässler, H.: phys. stat. sol. (a) *50* (1978) 95
45. Lochner, K., Bässler, H., Hinrichsen, T.: Ber. Bunsenges. Phys. Chem. *83* (1979) 899
46. Enkelmann, V., Wenz, G., Müller, M. A., Schmidt, M., Wegner, G.: Mol. Cryst. Liq. Cryst. *105* (1984) 11
47. Bässler, H.: in Ref. 4, p. 1
48. Enkelmann, V.: in Ref. 4, p. 91
49. Patel, G. N., Chance, R. R., Witt, J. D.: J. Chem. Phys. *70* (1979) 4387
50. Wenz, G., Wegner, G.: Mol. Cryst. Liq. Cryst. *96* (1983) 99
51. Wenz, G., Wegner, G.: Makromol. Chem. Rapid Comm. *3* (1982) 231
52. Plachetta, C., Rau, N. O., Hauck, A., Schulz, R. C.: Makromol. Chem. Rapid Comm. *3* (1982) 249
53. Wenz, G., Müller, M. A., Schmidt, M., Wegner, G.: Macromolecules *17* (1984) 837

54. Berlinsky, A. J., Wudl, F., Lim, K. C., Fincher, C. R., Heeger, A. J.: J. Polym. Sci. Polym. Phys. Ed. *22* (1984) 847
55. Sixl, H.: in Ref. 4, p. 49
56. Gross, H., Sixl, H.: Mol. Cryst. Liq. Cryst. *93* (1983) 261
57. Niederwald, H., Richter, K. H., Güttler, W., Schwoerer, M.: Mol. Cryst. Liq. Cryst. *93* (1983) 247
58. Niederwald, H., Schwoerer, M.: Z. Naturforsch. A, *38* (1983) 749
59. Day, D., Ringsdorf, H.: J. Polym. Sci. Polym. Lett. Ed. *16* (1978) 205
60. Day, D., Hub, H., Ringsdorf, H.: Isr. J. Chem. *18* (1979) 325
61. Day, D., Ringsdorf, H.: Makromol. Chem. *180* (1979) 1059
62. Tieke, B., Lieser, G., Weiss, K.: Thin Solid Films *99* (1983) 95
63. Tieke, B., Wegner, G.: in "Topics in Surface Chemistry", Plenum Press New York 1978, p. 121
64. Tieke, B., Lieser, G., Wegner, G.: J. Polym. Sci. Polym. Chem. Ed. *17* (1979) 1631
65. Olmstedt III, J., Strand, M.: J. Phys. Chem. *87* (1983) 4790
66. Schupp, H., Hupfer, B., van Wagenen, R. A., Andrade, J. D., Ringsdorf, H.: Coll. & Pol. Sci. *260* (1982) 262
67. Hub, H. H., Hupfer, B., Koch, H., Ringsdorf, H.: Angew. Chem. Int. Ed. Engl. *19* (1980) 938
68. Ruaudel-Teixier, A.: Mol. Cryst. Liq. Cryst. *96* (1983) 365
69. Tieke, B., Bloor, D., Young, R. J.: J. Mat. Sci. *17* (1982) 1156
70. Tieke, B., Lieser, G.: J. Coll. Interf. Sci. *88* (1982) 471
71. Kajzar, F., Messier, J.: Thin Solid Films *99* (1983) 109
72. Tieke, B., Bloor, D.: Makromol. Chem. *182* (1981) 173
73. Sarkar, M., Lando, J. B.: Thin Solid Films *99* (1983) 119
74. Tieke, B., Graf, H.-J., Wegner, G., Naegele, D., Ringsdorf, H., Banerjie, A., Day, D., Lando, J. B.: Coll. & Pol. Sci. *255* (1977) 521
75. Banerjie, A., Lando, J. B.: ACS Polymer Prepr. *19* (1978) 170
76. Kajzar, F., Messier, J.: J. Chem. Phys. *63* (1981) 123
77. Day, D., Lando, J. B.: Macromolecules *13* (1980) 1478
78. Day, D., Lando, J. B.: J. Appl. Polym. Sci. *26* (1981) 1605
79. Lieser, G., Tieke, B., Wegner, G.: Thin Solid Films *68* (1980) 77
80. Albrecht, O., Laschewsky, A., Ringsdorf, H.: Macromolecules *17* (1984) 937
81. Tieke, B., Enkelmann, V., Kapp, H., Lieser, G., Wegner, G.: ACS Org. Coat. Plast. Chem. Div. Prepr. *42*, 2 (1980) 396
82. Tieke, B., Enkelmann, V., Kapp, H., Lieser, G., Wegner, G.: J. Macromol. Sci. — Chem. A *15* (1981) 1045
83. Tieke, B., Wegner, G., Naegele, D., Ringsdorf, H.: Angew. Chem. Int. Ed. Engl. *15* (1976) 764
84. Tieke, B., Bloor, D.: Makromol. Chem. *180* (1979) 2275
85. Tieke, B., Weiss, K.: J. Coll. Interf. Sci. *101* (1984) 129
86. Tieke, B., Wegner, G.: Makromol. Chem. *179* (1978) 1639
87. Fouassier, J. P., Tieke, B., Wegner, G.: Isr. J. Chem. *18* (1979) 227
88. Lochner, K., Bässler, H., Tieke, B., Wegner, G.: phys. stat. sol. (b) *88* (1979) 653
89. Bloor, D., Hubble, C. L., Ando, J.: in "Molecular Metals", W. E. Hatfield ed., Plenum Press, New York 1979, p. 243
90. Tieke, B., Lieser, G.: J. Coll. Interf. Sci. *83* (1981) 230
91. Bubeck, C., Tieke, B., Wegner, G.: Ber. Bunsenges. Phys. Chem. *86* (1982) 499
92. Bubeck, C., Weiss, K., Tieke, B.: Thin Solid Films *99* (1983) 103
93. Bubeck, C., Tieke, B., Wegner, G.: Mol. Cryst. Liq. Cryst. *96* (1983) 109
94. Kajzar, F., Messier, J., Zyss, J.: J. Phys. Colloq. *C3* (1983) 709
95. Day, D., Lando, J. B.: Macromolecules *13* (1980) 1483
96. Tieke, B., Thesis, Ph. D.: University of Freiburg 1978
97. Bloor, D., Hubble, C. L.: Chem. Phys. Lett. **56** (1978) 89
98. Hupfer, B., Ringsdorf, H.: Chem. Phys. Lip. *33* (1983) 263
99. Wenz, G.: Diploma Thesis, University of Freiburg 1979
100. a) Sohn, J. E., Garito, A. F., Desai, K. N., Nurang, R. S., Kuzyk, M.: Makromol. Chem. *180* (1979) 2975
b) Garito, A. F.: Eur. Pat. Appl. EP 0022618 (1980)
101. Garito, A. F., Singer, K. D.: Laser Focus *80* (1982) 59 and references cited therein

102. Tieke, B., Weiss, K.: Coll. & Pol. Sci. 1985, to be published
103. Tieke, B.: Makromol. Chem. *185* (1984) 1455
104. Bader, H., Ringsdorf, H., Skura, J.: Angew. Chem. Int. Ed. Engl. *20* (1981) 91
105. Hub, H. H., Hupfer, B., Koch, H., Ringsdorf, H.: J. Macromol. Sci. — Chem. *A-15* (1981) 701
106. Hupfer, B., Ringsdorf, H., Schupp, H.: Chem. Phys. Lip. *33* (1983) 355
107. Tieke, B., Lieser, G.: Macromolecules *18* (1985) 327
108. Ostermayer, B., Vogt, W.: Makromol. Chem. Rapid Comm. *3* (1982) 563
109. Plachetta, C., Rau, N. O., Schulz, R. C.: Mol. Cryst. Liq. Cryst. *96* (1983) 141
110. Bader, H., Ringsdorf, H.: J. Polym. Sci. Polym. Chem. Ed. *20* (1982) 1623
111. Büschl, R., Hupfer, B., Ringsdorf, H.: Makromol. Chem. Rapid Comm. *3* (1982) 589
112. Lopez, E., O'Brien, D. F., Whitesides, T. H.: J. Am. Chem. Soc. *104* (1982) 305
113. Koch, H., Ringsdorf, H.: Makromol. Chem. *182* (1981) 255
114. Wagner, N., Dose, K., Koch, H., Ringsdorf, H.: FEBS Lett. *132* (1981) 313
115. Hupfer, B., Ringsdorf, H., Schupp, H.: Makromol. Chem. *182* (1981) 247
116. Johnston, D. S., Sanghera, S., Pons, M., Chapman, D.: Biochim. Biophys. Acta *602* (1980) 57
117. Albrecht, O., Johnston, D. S., Villaverde, C., Chapman, D.: Biochim. Biophys. Acta *687* (1982) 165
118. O'Brien, D. F., Whitesides, T. H., Klingbiel, R. T.: J. Polym. Sci. Polym. Lett. Ed. *19* (1981) 95
119. Pons, M., Johnston, D. S., Chapman, D.: J. Polym. Sci. Polym. Chem. Ed. *20* (1982) 513
120. Gros, L., Ringsdorf, H., Schupp, H.: Angew. Chem. Int. Ed. Engl. *20* (1981) 305
121. Lopez, E., O'Brien, D. F., Whitesides, T. H., Biochim. Biophys. Acta *693* (1982) 437
122. Ando, D. J., Bloor, D.: Polymer *20* (1979) 976
123. Iqbal, Z., Chance, R. R., Baughman, R. H.: J. Chem. Phys. *66* (1977) 5520
124. Chance, R. R., Patel, G. N., Witt, J. D.: J. Chem. Phys. *71* (1979) 206
125. Patel, G. N., Miller, G. G.: J. Macromol. Sci. Phys. B *20*, (1981) 111
126. Batchelder, D. N., Bloor, D.: J. Phys. *C 11* (1978) L 629
127. Plachetta, C., Schulz, R. C.: Makromol. Chem. *184* (1983) 613
128. Babbit, G. E., Patel, G. N.: Macromolecules *14* (1981) 554
129. Bubeck, C., Tieke, B., Wegner, G.: Ber. Bunsenges. Phys. Chem. *86* (1982) 495
130. Tieke, B., Bubeck, C.: unpublished results
131. Steiger, R., Hediger, H., Junod, P., Kuhn, H., Möbius, D.: Photogr. Sci. Eng. *24* (1980) 185
132. Weiss, K.: Diplomarbeit, Universität Freiburg i. Br., 1982
133. Baughman, R. H., Chance, R. R.: J. Polym. Sci. Polym. Phys. Ed. *14* (1976) 2037
134. Kopp, F., Fringeli, U. P., Mühlethaler, K., Günthard, H. H.: Biophys. Struct. Mechanism. *1* (1975) 75
135. Grunfeld, F., Pitt, C. W.: Thin Solid Films *99* (1983) 249
136. Steele, S. C., Wybourne, M. N., Möbius, D.: Thin Solid Films *99* (1983) 117
137. Sauteret, C., Hermann, J. D., Frey, R., Pradere, F., Ducuing, J., Baughman, R. H., Chance, R. R.: Phys. Rev. Lett. *36* (1976) 956
138. Shand, M. L., Chance, R. R.: in "Nonlinear Optical Properties of Organic and Polymeric Materials", D. J. Williams ed., ACS Symp. Ser. No. 233 (1983) p. 187
139. Kan, K. K., Roberts, G. G., Petty, M. C.: Thin Solid Films *99* (1983) 291
140. Kureka Chem. Ind. Co., Ltd., J. P. 58,111,029 (Chem. Abstr. *99* (1983) 149581 s)
141. Garito, A. F., Singer, K. D., Teng, C. C.: in "Nonlinear Optical Properties of Organic and Polymeric Materials", D. J. Williams ed., ACS Symp. Ser. No. 233 (1983) p. 1
142. Pitt, C. W., Walpita, L. M.: Thin Solid Films *68* (1980) 101
143. Carter, G. M., Chen, Y. J., Tripathy, S. K.: in "Nonlinear Optical Properties of Organic and Polymeric Materials", D. J. Williams ed., ACS Symp. Ser. No. 233 (1983) p. 213
144. Lalama, S. J., Singer, K. D., Garito, A. F., Desai, K. N.: Appl. Phys. Lett. *39* (1981) 940
145. Heckmann, K., Pfannemüller, B., Manecke, G., Ring, K., Ringsdorf, H.: Ger. Offen. DE 3,107,527 (Chem. Abstr. *98* (1983) 109377c)
146. Büschl, R., Folda, T., Ringsdorf, H.: Makromol. Chem. Suppl. *6* (1984) 245
147. Yager, P., Schoen, P. E.: Mol. Cryst. Liq. Cryst. *106* (1984) 371
148. Pabst, R., Ringsdorf, H., Koch, H., Dose, K.: FEBS Lett. *154* (1983) 5
149. Leaver, J., Alonso, A., Durrani, A. A., Chapman, D.: Biochim. Biophys. Acta *727* (1983) 327
150. Benz, R., Prass, W., Ringsdorf, H.: Angew. Chem. Suppl. *1982*, 869
151. Tieke, B., Wegner, G.: Makromol. Chem. *179* (1978) 2573

152. Braunschweig, F., Bässler, H.: Ber. Bunsenges. Phys. Chem. *84* (1980) 177
153. Mondong, K., Bässler, H.: Chem. Phys. Lett. *78* (1981) 371
154. Bubeck, C., Nguyen Xuan, T. H., Sixl, H., Tieke, B., Bloor, D.: Ber. Bunsenges. Phys. Chem. *87* (1983) 1149
155. de Jongh, L. J., Miedema, A. R.: Adv. Phys. *23* (1974) 1
156. Arend, H., v. Känel, H., Wachter, P.: phys. stat. sol. (b) *74* (1976) 151
157. a) Ledsham, R. D., Day, P.: J.C.S. Chem. Comm. *1981*, 921
b) Day, P., Ledsham, R. D.: Mol. Cryst. Liq. Cryst. *86* (1982) 163
158. Dunitz, J. D.: Acta Cryst. *10* (1957) 307
159. Gill, N. S., Nyholm, R. S., Barclay, G. A., Christie, T. I., Pauling, P. J.: J. Inorg. Nucl. Chem. *18* (1961) 88
160. Schmehl, R. H., Whitten, D. G.: J. Am. Chem. Soc. *102* (1980) 1938
161. Schmidt et al., G. M. J.: "Solid State Photochemistry", D. Ginsburg ed., Verlag Chemie Weinheim 1976
162. Lahav, M., Schmidt, G. M. J.: J. Chem. Soc. (B) *1967*, 312
163. Green, B. S., Lahav, M., Schmidt, G. M. J.: J. Chem. Soc. (B) *1971*, 1552
164. Ringsdorf, H., Schupp, H.: J. Macromol. Sci. — Chem. *A 15* (1981) 1015
165. Akimoto, A., Dorn, K., Gros, L., Ringsdorf, H., Schupp, H.: Angew. Chem. Int. Ed. Engl. *20* (1981) 90
166. Morawetz, H., Rubin, J. D.: J. Polym. Sci. *57* (1962) 687
167. Tieke, B.: J. Polym. Sci. Polym. Chem. Ed. *22* (1984) 391
168. Tieke, B., Wegner, G.: Angew. Chem. Int. Ed. Engl. *20* (1981) 687
169. Tieke, B., Wegner, G.: Makromol. Chem. Rapid Comm. *2* (1981) 543
170. Tieke, B.: IUPAC Macro 83, Bucharest Romania, Proceedings I (1983), p. 105
171. Tieke, B.: Coll. & Pol. Sci., 1985, to be published
172. Augenstein, M.: Diplomarbeit, Universität Freiburg 1984
173. Seeliger, J., Zagor, V., Blinc, R., Arend, H., Chapuis, G.: J. Chem. Phys. *78* (1983) 2661
174. Barraud, A., Rosilio, C., Ruaudel-Teixier, A.: Polymer Prepr. ACS Div. Polym. Chem. *19* (1978) 179
175. Regen, S. L., Czech, B., Singh, A.: J. Am. Chem. Soc. *102* (1980) 6638
176. Regen, S. L., Singh, A., Oehme, G., Singh, M.: J. Am. Chem. Soc. *104* (1982) 791
177. Tundo, P., Kippenberger, D. J., Politi, M. J., Klahn, P., Fendler, J. H.: J. Am. Chem. Soc. *104* (1982) 5352
178. Tundo, P., Kippenberger, D. J., Klahn, P. L., Prieto, N. E., Gao, T. C., Fendler, J. H.: J. Am. Chem. Soc. *104* (1982) 456
179. Reed, W., Gutermann, L., Tundo, P., Fendler, J. H.: J. Am. Chem. Soc. *106* (1984) 1897
180. Dorn, K., Klingbiel, R. T., Specht, D. P., Tyminski, P. N., Ringsdorf, H., O'Brien, D. F.: J. Am. Chem. Soc. *106* (1984) 1627
181. Chapela, V. M., Parry, G. S.: Nature *281* (1979) 134
182. Rebbah, H., Borch, M. M., Raveau, B.: Mat. Res. Bull. *15* (1981) 1930
183. Yamaka, S., Sakamoto, K., Hattori, M.: J. Phys. Chem. *85* (1981) 1930
184. Lagaly, G.: Angew. Chem. Int. Ed. Engl. *15* (1976) 575
185. Lagaly, G., Beneke, K.: J. Inorg. Nucl. Chem. *38* (1976) 1513
186. Weiss, A., Hilke, K. J.: Angew. Chem. Int. Ed. Engl. *4* (1965) 353
187. Subba Rao, G. V. and Shafer, M. W.: in "Intercalated Layer Materials". F. A. Levy ed., Reidel, Dordrecht 1979, p. 99
188. Yamanaka, S., Kobayashi, H., Tanaka, M.: Chem. Lett. *1976*, 329
189. Bornhart, D. M., Laughlan, C. N., Haque, M.-U.: Inorg. Chem. *8* (1969) 2768
190. Regen S. L., Shin, J. S., Yamaguchi, K.: J. Am. Chem. Soc. *106* (1984) 2446
191. Aliev, K. V., Ringsdorf, H., Schlarb, B., Leister, K. H.: Makromol. Chem. Rapid Comm. *5* (1984) 345

Kinetics and Mechanisms of Polyesterifications
II. Reactions of Diacids with Diepoxides*

Pierre-Jean Madec and Ernest Maréchal
Laboratoire de Synthèse Macromoléculaire, Université Pierre et Marie Curie, 12, rue Cuvier, 75005 Paris, France

This article is a critical review of most of the literature relative to non-catalyzed, base-catalyzed, and miscellaneous-catalyzed epoxy-carboxy esterifications and polyesterifications. Most kinetic data are relative to model esterifications; however, some polyesterification kinetics are analyzed. From the analysis of the results reported in the literature and found in our group it has been possible to compare and critize the various general mechanisms which have been proposed.

* Part I is published in Vol. 43 of this series

1 Introduction

The epoxy-carboxy reaction:

(1)

presents several important advantages:
— It can be carried out at moderate temperatures (80–120 °C) in the presence of basic catalysts. Moreover, catalysts other than amines and ammonium salts can be used; a better knowledge of this would allow a decrease of the reaction temperature.
— In addition to the ester linkages, hydroxy side groups are formed, which can be very useful for further chemical modifications of the polymer chains such as grafting or crosslinking.
— It is an addition reaction and no volatile compound is eliminated. This is an advantage since it prevents the formation of bubbles in the material. However, it can also be a disadvantage since the elimination of volatile compound is very often used, in laboratories and in industrial plants, to shift the equilibrium towards the formation of the ester.

In principle, at least three side reactions are possible:

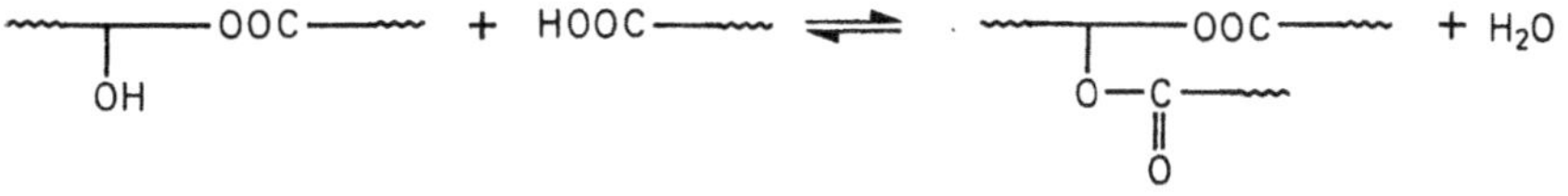

(2)

(3)

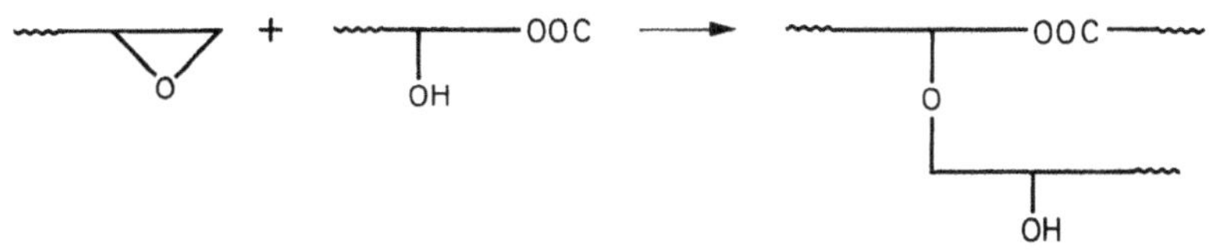

(4)

However, these reactions have a significant contribution only when the temperature is around 200 °C and, as already stated, many catalysts render a reaction betweeen 80 and 120 °C possible. In the section focussing on the kinetics and reaction mechanism we shall see that other, hardly avoidable side reactions can take place. The contribution of these side reactions depends not only on the catalyst and the temperature, but also on the nature of the reaction medium. When the reaction is carried out in the bulk, their contribution is generally higher than when a solvent is added.

There are few reviews [148,172] and, indeed, relatively few fundamental studies relative to the epoxy-carboxy reaction. Despite this fact, the reaction has many applications, mostly in polymer chemistry. Epoxy-carboxy reactions have been used not only to build polymer chains, but also to modify side or end groups of chains, to prepare monomers, and as a curing agent. The following listing of patents and applications is far from being exhaustive:

— Reaction of epoxy or carboxylic end (or side) groups of a chain with an unsaturated acid or an unsaturated epoxy compound to obtain oligomerizable or copolymerizable poly(β-hydroxy ester)s which are used as casting, curing, laminating, and photosensitive polymers [20, 22, 47, 176, 188, 190, 195, 204, 237, 259, 265].
— Synthesis of oligoesters with allyl side groups [237, 259].
— Modification of the end groups of a polymer to form an epoxy-terminated chain [78].
— Chemical modification of a chain bearing epoxy or carboxylic groups [37, 38, 78, 118, 151, 233].
— Epoxy resin intermediates and curing of epoxy resins by diacids or α,ω-dicarboxylic oligomers [17, 29, 44, 51–53, 72, 81, 89, 131, 151, 173, 193, 224, 225, 259, 293, 294].
— Polymer additives [154, 258].
— 2-hydroxyalkylesters of unsaturated acids, vinyl and allyl esters, and their polymers and copolymers [6, 41, 42, 49, 74–77, 80, 84, 87, 88, 90, 99, 101–106, 111, 112, 123, 125 127, 134, 135, 138, 158, 170, 171, 174, 184, 186, 196, 197, 200–202, 207–209, 213, 242, 257, 263, 270, 272, 274, 284, 290–292, 295].
— Various monomers (such as bishydroxyethyl or diglycidyl esters) [5, 28, 35, 36, 48, 61, 70, 71, 73, 93–97, 102, 108, 109, 119, 124, 126, 132, 136, 139, 164–166, 175, 177–180, 210, 217, 218, 221, 261, 262, 269, 275, 288].
— Various aliphatic or aromatic polyesters [2, 4, 18, 19, 32, 56–60, 63, 85, 91, 98, 120–122, 166, 181–183, 191, 205, 214, 220, 221, 287, 289].
— Polycondensation of oligomers [62, 155, 203].

2 Main Techniques Used in Kinetic Studies of Epoxy-Carboxy Reactions

2.1 Procedure and Apparatus

2.1.1 Single Run

This method is by far the most widely used as it is simple to perform in the case of homogeneous or non volatile-containing systems, and it gives a good picture of the batch reactions performed in industry. Reactions are carried out in a thermostated flask fitted with constant speed stirrer, inert gaz inlet, sampling device, thermometer, and condenser.

At definite times, samples are withdrawn and titrated by an appropriate method. The temperature is controlled either with a thermostated oil bath or a heating jacket and a P.I.D. regulation with a capter plunged in the reaction medium.

An inert atmosphere is required in all cases. No distillation column or vacuum line is necessary since there is no volatile to remove.

2.1.2 Separate Runs

Aliquots of the reaction mixture are placed in an inert atmosphere in sealed ampoules or flasks which are plunged in a thermostated oil bath. At different reaction times one ampoule is opened and its content is titrated.

Due to difficulty in carrying out all the experiments under exactly the same conditions, the experimental results show more discrepancies than in the case of a single run. This technique is used mainly for monoesterifications carried out on compounds which are volatile under reactional conditions (e.g., ethylene oxide).

2.1.3 Analytical Methods

The determination of reaction orders and of activation parameters requires accurate measurements of the concentrations of the different species present in the reaction medium, especially for high conversion.

If it can be accepted that the stoichiometry is known and that no side reactions take place, the concentration of only one of the products or of the reactants needs to be measured.

Two methods can be used: discontinuous measurements of samples and continuous measurements following a physical parameter which, after standardization, can be related to a parameter such as concentration, degree of polymerization, or conversion.

2.1.3.1 Discontinuous Measurements

— Determination of carboxy groups. Esterification or polyesterification kinetics is usually followed by this titration, which is both easy and accurate. Each sample is dissolved in a solvent or a mixture of solvents ($CHCl_3$, C_6H_6/EtOH or MeOH, toluene/EtOH or MeOH, . . .) and then titrated with alcoholic KOH. The end point is determined either with an indicator (in most cases phenolphthalein) or with a pH-meter. An accuracy of about 0.1–1% can generally be achieved.

— Determinations of epoxy groups. Epoxy groups are usually titrated by Durbetaki's method [50] or by Jay's derivative process [110]. In the latter, which today is the most commonly used, epoxy groups are reacted with HBr resulting from the reaction of perchloric acid with tetraethylammonium bromide. This reaction, which is stoichiometric, is carried out in the presence of crystal violet as indicator. When all epoxy groups have reacted, the excess of HBr changes the colour of the indicator.

Various well-established procedures have been reviewed [172]. Some of them have been used for epoxide titration in epoxy-carboxy kinetic studies:

— Direct infrared measurements [15,169].

— Iodometry [232], especially for ethylene oxide [11].

— Hydrochloric acid/methyl ethyl ketone procedure [113].

— Determination of low-molecular-weight compounds. The kinetics of epoxy-carboxy esterification can be followed by chromatographic methods. The concentration of each of the species present in the medium can be determined, at least for low-molecular-weight compounds.

Gas-phase chromatography is mainly used for the monoesterification of low-molecular-weight compounds [91,166]. Gas-liquid chromatography allows the characterization of free 1,2-ethane diol (or dimer) in the synthesis of bis(β-oxyethyl)terephthalate from terephthalic acid and ethylene oxide [218], thin-layer chromatography is used for diesterifications [166,167] or oligomer characterization [218], and size-exclusion chromatography in most general cases [11,20,169].

Monoesters and diesters of terephthalic acid and ethylene oxide can be determined separately by polarography [133,286].

2.1.3.2 Continuous Monitoring

— Infrared spectroscopy. Due to experimental difficulties, infrared spectroscopy is used infrequently in these kinetic studies. However, continuous measurements have been carried out by Fellers [63] in the case of polyacid-epoxy condensation reaction for gel point determination.

— Nuclear magnetic resonance spectroscopy. NMR spectroscopy has been used for kinetic investigation of the stearic acid-ethylene oxide reaction [13]. On the other hand, the use of the NMR spin-echo method [133] seems to be convenient for continuous monitoring of 1,2-epoxy-3-phenoxypropane and hexanoic acid. This is based on the relation between the change in the nuclear spin-spin relaxation time during the reaction and the change in the number average molecular weight of the reaction medium.

— Miscellaneous methods. Some other techniques, such as dielectric constant measurements [267], conductimetry [268], mass spectrometry [166,169], and electronic spectroscopy [214] are used less frequently but can also be useful. Change in vapor pressure determined with a manometric device can be used when the reaction mixture contains a volatile compound such as ethylene oxide [145,162,231].

2.2 Studies of Side Reactions

Before studying epoxy-carboxy esterification kinetics, it must be remembered that side reactions can interfere with the main reaction. These must either be avoided by well-defined experimental conditions or taken into account in kinetic calculations.

As mentioned in Sect. 1, at least three side reactions are possible [222,223]: esterification (*2*), etherification (*4*) of secondary hydroxyl groups, and hydrolysis of epoxide (*3*). However, it is generally admitted that under well-defined conditions (solution, moderate temperature, basic catalyst) their contribution can be neglected, as shown by simultaneous titration of carboxyl and epoxide groups [64,100,116,156–158,266]. Nonetheless, when the reaction is carried out in the bulk, the contribution of epoxide side reactions seems higher [159,169]. In addition to the side reactions already mentioned, several authors [159,203,276] have pointed out the contribution of the polymerization of epoxide functions initiated by tertiary amines; some authors have been able to take these reactions into account in their kinetic relations [159].

2.3 General Treatment of Experimental Data

Several textbooks deal with this problem [9,30,31]. The general rate equation relative to the reaction of acid A with epoxy compound E, in the absence of catalyst, is:

$$-\frac{d[A]}{dt} = k_m[A]^a[E]^e \tag{1}$$

where a and e are the orders of acid and epoxy compounds and $m = a + e$.

When a basic catalyst (B) is present, Eq. (1) becomes:

$$-\frac{d[A]}{dt} = k_n[A]^a[E]^e[B]^b \tag{2}$$

where b is the order of the catalyst and $n = a + e + b$.

With $r = \frac{[E]_0}{[A]_0}$, $[A]_0 = x_0$, $[A] = x$, and $\alpha = x/x_0$, Eq. (1) becomes:

$$-\frac{d\alpha}{dt} = k_m x_0^{a+e-1} \alpha^a (\alpha + r - 1)^e \tag{3}$$

If the study is performed at stoichiometry ($r = 1$):

$$-\frac{d\alpha}{dt} = k_m x_0^{m-1} \alpha^m \tag{4}$$

A method allowing the determination of m from the curve $\alpha = f[\log (t - t_0)]$ is described by Bamford and Tipper [9]. However, the general evaluation of the kinetics is given by the plot of the integrated form $F(\alpha)$, with the order obtained by graphical determination, against $t - t_0$. The rate constant is determined by linear regression according to the least squares method, from the system of values $[F(\alpha_i), t_i - t_0]$. The validity of the integrated form is tested by determination of the correlation coefficient.

Integration of Eq. (4) leads, for each pair $(\alpha_i, t_i - t_0)$, to:

$$F(\alpha_i) = \frac{x_0^{1-m}}{1 - m} \alpha_i^{1-m} \qquad m \neq 1 \tag{5}$$

$$F(\alpha_i) = -x_0^{1-m} \ln \alpha_i \qquad m = 1 \tag{6}$$

When the study is not performed at stoichiometry ($r \neq 1$), the graphical method cannot be used, due to the high number of curves which are required and the difficulty in distinguishing them. After the global order m in reactants had been obtained from the study at stoichiometry, a numerical integration of Eq. (3) for partial orders between zero and m with 0.5 step width is carried out.

Remark: In the case of catalyzed reactions, the contribution of non-catalyzed reaction is not taken into account (below 1 %).

2.4 Discrimination Between Possible Values of Reaction Orders

Unambiguous results can be obtained only if some essential conditions are observed:

— If only narrow ranges of conversion are studied, it is impossible to discriminate between close values of reaction orders unless very accurate experimental data have been obtained. For most common accuracies (0.5–1 %), the reaction must be controlled until conversion is 0.55 with at least 10 experimental points.

— If the balance between reactants is stoichiometric, only overall orders can be obtained. However, the determination of orders in acid and in epoxide (obtained by non-stoichiometric studies) is highly desirable since it confirms the overall order and provides valuable information on the mechanism.

2.5 Determination of Activation Parameters

The dependence of the rate constant on temperature can be obtained either by the classical Arrhenius equation:

$$k = A \exp(-E/RT) \tag{7}$$

where E and A are the activation energy and the preexponential term, respectively, or by the transition state theory (see, for instance, Ref. [9]).

$$k = \frac{\varkappa T}{h} \exp(\Delta^{\neq}/R) \cdot \exp(-\Delta H^{\neq}/RT) \cdot (C^{\neq})^{1-d} \tag{8}$$

where $\Delta H^{\neq}$ and $\Delta S^{\neq}$ are the activation enthalpy and activation entropy, respectively, and $C^{\neq}$ is a homogeneity term (equal to 1 unit of concentration).

More and more authors use Eq. (8) which gives $\Delta H^{\neq}$ and $\Delta S^{\neq}$ a more theoretical base than Eq. (7). However, it must be remembered that relation (8) has been established for reactions in the gaseous phase or at the best in highly diluted phase. This is not the case with many of the polyesterification systems studied where concentrations are high and expressed in term of mole per kilogram and not, as normally, in mole per liter. The consequences of these approximations are not clear and it is obvious that a considerable amount of additional work must be carried out on this subject.

2.6 Description of a Method Allowing the Evaluation of Two Rate-Determining Steps

A global analysis of the kinetic results relative to the model reactions of benzoic acid with 1,2-epoxy-3-phenoxypropane in solution [156–158] and of octadecanoïc acid with 1,2-epoxy-3-dodecyloxypropane in the bulk [159], both carried out in the presence of N,N-dimethyldodecylamine as catalyst, showed that two competitive reactions take place. A mathematical treatment allowed the determination of the rate constants of each reaction.

2.6.1 Reaction under stoichiometric conditions

The behaviour of the system in solvents or in the bulk corresponds to a global order of 1.5 in reactants, which can be described by the simultaneous contribution of two competitive reactions:

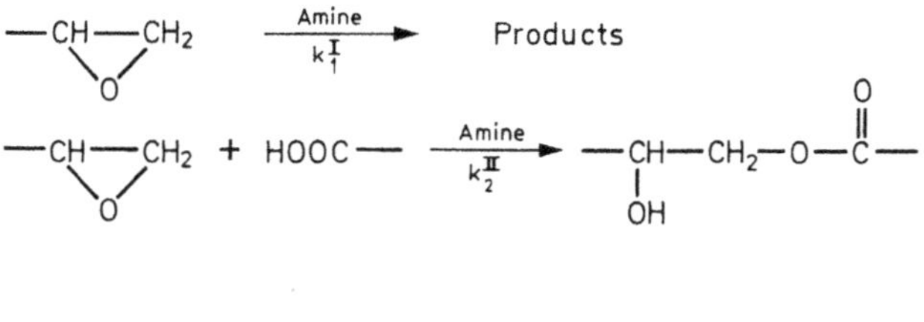

(5)

(6)

Superscripts I and II relate the rate constants to Eq. (5) and Eq. (6) and subscripts 1 and 2 to the global order in reactants, respectively. The contribution of Eq. (5) and (6) can be described as follows:

$$-\frac{dx}{dt} = k_1^I x + k_2^{II} x^2 \tag{9}$$

or

$$-\frac{d\alpha}{dt} = k_1^I \alpha + k_2^{II} x_0 \alpha^2 \tag{10}$$

Eq. (10) must be equivalent to Eq. (11) describing the global phenomenon (see Eq. (8) [156]):

$$-\frac{d\alpha}{dt} = k_{1.5} x_0^{0.5} \alpha^{1.5} \tag{11}$$

It is obvious that integration of these equations does not allow the exact determination of k_1^I and k_2^{II}. But, if the pair (k_1^I, k_2^{II}) exists, a value of k_1^I corresponds to each value of the ratio k_2^{II}/k_1^I and this is true for any value of the pair $[\alpha_i, (t_i - t_0)]$. α_i is the conversion at time t_i, and t_0 the initial time for the interval under consideration. From Eq. (10) it follows:

$$-\frac{d\alpha}{\alpha\left(1 + \frac{k_2^{II}}{k_1^I} x_0 \alpha\right)} = k_1^I \, dt \tag{12}$$

and, after integration:

$$\ln\left[\frac{1 + \frac{k_2^{II}}{k_1^I} x_0 \alpha}{\alpha\left(1 + \frac{k_2^{II}}{k_1^I} x_0\right)}\right]_1^{\alpha} = k_1^I (t - t_0) \tag{13}$$

which is equivalent to:

$$\ln \frac{1 + \frac{k_2^{II}}{k_1^I} x_0 \alpha}{\alpha\left(1 + \frac{k_2^{II}}{k_1^I} x_0\right)} = k_1^I (t - t_0) \tag{14}$$

Equation (14) corresponds to a 1st order rate equation. However, it is necessary to know the value of k_2^{II}/k_1^I which allows the calculation of the rate constant k_1^I and, hence, that of k_2^{II}, for the widest range of conversion possible.

The calculation process is the following: an arbitrarily chosen value is given to

k_2^{II}/k_1^I, then the value of k_1^I corresponding to each pair $[\alpha_i, (t_i - t_0)]$ is computer-calculated; an iteration process allows variations of k_2^{II}/k_1^I.

Let

$$G\left[\alpha_i, \left(\frac{k_2^{II}}{k_1^I}\right)_j\right] \quad \text{be} \quad \ln \frac{1 + \left(\frac{k_2^{II}}{k_1^I}\right)_j x_0 \alpha_i}{\alpha_i \left[1 + \left(\frac{k_2^{II}}{k_1^I}\right)_j x_0\right]}$$

A linear regression by the least-squares method is applied to couples $[G[\alpha_i, (k_2^{II}/k_1^I)_j]$, $(t_i - t_0)]$. The correlation coefficient I is determined in each case. Convergence towards a value of k_1^I, which remains constant during the whole reaction, takes place for the maximum value of $I(I \simeq 1)$. k_2^{II} can then be calculated.

2.6.2 Reactions Under Non-Stoichiometric Conditions

Let c be $r - 1$ with $r = [E]_0/[A]_0$.

Then

$$-\frac{d\alpha}{dt} = k_m x_0^{m-1} \alpha^a (\alpha + c)^e \tag{15}$$

This relation has been established elsewhere [9]. It must be equivalent to:

$$-\frac{d\alpha}{dt} = k_1^I(c + \alpha) + k_2^{II} x_0 \alpha(\alpha + c) \tag{16}$$

The integrated form is:

$$\frac{1}{1 - c\frac{k_2^{II}}{k_1^I} x_0} \ln \frac{(c + 1)\left(1 + \frac{k_2^{II}}{k_1^I} x_0 \alpha\right)}{(c + \alpha)\left(1 + \frac{k_2^{II}}{k_1^I} x_0\right)} = k_1^I(t - t_0)$$

The principle of calculation used in the case of stoichiometric study also can be applied to Eq. (17); it is merely necessary to take into account the value of c.

2.7 The Problem of the Diffusion of a Partially Soluble Reactant

Several articles relate to the reaction of terephthalic acid with ethylene oxide, catalyzed either by tertiary amines [18,120,164,218] or quaternary ammonium salts [4,231]. In most cases, the reaction was carried out in a primary alcohol such as *n*-butanol [18,164] or *n*-pentanol [231]. Anfinogentov et al. [4] used various dialkyl esters as solvents; according to these authors, the use of dialkyl esters is especially convenient and advantageous, because the resultant mixture of monomers can be converted directly into the polymer, which means that the usual stages of separation and solvent recovery are not necessary.

The solubility of terephthalic acid in the above-mentioned solvents is very low, which means that the acid must diffuse continuously from the solid particules to the solution where the reaction takes place. In such a case, the first question which arises is: does the diffusion control the kinetics of the overall process? In all cases, the authors claimed that the reaction rate is never affected by the amount of undissolved terephthalic acid and that the reaction proceeds through a chemical kinetic control. Under the experimental conditions used by Bhatia et al. [18], the diffusion rate of terephthalic acid from the solid particles to the solution is 9.5×10^{-5} mol cm^{-3} s^{-1} at 100 °C and that of ethylene oxide from the gas phase to the liquid is 19.4×10^{-4} mol $\cdot$ cm^{-3} s^{-1}. These values are far above the rate of formation of the diester(bishydroxy-ethylterephthalate), as this is only 5.84×10^{-8} mol cm^{-3} s^{-1}. Moreover, the independence of the reaction rate on the mass transfer effects was shown by varying the values of some parameters (e.g., ethylene oxide flow-rate, stirrer-speed, particule size, terephthalic acid charge) in a large range.

3 Inventory of the Main Catalysts Used in Epoxy-Carboxy Esterifications and Polyesterifications

The main catalysts used in esterification or polyesterification are listed in Table 1. We reported each catalyst as it is defined in the corresponding reference, without any consideration about its behaviour and any assumption on the effective catalytic species. This will be discussed in Sect. 4.

Table 1. Catalysts used in epoxy-carboxy esterifications and polyesterifications

Catalyst	Epoxy	Acid	Ref.
Group I_A			
Li:			
Li OAC	Alkylene oxide	Acetic	92)
Li-methacrylate	Propylene oxide	Methacrylic	270)
Na:			
Na OH	Ethylene oxide	Stearic	11, 279)
	1,2-epoxy-3-hydroxy-propane	Various carboxylic acids	129)
Na OMe	Monosubstituted ethylene oxides	Various carboxylic acids	281)
Na OAC	Alkylene oxide	Acetic	92)
	Propylene oxide	Acetic	260)
	1,2-epoxy-3-allyloxy-propane	Diacids	237)
	1,2-epoxy-3-phenoxy-propane	Hexanoic	235)

Contin. p. 164

Table 1. (continued)

Catalyst	Epoxy	Acid	Ref.
K:			
KOH	Ethylene oxide	Chloracetic	[142]
	Ethylene oxide	Stearic	[229, 279]
	1,2-epoxy-3-phenoxy-propane	Hexanoic	[222, 235]
	Epoxy resin	Acetic, nonanedioic	[2]
K_2O	Epoxy resin	Carboxyl-contg. oligodiene	[81]
Various alkali metal derivatives:			
Alkali metal carbonate	Ethylene oxide	Fatty acids	[216]
	Propylene oxide	Fatty acids	[191]
	1,2-epoxy-3-chloropropane	Methacrylic	[21]
	Alicylic monoepoxide	Various carboxylic acids	[239]
	1,2-epoxy-3-phenoxy-propane	Hexanoic	[235]
Alkali metal hexadecanoate	1,2-epoxy-3-hydroxy-propane	Fatty acids	[166]
Alkali metal iodide	3,4-epoxy-1-butene	Fatty acids	[130]
Alkali metal terephthalate	Ethylene oxide	Terephthalic	[230]
F_3C COOK	Cyclohexane diepoxy derivative	Dicarboxylic polyester	[78]
Group II_A			
Mg:			
Mg O	Ethylene oxide	Acrylic	[88]
	Epoxy resin	Carboxyl-contg. oligodiene	[81]
Ca:			
Ca O	Epoxy resin	Carboxyl-contg. oligodiene	[81]
Ca octadecanoate	1,2-epoxy-3-hydroxy-propane	Fatty acids	[177]
Group II_B			
Zn:			
$ZnCl_2$	Alkylene oxide	Acrylic or Methacrylic	[103]
	1,2-epoxy-3-allyloxy-propane	Acetic	[199]
Group III_A			
B:			
BF_3	Propylene oxide	Acrylic or methacrylic	[287]
Al:			
$AlCl_3$	Alkylene oxide	Acrylic or methacrylic	[213, 287]
	1,2-epoxy-3-allyloxy-propane	Acetic	[199]
Various Al organic derivatives	Ethylene oxide	Terephthalic	[93]
Group IV_A			
Ge:			
GeO_2	Ethylene oxide	Terephthalic	[28]

Table 1. (continued)

Catalyst	Epoxy	Acid	Ref.
Sn:			
$SnCl_4$	Propylene oxide	Acrylic or methacrylic	287)
	Ethylene oxide	Stearic acid	216)
Various Sn organic derivatives	Glycidyl methacrylate copolymer	Various dicarboxylic	233)
	Epoxy resin	Various dicarboxylic	225)
Various Sn halides associated with amines	Ethylene oxide	Terephthalic	94)
Group IV_B			
Ti:			
Ti			
Bis(cyclopentadienyl) titanium dihalide	Ethylene oxide	Terephthalic	60)
Monobutoxy amino titanate	1,2-epoxy-3-phenoxy-propane	Hexanoic	133)
Group V_A			
N:			
aliphatic tertiary amines			
Me_3N	1,2-epoxy-3-chloropropane	Fatty acids	258)
	1,2-epoxy-3-phenoxy-propane	Substituted benzoic	215)
N,N-dimethyldodecyl amine	1,2-epoxy-3-phenoxy-propane	Benzoic	116, 156–158)
	1,2-epoxy-3-dodecyloxy-propane	Octadecanoic	159)
	1,2-epoxy-3-dodecyloxy-propane	2,4-hexadienoic	203)
N,N-dimethylhexyl amine	1,2-epoxy-3-phenoxy-propane	Various carboxylic	245)
N,N-dimethylbenzyl amine	1,2-epoxy-3-butoxy-propane	Methacrylic	242)
	Alkylene oxide	Methacrylic	208)
	Epoxy resin	Various dicarboxylic	44, 225)
	1,2-epoxy-3-benzoate-propane	Benzoic	64)
	1,2-epoxy-3-phenoxy-propane	Hexanoic	222)
N,N-dimethylphenyl amine	Epoxy resin	Poly(acrylic acid-styrene-methacrylate)	293)
	Epoxy resin	Methacrylic	131)
	1,2-epoxy-3-phenoxy-propane	Various carboxylic	245)
Et_3N	Alkylene oxide	Terephthalic	18, 56–58, 93, 98, 107, 120, 181, 217, 218)
	Alkylene oxide	Acrylic or methacrylic	126, 263, 287)
	Monosubstituted ethylene oxides	Various carboxylic	281)
	1,2-epoxy-heptane	Methacrylic	106)

Contin. p. 166

Table 1. (continued)

Catalyst	Epoxy	Acid	Ref.
	1,2-epoxy-3-butoxypropane	Methacrylic	77)
	1,2:2,3-diepoxy butane	Acrylic	138)
	1,2-epoxy-3-hydroxy-propane	Fatty acids	166)
	1,2-epoxy-3-methacrylate-propane	Acetic	43)
	1,2-epoxy-3-cinnamate-propane	Poly(acrylic acid)	195)
	Glycidyl grafted copolymer	Polycarboxylic	224)
	Epoxy resin	Various dicarboxylic	225)
Bu_3N	1,2-epoxy-3-hydroxy-propane	Various mono- or dicarboxylic	247, 251, 252)
	1,2-epoxy-3-phenoxy-propane	Acetic	243)
	1,2-epoxy-3-phenoxy-propane	Substituted benzoic	100)
	Ethylene oxide	Acetic and chlorinated derivatives	91)
	Epoxy resin	Acetic, nonanedioic	2)
Tri(*n*-hexyl)amine	1,2-epoxy-3-chloropropane	Hexanoic	236)
Tri(*i*-amyl)amine	Ethylene oxide	Terephthalic	183)
Various NR_3 (R: alkyl or aryl)	Ethylene oxide	Terephthalic	32, 93, 147, 162, 164, 165, 230, 269)
	Ethylene or propylene oxide	Tere- or isophthalic	107)
	Ethylene oxide	Acetic and chlorinated derivatives	91)
	2-transbutene oxide	Acetic	227)
	Various α-epoxides	Various carboxylic	198)
	1,2-epoxy-3-chloropropane	Dehydroabietic	194)
	1,2-epoxy-3-phenoxy-propane	Acetic	243, 244)
	1,2-epoxy-3-phenoxy-propane	Benzoic	244, 276)
	1,2-epoxy-3-phenoxy-propane	Various carboxylic	234, 240, 249)
	Epoxy resin	Phthalic	246)
	Epoxy resin	Acrylic or methacrylic	176, 190)
	Epoxy resin	Acetic	2)
	Epoxy resin	Carboxy-contg. rubber	51)
	Epoxy resin	COOH-terminated oligoisobutene	294)
	Diepoxide	Polymeric carboxylic acid	63)
N,N-dimethyl coconut oil; polymeric amines	Ethylene oxide	Fatty acids	20)
Primary and secondary amines or salts	Ethylene oxide	Terephthalic	59, 93)
	Ethylene oxide	Acetic or chlorinated derivatives	91)
	Alkylene oxide	Acrylic or methacrylic	263)
	Epoxy resin	Methacrylic	265)
	Glycidyl methacrylate copolymer	Various dicarboxylic	136)

Table 1. (continued)

Catalyst	Epoxy	Acid	Ref.
Aliphatic quaternary ammonium salts			
Tetramethylammonium chloride	1,2-epoxy-3-chloropropane	Various dicarboxylic	48)
Tetraethylammonium bromide	Alkylene oxide	Aromatic dicarboxylic	107)
Tetrabutylammonium bromide	Glycidyl methacrylate copolymer	Dicarboxylic	134, 135, 137)
Trimethylbenzyl ammonium chloride	1,2-epoxy-3-chloropropane	Methacrylic	87, 123)
		Various carboxylic	160)
		isophthalic	262)
		4-methyl isophthalic	261)
	1,2-epoxy-3-hydroxy-propane	Fatty acids	166, 167)
	1,2-epoxy-3-phenoxy-propane	Hexanoic	133)
	Epoxy mono (or poly)-acrylate (or methacrylate)	Methacrylic	274)
	α-glycidyl-ω-hydroxy polyoxyethylene	Hexadecanoic	194)
Triethylbenzylammonium chloride	Ethylene oxide	Terephthalic	60)
	Glycidyl methacrylate	Octanoic	114)
	Glycidyl methacrylate copolymer	Methacrylic	204)
		Cinnamic	161)
Tripropylbenzyl ammonium chloride	Ethylene oxide	Terephthalic	60)
Various quaternary ammoniums salts	Ethylene oxide	Terephthalic	182, 231)
	1,2-epoxy-3-chloropropane	Methacrylic	101)
	1,2-epoxy-3-chloropropane	Various carboxylic	175, 187)
	1,2-epoxy-3-phenoxy-propane	Benzoic	276)
	Epoxy resin	Acrylic	190)
		Methacrylic	176)
		Terephthalic	33)
		Hexanedioic	89)
Ammonium salts of carboxylic acids or phenol			
Triethylamine monosalt of terephthalic acid	Ethylene oxide	Terephthalic	61)
Various quaternary ammonium carboxylate	Ethylene oxide	Terephthalic	4, 121, 277)
	Alkylene oxide	Aromatic dicarboxylic	36, 119)
Triamyl ammonium phenolate	Epoxy resin	α,ω-dicarboxy polyester	220)
Quaternary ammonium anion exchange resin in the OH form	Alkylene oxide	Terephthalic	139)

Contin. p. 168

Table 1. (continued)

Catalyst	Epoxy	Acid	Ref.
Quaternized poly-(dimethylamino ethylmethacrylate)	1,2-epoxy-3-chloropropane	Acrylic	[196]
Aromatic amines			
Pyridine	Alkylene oxide	Tere- or isophthalic	[107]
	1,2-epoxy-3-phenoxy-propane	Benzoic	[117,267,268]
	Poly(glycidyl methacrylate)	Acetic	[192]
Substituted			
Pyridines or quinoleines	1,2-epoxy-3-phenoxy-propane	Benzoic	[266]
N-phenyl α-naphthyl amine	1,2-epoxy-3-chloropropane	Methacrylic	[87]
Various aromatic amines	1,2-epoxy-3-phenoxy-propane	Various carboxylic	[240]
Pyridine-contg. anion exchange resin	Various epoxy compounds	Methacrylic	[200]
Aromatic quaternary ammonium salts			
N-methylpyridinium chloride	Alkylene oxide	Terephthalic	[272]
Various quaternary pyridinium salts	Alkylene oxide	Acrylic or methacrylic	[75,99]
Amine oxides	Ethylene oxide	Various mono- or dicarboxylic	[278]
Amino alcohols			
N,N-dimethyl ethanol amine	Propylene oxide	Methacrylic	[80]
Triethanol amine	Epoxy resin	Carboxy-contg. rubber	[53]
Triisopropanol amine	Epoxy resin	Acetic, nonanedioic	[2]
Esters of amino alcohols			
Dimethylamino acrylate	Alkylene oxide	Methacrylic	[292]
Dimethylamino methacrylate	Glycidyloxyalkyl-alkoxysilanes	α,β-unsaturated monocarboxylic	[127]
t-butylaminoethyl-methacrylate	Epoxy resin	Various dicarboxylic	[259]
Tertiary amine or quaternary ammonium salt + metallic salt			
Quaternary ammonium salt or amines + Cr (halide, acetylacetonate, oxide)	Ethylene oxide	Acrylic	[84]

Table 1. (continued)

Catalyst	Epoxy	Acid	Ref.
Various amides	Alkylene oxide	Terephthalic	179)
Various alkylated amides	1,2-epoxy-3-chloropropane	Various carboxylic	212, 275)
	Alkylene oxide	Methacyclic	284)
	Ethylene oxide	p-$C_6H_4(CH_2CH_2COOH)_2$	34)
	Epoxy resin	Acetic, nonanedioic	2)
Polyamides	Propylene oxide	Isophthalic	210)
	Epoxy resin	Carboxy-contg. rubber	53)
P:			
Ph_3P	Various epoxides	Acrylic acid	47)
	Ethylene oxide	Terephthalic acid	70, 108, 122)
	Resorcinol diglycidyl ether	Methacrylic acid	74, 112)
Bu_3P	1,2-epoxy-3-chloropropane	Acrylic acid Methacrylic acid	101)
	Ethylene oxide	Terephthalic acid	122)
$(CH_2OH)_3P$	Poly(epoxyurethane)	α,ω-diCOOH oligobutadiene	193)
Various P-organic derivatives	Alkylene oxide	Acrylic	45, 273)
As:			
Ph_3As	Ethylene oxide	Terephthalic acid	122)
Sb:			
Ph_3Sb	Ethylene oxide	Terephthalic acid	122)
	Various epoxides	Acrylic acid	47)
Sb_2O_3	Ethylene oxide	Terephthalic acid	28)
$SbCl_5$	Ethylene oxide	Stearic acid	216)
Group V_B			
V:			
VCl_3	Alkylene oxide	Acrylic	174)
Organic derivative associated with tertiary amines	Ethylene oxide	Terephthalic	96)
	1,2-epoxy-3-phenoxy-propane	Carboxy polyisobutylene	62)
Group VI_A			
S:			
Phenothiazine	Alkylene oxide	Acrylic or methacrylic	125, 291)
	1,2-epoxy-3-chloropropane	Acrylic or methacrylic	6, 102)
Various organic derivatives	Alkylene oxide	Various dicarboxylic	97, 209)
	1,2-epoxy-3-chloropropane	Acrylic or methacrylic	102)
Group VI_B			
Cr:			
Cr oxide (CrO_3)	Alkylene oxide; 1,2-epoxy-3-chloropropane	Acrylic or methacrylic	5)

Contin. p. 170

Table 1. (continued)

Catalyst	Epoxy	Acid	Ref.
Carboxylic acid derivatives	Alkylene oxide	Acrylic or methacrylic	76, 186, 205)
	Alkylene oxide	Terephthalic	73, 124, 180)
	1,2-epoxy butane	2-ethylhexanoic	253, 255)
	Ethylene oxide	Hexanedioic	254)
	Alkylene oxide	Acetic or propanoic	206)
	1,2-epoxy-3-phenoxy-propane	Hexanoic	150, 169)
Organic derivatives	Ethylene oxide	Acrylic	290)
		Terephthalic	73, 180)
Mineral derivatives associated with other metallic derivatives	Alkylene oxide	Acrylic or methacrylic	184)
Cr chelates	Epoxy resin	Carboxylated poly-butadiene	17)
	Epoxy resin	Oxalic; citric	280)
Cr compound + organic derivatives (e.g.: amine,	Ethylene oxide	Terephthalic	83, 95, 295)
phosphine thioether, etc.)	Alkylene oxide	Acrylic or methacrylic	46, 83, 84, 177)
Group VIII			
Fe:			
$FeCl_3$	Alkylene oxide	Acrylic or methacrylic	214, 257, 287)
	Alkylene oxide	Tere- or isophthalic	71, 132)
	1,2-epoxy-3-chloropropane	Acrylic	105)
$FeCl_3$ associated with a tertiary amine	Ethylene oxide	Terephthalic	82)
$Fe(OH)_3$	Propylene oxide	Acrylic or methacrylic	170)
Ru:			
$RuCl_3$, H_2O	Propylene oxide	Acrylic	90)
Co:			
halides	Ethylene oxide	Terephthalic	71)
Co-organic derivatives	1,2-epoxy-3-phenoxy-propane	Carboxy polyisobutylene	62)
Ni:			
Halides	Ethylene oxide	Terephthalic	71)
Other catalytic systems			
Anion exchange resin	Alkylene oxide	Acrylic or methacrylic	111, 197)
	Various epoxides	Acrylic or methacrylic	202)
	Ethylene oxide	Terephthalic	288)
	1,2-epoxy-3-allyloxy-propane	Acetic	199)
	1,2-epoxy-3-chloro-propane	Acrylic or methacrylic	49, 201)
Pyridine-contg. anion exchange resin	Various epoxides	Methacrylic	200)
γ-radiations	1,2-epoxy-3-phenoxy-propane	Various carboxylic	283)

4 Kinetic Relations and Mechanisms

4.1 Non-Catalyzed Reactions

The literature on the basic study of the non-catalyzed epoxy-carboxy reaction is, surprisingly, rather abundant in number and volume, coming just after the amine- or the ammonium salt-catalyzed reaction. This is, in fact, surprising since the non-catalyzed reaction is generally very slow and of little interest for applications.

Several mechanisms have been proposed, but very few authors have shown that the orders in reactants depend on the nature of the reaction medium, although Madec and Maréchal, studying the benzoic/1,2-epoxy-3-phenoxypropane system found that the overall order is 2 in xylene, chlorobenzene, and orthodichlorobenzene as solvent [156]; the authors suggested a bimolecular process. However, compared to the amine-catalyzed reaction, the reaction is very slow.

On the other hand, in nitrobenzene [157] the overall order is 1, which can be reasonably explained by the formation of a complex between the solvent and the epoxy compound:

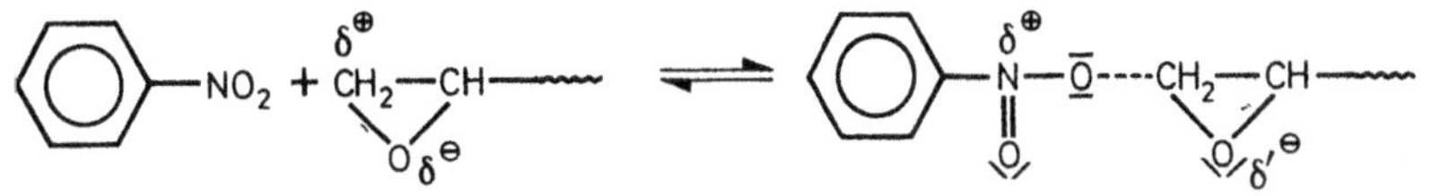
(7)

This complex has a nucleophilic power much higher than that of the epoxy Several authors [142, 236, 264, 282] have suggested formation of a cyclic transition state and the *cis*-opening of the oxide ring as the rate-determining step:

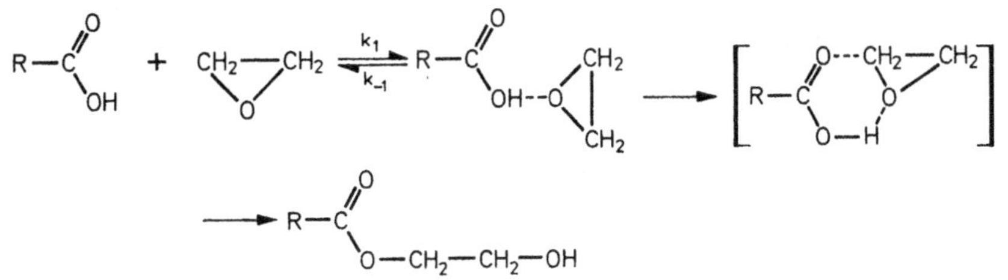
(8)

This mechanism is confirmed by the second order of the reaction and the generally accepted fact that the acid-oxide complex is formed through hydrogen bonds [142]. However, a detailed investigation of the stereochemistry of the reaction [228] showed that the ring-opening is *trans* which is absolutely incompatible with the mechanism reported in reaction (9). To make this point clear, Shvets et al. [228] completely reinvestigated the kinetics of the reaction and showed that its order depends on the nature of the reactants. It can be 1st or zero order in oxide, depending on the basicity of the oxide, the acidity of the acid, or the acid-oxide molar ratio. However, these additional kinetic data are in complete agreement with scheme (8), involving the rapid equilibrium formation of a complex between the oxide and the acid, followed by the slow monomolecular conversion of this complex into the reaction product. It would appear as a contradiction between kinetics and stereochemistry. Another mechanism has been proposed [115], which fits stereochemistry but does not explain pseudo-zero order observed by Shvets [228] when oxide is in excess:

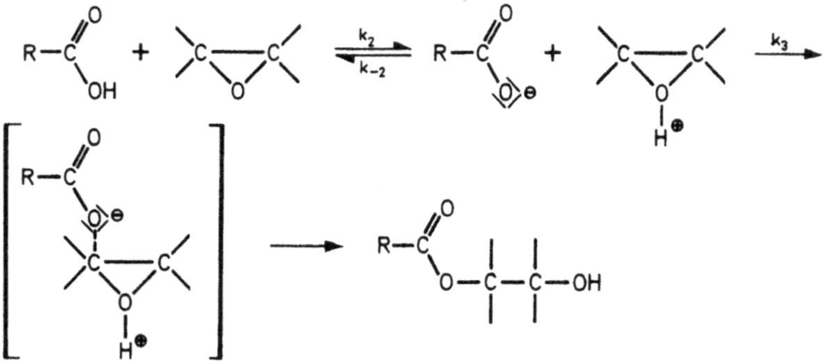

(9)

This mechanism can fit zero order only if, in oxide, the acid is completely dissociated, which is not the case. However, the authors [228] made this reaction scheme compatible with kinetic observations by assuming that:

$$k_2 \ll k_{-2} \qquad \text{and} \qquad k_2 \ll k_3$$

In this case, the formation of the products results from the slow ionization of the complex. The rapid subsequent interaction of the ions gives products through the *trans*-opening of the oxide or leads to the regeneration of the original complex.

A mechanism analogous to that described by scheme (8) has been proposed by Batog et al. [15] for the reaction of norbornene-derived carboxylic acid with alicyclic monoepoxides. The reaction, which is carried out in the bulk, is second order.

Rokaszewski [211,212] studied the mechanism of the reaction of acetic acid with 1,2-epoxy-3-chloropropane in the presence of N,N-dimethylformamide (S) [114]; in most experiments the concentration of S is higher than that of each reactant. This author proposed an autocatalytic mechanism:

$$A|^{\ominus} + E|\ HA \xrightarrow{k_1} AEH + A^{\ominus} \qquad (10)$$

and

$$A|^{\ominus} + E|\ HEA \xrightarrow{k_2} AEH + AE|^{\ominus} \qquad (11)$$

where $A|^{\ominus}$ is probably a mixture of anion $A|^{\ominus}$ and of ion pair $A^{\ominus}/HS^{\oplus}$, and determined the ratio $r = k_2/k_1$, finding that this depends on the molar ratio of the reactants. This is in opposition to theoretical calculations carried out from the kinetic model and indicates that r should not depend on this ratio.

4.2 Amine- and Ammonium Salt-Catalyzed Reactions

This class of catalysts is the most widely used and studied. However, few kinetic studies are really reliable, and many authors have not considered the possible contribution of side reactions or have given only qualitative information. This is a serious deficiency since the contribution of such reactions can be relatively important, particularly when the esterification is carried out in the bulk [159,203].

Whatever the catalyst and the mechanism, it is important to determine the various side reactions which accompany the main esterification. Thus reactions (13) to (16):

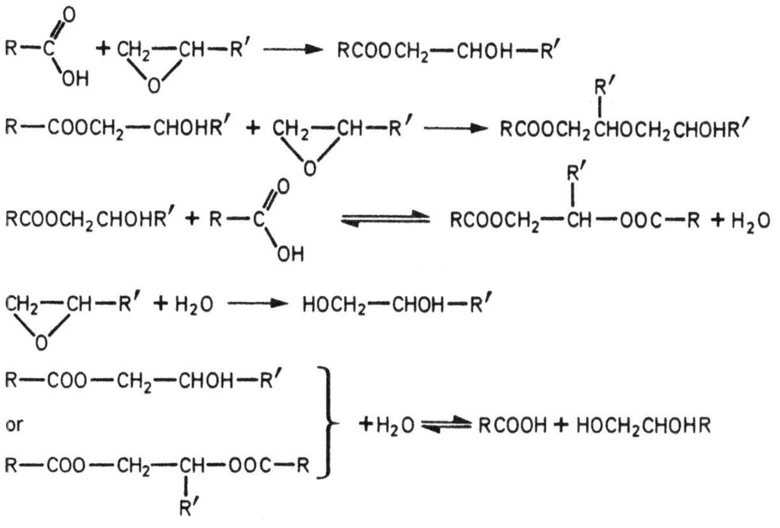

(12) (13) (14) (15) (16)

will take place or not, depending on the temperature, the nature of the catalyst, and on the fact that the reaction is carried out in the bulk or in the presence of a solvent. Unfortunately, the possibility of their contribution is very often not considered.

Many other side reactions can take place, even under mild conditions; one of the most important side reactions is the opening of the epoxy ring by the catalyst and the subsequent polymerization of the epoxy compound. This has been quantitatively analyzed by Madec and Maréchal [156–159] and applied to the synthesis of polymers by block-polycondensation [37, 38] and will be discussed later.

Surprisingly, almost no authors have paid attention to the fact that, in principle, the reaction can take place either on the α or on the β position of the epoxy ring. Martinez Utrilla and Olano Villen [166] discussed this problem when studying the reaction of fatty acids with 1,2-epoxy-3-hydroxypropane in the presence of various catalysts, of which benzyltrimethylammonium chloride is the most active. They showed that in some cases an interchange reaction occurs:

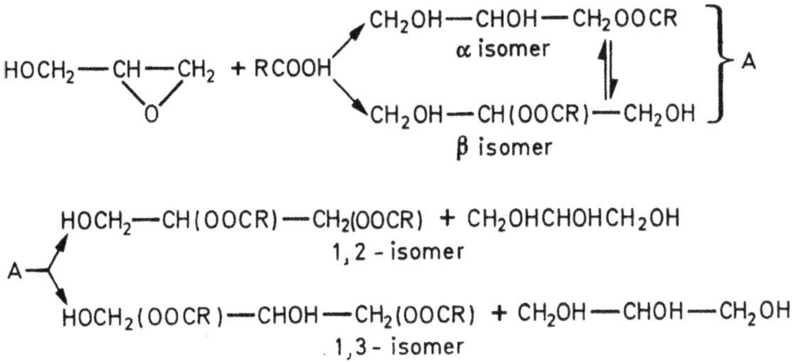

(17) (18)

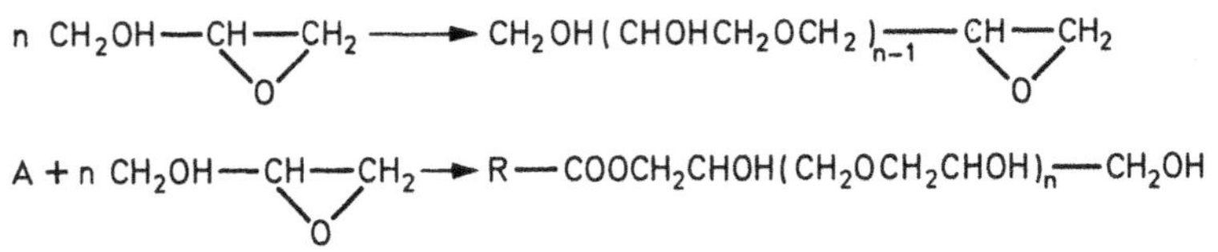

(20)

The analysis of the reactional medium showed that reaction (17) is widely predominent.

Many mechanisms have been proposed, but these can be grouped into two main classes.

4.2.1 Mechanisms which do not Involve the Formation of a Cyclic Complex

Several mechanisms involve the formation of a more or less dissociated ion-pair which reacts with the epoxy compound to form the ester without any contribution of a cyclic species.

The hypothesis of the existence of a pure ionic mechanism, involving species such as $C_6H_5O^{\ominus}$ or $C_6H_5COO^{\ominus}$ in the reaction of epoxy with either phenol or benzoic acid, has been mentioned in the earliest publications relating to this reaction [23,24,222]. Shechter et al. [222], studying the reaction between 1,2-epoxy-3-phenoxypropane and octanoic acid in the presence of a basic catalyst (KOH or benzyldimethylamine), proposed an ionic mechanism with the carboxylate ion as active species. However, their hypothesis has no fundamental kinetic grounds.

Such pure ionic mechanisms have been severely criticized by several authors. Thus, Ishii et al. [100], studying the reaction of various substituted benzoic acids with 1,2-epoxy-3-phenoxypropane in the presence of NMe_3, determined values of Hammett's plot. They observed that ϱ is positive and decreases with increasing reaction temperature. According to these authors, this shows that ionic dissociation of benzoic acid is not the rate-controlling factor, since ϱ values relative to the dissociation constant of substituted benzoic acid do not change with temperature [25]. Even if Ishii's comments about the values of ϱ are not clear, his experimental observations fit those of Kakiuchi and Tanaka [116].

Many of the studies which have proposed mechanisms without the formation of a cyclic complex were carried out with ethylene oxide. In many cases, and at least within the limit of experimental conditions, the authors found a pseudo-zero order in acid and first order in epoxy and catalyst.

On the other hand, studying the tertiary amine-catalyzed reaction of acetic acid with ethylene oxide in *n*-butanol, Bazant et al. [91] found a pseudo-zero order for the consumption of ethylene oxide:

$$-\frac{d[E]}{dt} = k_0 \qquad (18)$$

k_0 depends linearly on the acid and catalyst concentrations:

$$k_0' = k_1[RCOOH] + k_1'[R_3N] \qquad (19)$$

The amine-catalyzed reaction is bimolecular:

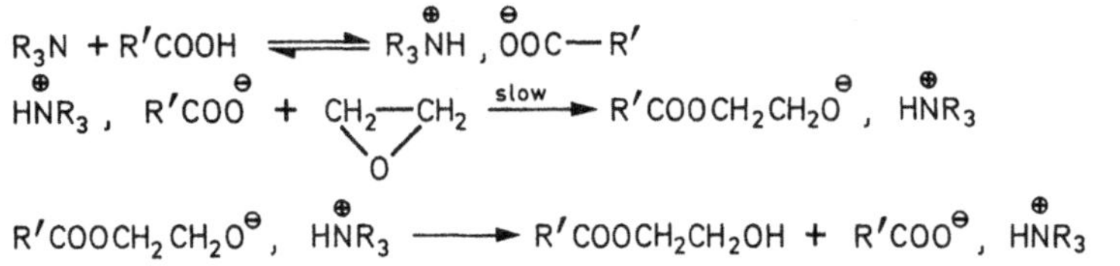

(21)

(22)

(23)

The initial formation of an ammonium salt has been suggested by several authors [164, 218, 230].

Thus, Samsonova et al. [218], studying the kinetics of the triethylamine-catalyzed reaction of terephthalic acid, in the bulk under a pressure of ethylene oxide, found a third-order relation for the consumption of ethylene oxide:

$$-\frac{d[E]}{dt} = k_3[E][A][Cat] \tag{20}$$

The reaction rate varies only very slightly with the basicity of the amine, which, according to these authors, cannot be explained by the formation of a cyclic complex. Like other authors [164, 230], they propose a mechanism in which the first step is the formation of β-ethoxytriethylammonium terephthalate. They isolated 80% of the theoretical amount of this salt which was identified by chemical analysis and ^{1}H NMR. When used as a catalyst, it exhibited the same activity as triethylamine. Their results fit Shvets and Romashkin's mechanism [230]:

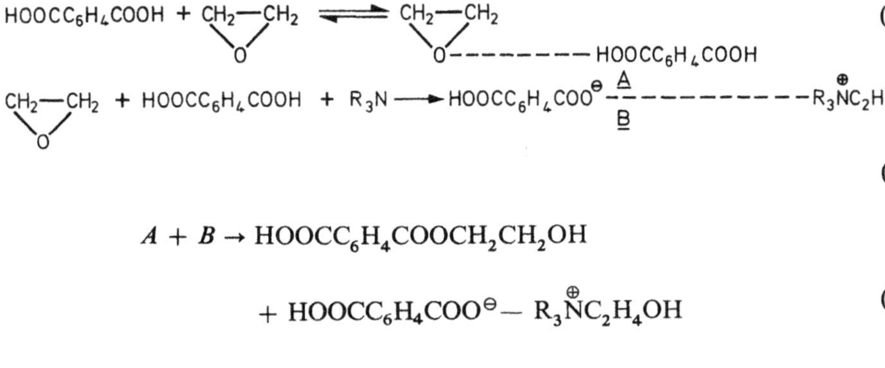

(24)

(25)

$$A + B \rightarrow HOOCC_6H_4COOCH_2CH_2OH$$

$$+ HOOCC_6H_4COO^{\ominus} — R_3\overset{\oplus}{N}C_2H_4OH \tag{26}$$

Samsonova et al. [218] showed that both the ethoxylation of the hydroxy groups into hydroxyethylterephthalate and bis(hydroxyethyl)terephthalate, and the oligomerization of ethylene oxide, are sufficiently slow to consider that the reaction of terephthalic acid with ethylene oxide is highly selective. The problem of the relative contribution of the second carboxylic group of terephthalic acid to the kinetics will be discussed later.

Although it is described by another reaction scheme, the mechanism proposed by Bathia et al. [18] and by Mares et al. [164] for the reaction of ethylene oxide with terephthalic acid in *n*-butanol involves the same active species:

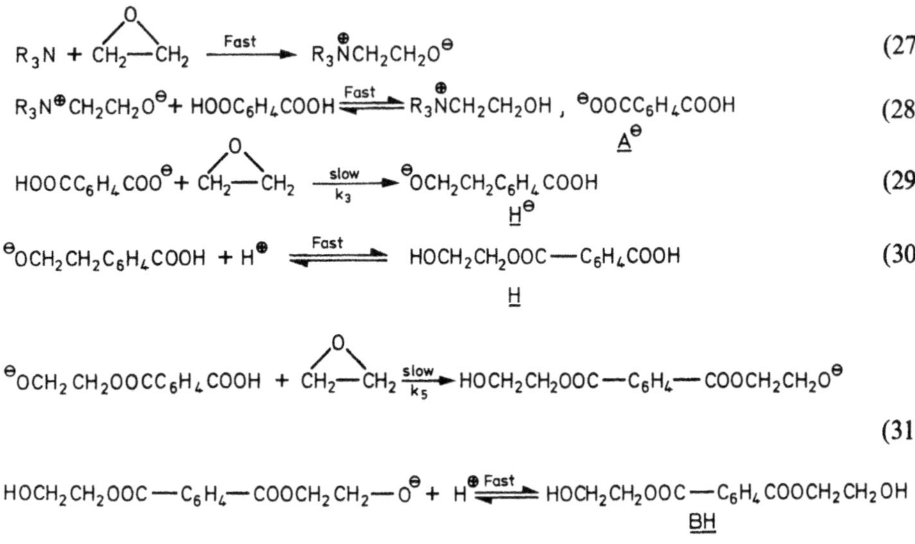

(27) (28) (29) (30) (31) (32)

Few authors have taken into account the fact that, when the acid molecule contains two carboxylic groups, interferences between the corresponding kinetics are possible, although this problem has been analyzed [18, 164]. The kinetic relations obtained by Mares et al. [164] and by Bhatia et al. [18] are the same. In both articles, these were established taking into account the fact that the ratio of the dissociation constants of *H* and *A* is 0.5, which introduces a factor 2 in the relations. In the following, [A] is the concentration of terephthalic acid; that of carboxylic groups is obviously:

$$2[A] + [H] = 2[A]_0 - 2[BH] - [H] \quad (21)$$

$$-\frac{d[E]}{dt} = [E]\,(k_3[A^\ominus] + k_5[H^\ominus]) \quad (22)$$

$$[H^\ominus] = [H]\,[A^\ominus]/2[A] \quad (23)$$

$$[A^\ominus] = 2[Cat]\,[A]/(2[A] + [H]) \quad (24)$$

$$-\frac{d[E]}{dt} = [E]\,[Cat]\,\{2k_3[A]/(2[A] + [H]) + k_5[H]/(2[A] + [H])\} \quad (25)$$

$$-\frac{d[A]}{dt} = [E]\,[Cat]\,\{2k_3[A]/(2[A] + [H])\} \quad (26)$$

$$\frac{d[H]}{dt} = [E]\,[Cat]\,\{2k_3[A]/(2[A] + [H]) - k_5[H]/(2[A] + [H])\} \quad (27)$$

$$\frac{d[BH]}{dt} = [E]\,[Cat]\,\{k_5[H]/(2[A] + [H])\} \quad (28)$$

From here on, the kinetic problem is treated by Bathia and Mares in two different ways.

Mares et al. [164] found that the consumption of carboxylic groups or of ethylene oxide is described by a pseudo-zero order in acid and a first order in ethylene oxide and in catalyst. From Eqs. (25) or (26) and (27) it can be inferred that the consumption of carboxylic end groups is pseudo-zero order at the constant concentration of *E* and *Cat*, only if k_3 and k_5 have the same (or nearly the same) value, since in that case (25) becomes:

$$-\frac{d[E]}{dt} = k[E]\,[Cat] \qquad \text{where } k_3 = k_5 = k \tag{29}$$

Since $k_3/k_5 = \frac{[H]_{max}}{2[A]}$, it can be determined by following the concentration of individual components of the reaction mixture with respect to time; Mares et al., found $k_3/k_5 = 1.7$, which strongly supports their hypothesis.

Bathia et al. [18] obtained completely different results due to the fact that they applied the pseudo-zero order to the consumption of terephthalic acid ([A]) and not to that of all carboxylic groups (2[A] + [H]). Using Capellos method [31], they solved the system of Eqs. (25) to (27) and found:

$$[H] = \frac{[A]}{a-1}\left[1 - \left(\frac{[A]}{[A]_0}\right)^{a-1}\right] \tag{30}$$

$$[BH] = [A]_0 - [A] - [H] \tag{31}$$

where $a = \frac{k_5}{2k_3}$ and $[A]_0$ is the initial concentration of terephthalic acid.

Determination of k_3 and k_5 from a ratio gave values which vary very little with time; k_5/k_3 is in the range of 14.3 to 22.8.

According to Refs. [18, 164], kinetic determinations were carried out with triethylamine as catalyst. However, Mares et al. [164] determined k_0 in:

$$-\frac{d[E]}{dt} = k_0 \tag{32}$$

for different amines and found that it was almost independent of their pK, as shown in Table 2.

Table 2. Dependence of k_0 on the pK of the catalyst in the amine-catalyzed reaction of terephthalic acid with ethylene oxide in *n*-butanol

Amine	pK	k_0 (mol kg^{-1} h^{-1})
Dimethylaniline	8.94	0.228
Pyridine	5.23	0.231
Tributylamine	4.07	0.242

In the same way, Kamatani [120], studying the reaction of terephthalic acid with ethylene oxide in solvents such as THF and nitrobenzene and in the presence of amines, showed that the real catalyst is not the amine itself but the quaternary ammonium salt formed from TPA, EO, and amine: $R'COO^{\ominus}$, $R_3N^{\oplus}CH_2CH_2OH$.

These quaternary ammonium terephthalates were prepared separately and used as catalyst for the reaction of TPA with EO. It was found [120] that the different ammonium salts have almost the same catalytic activity, showing that the formation of the catalyst from amine, ethylene oxide, and TPA is the rate-determining step.

Although not implying the formation of a cyclic complex, the mechanism proposed by Shvets et al. [231] for the reaction of terephthalic acid with ethylene oxide in the presence of tetraalkylammonium halides and in *n*-pentanol is more complex than that we have just described. According to these authors, ethylene halohydrin and tetraalkylammonium terephthalate are formed:

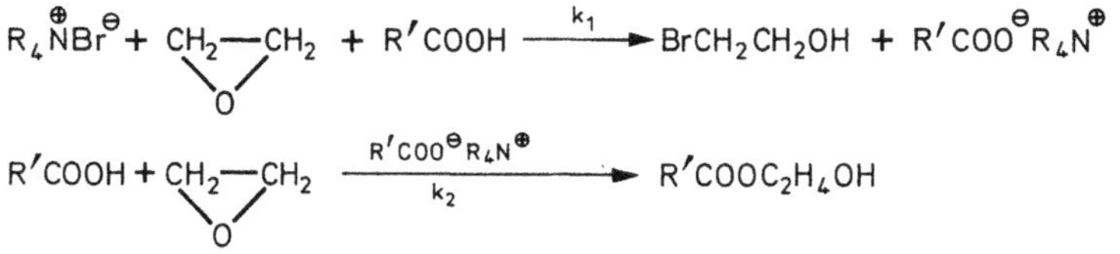

(33)

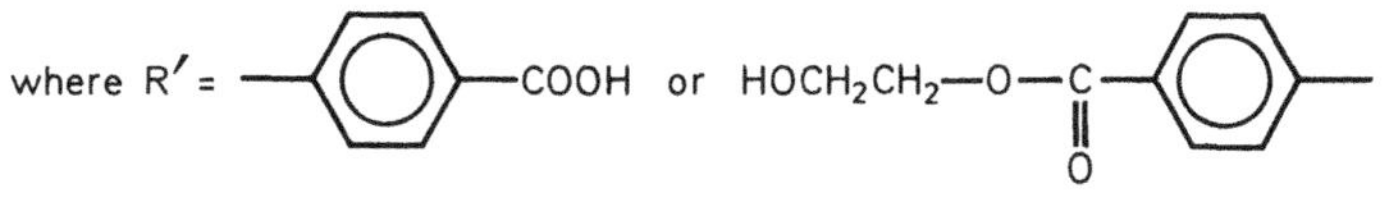

(34)

The content in the halohydrin was determined by gas chromatography; at the beginning of the reaction, the halide concentration rapidly decreases with a simultaneous increase in the concentration of the halohydrin, the sum of these two quantities remaining constant. However, the concentration of the halide does not fall to zero, and after a certain time it becomes steady at a value which depends on its nature and on its initial concentration. Moreover, when the supply of ethylene oxide is stopped, the concentration of the halide ion begins to rise and reaches its initial value. This shows that another reaction takes place:

$$R'COO^{\ominus}\overset{\oplus}{N}R_4 + BrCH_2CH_2OH \xrightarrow{k_3}$$
$$\xrightarrow{k_3} R'COOC_2H_4OH + R_4N^{\oplus}Br^{\ominus} \qquad (35)$$

Experimental study of the system showed that reaction (33) is much more rapid than (34) and (35), and it was established that the consumption of ethylene oxide obeys a second-order law:

$$-\frac{d[E]}{dt} = k[E][Cat] \qquad (33)$$

Kinetic treatment of this system allowed the determination of k_1, k_2, k_3 by an approximation method.

Comparison of the rate constants for reaction (33), (34), and (35) for various halides and tetralkylammonium cations shows that the rates of all reactions rise in the sequence:

$$(C_2H_5)_4NCl < (C_2H_5)_4NBr < (C_2H_5)_4NI$$

This corresponds to the rise in nucleophilicity of the halogen in reaction (33) and (34) and with the decrase in strength of the carbon halogen bond in reaction (35).

The rate of these reactions also increases with the size of the cation: $(C_4H_9)_4N^{\oplus} > (C_2H_5)_4N^{\oplus} > (CH_3)_4N^{\oplus}$. According to the authors [231], this is connected with a weakening of the association of the ion pair which increases the nucleophilic activity of the corresponding anion in all reactions; lastly, the tetraethyl and tetrabutylammonium bromide and iodide, respectively, are twice to six times as efficient as the corresponding tertiary amines. This is connected with the higher catalytic activity of tetraethyl and tetrabutyl ammonium terephthalates as compared with that of β-hydroxyethyltriethylammonium terephthalate which is considered the active species of the systems catalyzed by triethylamine, as reported above.

Bleha et al. [20] studying the reaction of octadecanoic acid with ethylene oxide, supposed that the real catalyst was a quaternary ammonium hydroxide and not the corresponding octadecanoate:

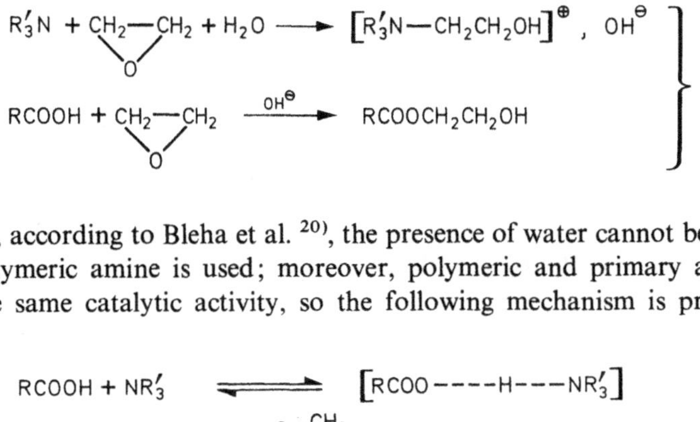

(36)

However, according to Bleha et al. [20], the presence of water cannot be considered when a polymeric amine is used; moreover, polymeric and primary amines have roughly the same catalytic activity, so the following mechanism is proposed [162]:

$$RCOOH + NR'_3 \rightleftharpoons [RCOO{-}{-}{-}{-}H{-}{-}{-}NR'_3] \qquad (37)$$

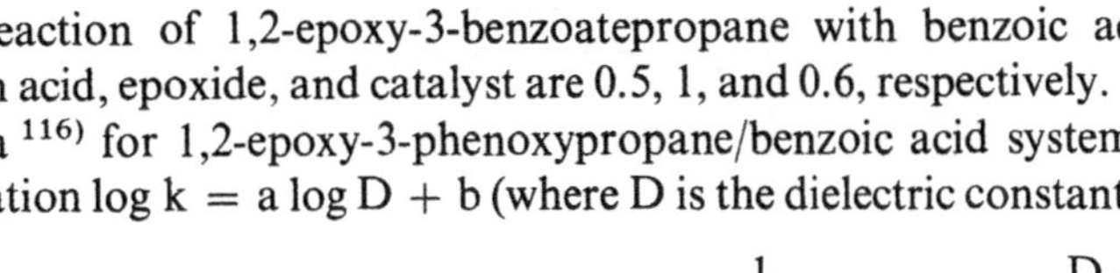

(38)

Some authors have denied the existence of mechanisms involving only ionic or only complex forms. Thus, Fiala and Lidarik [64] studied the benzyldimethylamine-catalyzed reaction of 1,2-epoxy-3-benzoatepropane with benzoic acid in xylene. The order in acid, epoxide, and catalyst are 0.5, 1, and 0.6, respectively. Like Kakiuchi and Tanaka [116] for 1,2-epoxy-3-phenoxypropane/benzoic acid systems, they found that the relation $\log k = a \log D + b$ (where D is the dielectric constant of the system) fits the results better than the relations $\log k = \frac{1}{D}$ or $\log k = \frac{D-1}{2D+1}$ do. They conclude that the real mechanism is a combination of both ionic and complex-forming reactions. The 0.6 order in catalyst would result from the formation of an acid-base complex in xylene:

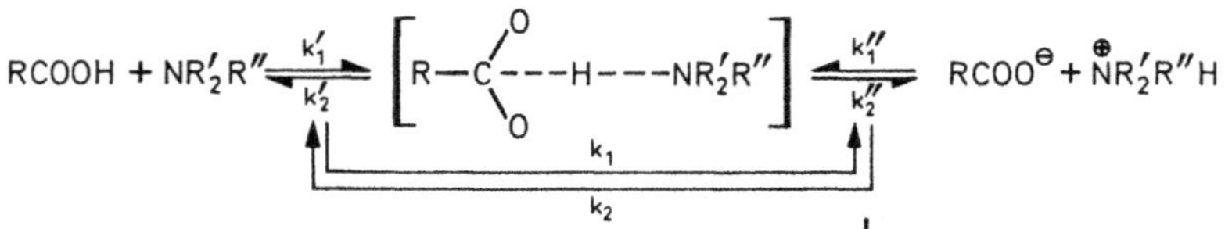
(39)

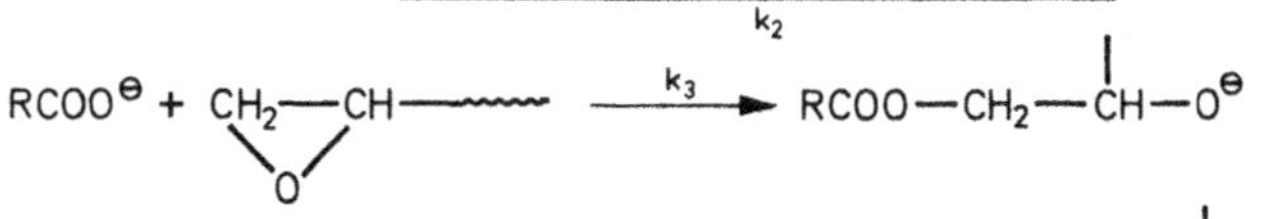
(40)

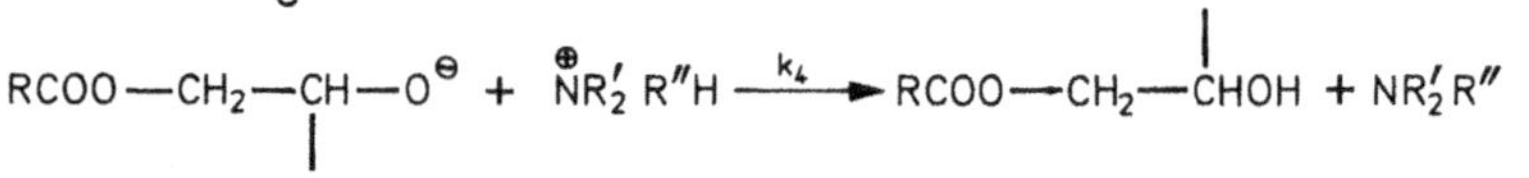
(41)

From this scheme, and accepting some assumptions, Fiala and Lidarik [64] established a complex reaction giving orders 1, 0.5, and 0.5 in epoxide, acid, and catalyst, respectively. Unfortunately, this relation involves the concentration of intermediary species which were not experimentally measured.

4.2.2 Mechanisms which Involve the Formation of a Cyclic Complex

Several authors, particularly in most recent publications, have proposed mechanisms involving the formation of a cyclic complex, of which the formation is the rate-controlling step.

Sorokin and Gershanova [236] carried out a kinetic study of hexanoic acid with various epoxy compounds: 1,2-epoxy-3-phenoxypropane and various substituted derivatives, 1,2-epoxy-3-allyloxypropane, 1,2-epoxy-3-hydroxypropane. They found a zero order in acid and first order in epoxy and catalyst, respectively.

They assumed that the reaction begins by the formation of a complex:

$$\underset{}{RCOOH + |NR_3 \rightleftharpoons \underset{A}{RCOOH{-}|NR_3}} \tag{42}$$

Due to the large excess of the acid compared to amine (50 to 100 times the catalyst concentration), this equilibrium is shifted to the right and the concentration of the complex is mainly determined by the tertiary amine concentration.

Step 1 is followed by the formation of a six-centered cyclic transition complex, in which a synchronized transfer of an electron pair takes place:

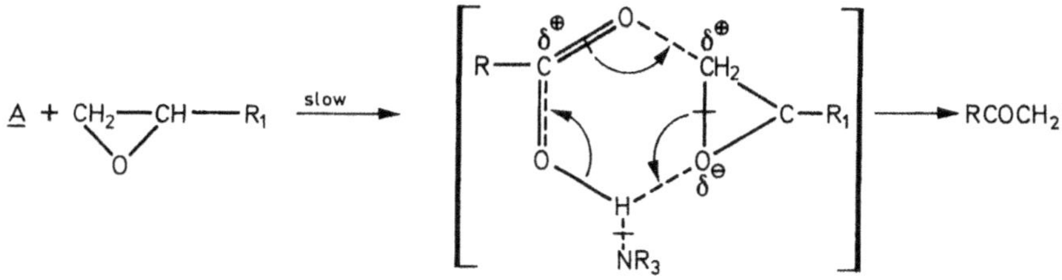
(43)

This mechanism does not differ from that of Tanaka, but it does not explain orders 0, 1, 1 in acid, epoxide, and catalyst, respectively [236].

Sakai et al. [215] proposed the following mechanism for the base-catalyzed reaction of 1,2-epoxy-3-phenoxypropane with benzoic acid in non-polar solvents:

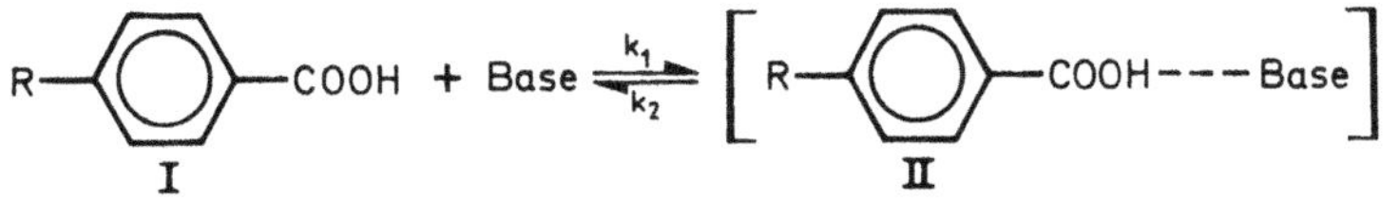
(44)

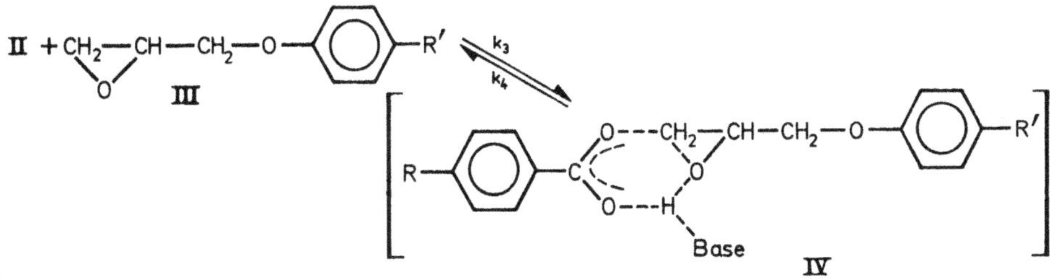
(45)

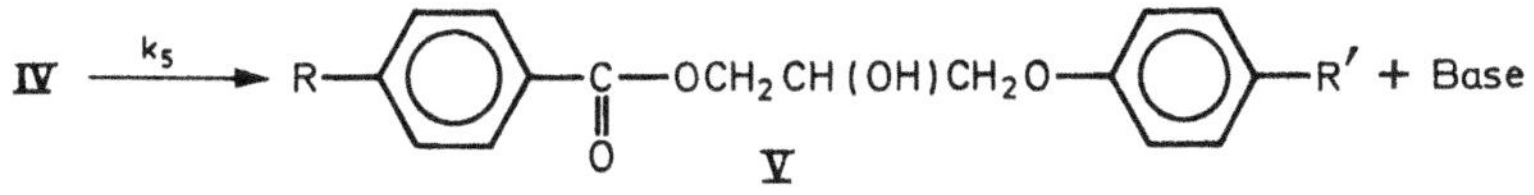
(46)

Kakiuchi and Tanaka [116] extended this study to several substituted acids and 1,2-epoxy-3-phenoxypropane. In addition to reactions (44) to (46), they took into consideration the following:

$$\mathrm{V + III \rightarrow C_6H_5COOCH_2\underset{\underset{R'}{|}}{C}HOCH_2CHOHR''} \tag{47}$$

VI

$$\mathrm{I + V \rightarrow C_6H_5COOCH_2\underset{\underset{R''}{|}}{C}HOOC{-}C_6H_5 + H_2O} \tag{48}$$

$$\mathrm{III + H_2O \rightarrow HOCH_2CHOHR'} \tag{49}$$

$$\mathrm{V + VII + H_2O \rightarrow I + VIII} \tag{50}$$

$\mathrm{R'': {-}CH_2OC_6H_4{-}R'}$

However, Kakiuchi and Tanaka showed that reaction rates hardly change with the nature of the solvent (xylene, mono or dichlorobenzene, nitrobenzene) and, that disappearance of benzoic acid is completely accounted for by the appearance of ester and hydroxy groups. Reactions (47) to (50) can thus be neglected, which is contrary to the observations of Malkemus and Swan [163] and Wrighley et al. [286], who, studying the reaction of fatty acids with ethylene oxide, found that reactions (47) to (50) take place.

Table 3. Kinetic parameters of epoxy-carboxy esterification and polyesterification reactions – Table is in two parts: Part 1 (page 182 to 198 for experimental conditions, Part 2 (page 199 to 208 for results and references. r = initial molar ratio of epoxide (or diepoxide) to acid (or diacid); when r is given (third column), no solvent is present. $[A]_0$ and $[E]_0$ are acid and epoxide initial concentrations, respectively

(Table 3. Part 1)

No.	Epoxide	Acid	Solvent	Catalyst	Temperature Range (°C)
1	1,2-Epoxyethane	Acetic	1-Butanol $[A]_0 = 1.8$ mol l^{-1} Partial pressure of E: 680 torr (p_E)	No	80
2	1,2-Epoxyethane	Acetic	1-Butanol $[A]_0 = 0.5$ mol l^{-1} Partial pressure of E: 680 torr (p_E)	Pyridine 0.13 mol l^{-1}	80
3	1,2-Epoxyethane	Acetic	1-Butanol $[A]_0 = 2.1$ mol l^{-1} p_E as in 1	Pyridine 0.13 mol l^{-1}	80
4	1,2-Epoxyethane	Chloracetic	1-Butanol $[A]_0 = 0.9$ mol l^{-1} p_E as in 1	No	80
5	1,2-Epoxyethane	Chloracetic	1-Butanol $[A]_0 = 0.9$ mol l^{-1} p_E as in 1	Dimethylcyclohexylamine 0.075 mol l^{-1}	80
6	1,2-Epoxyethane	Chloracetic	1-Butanol $[A]_0$ and p_E as in 5	Dimethylcyclohexylamine 0.300 mol l^{-1}	80
7	1,2-Epoxyethane	Dichloracetic	1-Butanol $[A]_0 = 0.25$ mol l^{-1} p_E as in 1	No	80
8	1,2-Epoxyethane	Dichloracetic	1-Butanol $[A]_0 = 0.75$ mol l^{-1} p_E as in 1	Dimethylcyclohexylamine 0.075 mol l^{-1}	80

9	1,2-Epoxyethane	Terephthalic	1-Butanol $[A]_0 = 0.743$ mol kg^{-1} $[E]_0 = 0.841$ mol kg^{-1}	Dimethylaniline 0.156 mol kg^{-1}	80
10	1,2-Epoxyethane	Terephthalic	1-Butanol $[A]_0$ and $[E]_0$ as in 9	Pyridine as in 9	80
11	1,2-Epoxyethane	Terephthalic	1-Butanol $[A]_0$ and $[E]_0$ as in 9	Tributylamine as in 9	80
12	1,2-Epoxyethane	Terephthalic	1-Butanol $[A]_0 = 0.985$ mol l^{-1} $[E]_0 = 0.306$ mol l^{-1}	Triethylamine 0.0932 mol l^{-1}	60–120
13	1,2-Epoxyethane	Terephthalic	Dimethylterephthalate $[A]_0$ not indicated $[E]_0$ not indicated	β-Hydroxybutylammonium terephthalate	145–165
14	1,2-Epoxyethane	Terephthalic	Nitrobenzene $[A]_0 = 1.16$ mol kg^{-1} $[E]_0 = 2.33$ mol kg^{-1}	Bu_3P 0.0116 mol kg^{-1}	120
15	1,2-Epoxyethane	Terephthalic	as in 14	Ph_3P as in 14	120
16	1,2-Epoxyethane	Terephthalic	as in 14	Ph_3As as in 14	120
17	1,2-Epoxyethane	Terephthalic	as in 14	Et_3N as in 14	120
19	1,2-Epoxyethane	Terephthalic	as in 14	Pr_3N as in 14	120
20	1,2-Epoxyethane	Terephthalic	as in 14	$Et_2NC_2H_4OH$ as in 14	120
21	1,2-Epoxyethane	Terephthalic	as in 14	$EtN(C_2H_4OH)_2$ as in 14	120
22	1,2-Epoxyethane	Terephthalic	as in 14	EtN(cyclohexyl)$_2$ as in 14	120
23	1,2-Epoxyethane	Terephthalic	as in 14	$(HOC_2H_4)_3N$ as in 14	120
24	1,2-Epoxyethane	Terephthalic	as in 14	Pr_2NH as in 14	120
25	1,2-Epoxyethane	Terephthalic	as in 14	Bu_2NH as in 14	120
26	1,2-Epoxyethane	Terephthalic	as in 14	$EtNH_2$ as in 14	120
27	1,2-Epoxyethane	Terephthalic	as in 14	$PrNH_2$ as in 14	120
28	1,2-Epoxyethane	Terephthalic	as in 14	$BuNH_2$ as in 14	120
29	1,2-Epoxyethane	Terephthalic	1-Pentanol $[A]_0 = 0.595$ mol l^{-1} $[E]_0 = 0.242$ mol l^{-1}	Et_4NBr 0.060 mol l^{-1}	110
30	1,2-Epoxyethane	Terephthalic	as in 29	Et_4NI as in 29	110
31	1,2-Epoxyethane	Terephthalic	as in 29	Et_4NCl as in 29	110
32	1,2-Epoxyethane	Terephthalic	as in 29	Bu_4NI as in 29	110

Contin. p. 184

Table 3. (continued)

No.	Epoxide	Acid	Solvent	Catalyst	Temperature Range (°C)
33	1,2-Epoxyethane	Terephthalic	as in 29	Me_4NI as in 29	110
34	1,2-Epoxyethane	Terephthalic	No $[A]_0 = 0.03$ mol $[E]_0 = 1.2$ mol	Et_4N^+ terephthalate $3 \cdot 10^{-5}$ mol	92–127
35	1,2-Epoxyethane	Terephthalic	No $[A]_0 = 0.06$ mol l^{-1} $[E]_0 = 0.48$ mol l^{-1}	Et_3N 0.75×10^{-2} mol l^{-1}	120–145
36	1,2-Epoxyethane	Terephthalic	No $[A]_0$ and $[E]_0$ as in 35	(β-ethoxy)triethylammonium terephthalate 0.44×10^{-2} mol l^{-1}	50
37	1,2-Epoxyethane	Octadecanoic	No r = 1	K octadecanoate (K) K/acid: 0.05	120
38	1,2-Epoxyethane	Octadecanoic	as in 37	K/acid: 0.01	120
39	1,2-Epoxyethane	Octadecanoic	as in 37	K/acid: 0.005	120
40	1,2-Epoxyethane	1,3,5-tri(2-carboxy-ethyl)isocyanurate	Dimethylformamide $[A]_0 = 3$ mol l^{-1} $[E]_0 = 3$ mol l^{-1}	No	65.2–75.2
41	1,2-Epoxypropane	Acetic	*o*-Xylene $[A]_0 = 0.228$ mol l^{-1} $[E]_0 = 6.2$ mol l^{-1}	No	161.5
42	1,2-Epoxypropane	Trichloracetic	*o*-Xylene $[A]_0 = 0.213$ mol l^{-1} $[E]_0 = 2.63$ mol l^{-1}	No	25.2
43	1,2-Epoxypropane	1,3,5-tri(2-carboxy-ethyl)isocyanurate	as in 40	No	65.2–75.2
44	1,2-Epoxybutane	Trichloracetic	*o*-Xylene $[A]_0 = 0.6$ mol l^{-1} $[E]_0 = 3.0$ mol l^{-1}	No	15
45	(3,4-Epoxycyclohexyl)-acetoxymethane	3-Cyclohexenecarboxylic	No r = 1	No	100–160

46	(3,4-Epoxycyclohexyl)-benzoyloxymethane	3-Cyclohexenecarboxylic	No	No	100–160
47	(3,4-Epoxycyclohexyl)-benzoyloxymethane	Monocyclohexyl ester of 4-cyclohexene-1,2-dicarboxylic	No r = 1	No	100–160
48	(3,4-Epoxycyclohexyl)-benzoyloxymethane	Monocyclohexyl ester of 5-norbornene-2,3-dicarboxylic	No r = 1	No	100–160
49	(3,4-Epoxycyclohexyl)-ethylcarboxylate	3-Cyclohexenecarboxylic	No r = 1	No	120–160
50	4,5-Epoxy-2′-furylcyclo-hexanespiro-5′-(1′,3′-dioxane)	3-Cyclohexenecarboxylic	No r = 1	No	100–160
51	4,5-Epoxycyclohexane-spiro-5′-(1_1,3_3-dioxane)-2′-spirocyclo-hexane	3-Cyclohexenecarboxylic	No r = 1	No	100–160
52	4,5-Epoxycyclohexane-spiro-5′-(1_1,3_3-dioxane)-2′-spirocyclo-hexane	Monocyclohexylester of 4-cyclohexene-1,2-dicarboxylic	No r = 1	No	100–160
53	4,5-Epoxycyclohexane-spiro-5′-(1_1,3_3-dioxane)-2′-spirocyclo-hexane	Monocyclohexylester of 5-norbornene-2,3-dicarboxylic	No r = 1	No	100–160
54	(4,5-Epoxycyclohexyl)-1,2-dimethyldicarboxylate	3-Cyclohexenecarboxylic	No r = 1	No	120
55	1,2-Epoxy-3-chloropropane	Acetic	Dimethylformamide $[A]_0 = 3.22$ mol l^{-1} $[E]_0 = 5.36$ mol l^{-1}	Dimethylformamide 5.24 mol l^{-1}	80.2
56	1,2-Epoxy-3-chloropropane	Hexanoic	not indicated	No	80–110
57	1,2-Epoxy-3-chloropropane	Hexanoic	not indicated	Trihexylamine concentration non indicated	80–110
58	1,2-Epoxy-3-chloropropane	1,3,5-tri(2-carboxyethyl)-isocyanurate	Dimethylformamide $[A]_0 = 3$ mol l^{-1} $[E]_0 = 3$ mol l^{-1}	No	75.2–80.2
59	1,2-Epoxy-3-hydroxypropane	Hexadecanoic	Dioxane $[A]_0 = 0.324$ mol l^{-1} $[E]_0 = 0.809$ mol l^{-1}	Benzyltrimethylammonium chloride 0.016 mol l^{-1}	81–96.5

Contin. p. 186

Table 3. (continued)

No.	Epoxide	Acid	Solvent	Catalyst	Temperature Range (°C)
60	1,2-Epoxy-3-hydroxypropane	Hexadecanoic	Dimethylsulfoxide $[A]_0$ and $[E]_0$ as in 59	as in 59	85–100.5
61	1,2-Epoxy-3-hydroxypropane	Hexadecanoic	Xylene $[A]_0 = 0.260$ mol l^{-1} $[E]_0 = 0.286$ mol l^{-1}	Benzyltrimethylammonium chloride 0.01115 mol l^{-1}	107.9
62	1,2-Epoxy-3-phenoxypropane	Benzoic	Xylene $[A]_0 = 0.5$ mol kg^{-1} $[E]_0 = 0.5$ mol kg^{-1}	Trimethylamine 0.5×10^{-2} mol kg^{-1}	95–125
63	1,2-Epoxy-3-phenoxypropane	Benzoic	Chlorobenzene $[A]_0$ and $[E]_0$ as in 62	as in 62	95–125
64	1,2-Epoxy-3-phenoxypropane	Benzoic	Nitrobenzene $[A]_0$ and $[E]_0$ as in 62	as in 62	95–125
65	1,2-Epoxy-3-phenoxypropane	Benzoic	Xylene $[A]_0 = 0.27$ mol l^{-1} $[E]_0 = 0.27$ mol l^{-1}	Dimethyldodecylamine 0.023 mol l^{-1}	80–130
66	1,2-Epoxy-3-phenoxypropane	Benzoic	Chlorobenzene $[A]_0$ and $[E]_0$ as in 65	as in 65	80–130
67	1,2-Epoxy-3-phenoxypropane	Benzoic	*o*-Dichlorobenzene $[A]_0$ and $[E]_0$ as in 65	as in 65	80–130
68	1,2-Epoxy-3-phenoxypropane	Benzoic	Nitrobenzene $[A]_0$ and $[E]_0$ as in 65	as in 65	80–130
69	1,2-Epoxy-3-phenoxypropane	Benzoic	Xylene $[A]_0 = 0.31$ eq kg^{-1} $[E]_0 = 0.31$ eq kg^{-1}	Dimethyldodecylamine 2.12×10^{-2} eq kg^{-1}	86.6–115.5
70	1,2-Epoxy-3-phenoxypropane	Benzoic	Chlorobenzene $[A]_0 = 0.23$ eq kg^{-1} $[E]_0 = 0.23$ eq kg^{-1}	as in 69 1.96×10^{-2} eq kg^{-1}	86.6–115.5
71	1,2-Epoxy-3-phenoxypropane	Benzoic	*o*-Dichlorobenzene $[A]_0 = 0.21$ eq kg^{-1} $[E]_0 = 0.21$ eq kg^{-1}	as in 69 2.14×10^{-2} eq kg^{-1}	86.6–115.5

72	1,2-Epoxy-3-phenoxypropane	Benzoic	Nitrobenzene $[A]_0 = 0.23$ eq kg^{-1} $[E]_0 = 0.23$ eq kg^{-1}	as in 69 2.03×10^{-2} eq kg^{-1}	86.6–115.5
73	1,2-Epoxy-3-phenoxypropane	Benzoic	Xylene $[A]_0 = [E]_0 = 0.5$ mol l^{-1}	Isoquinoline 0.02 mol l^{-1}	96–118
74	1,2-Epoxy-3-phenoxypropane	Benzoic	as in 73	3-Picoline as in 73	96–118
75	1,2-Epoxy-3-phenoxypropane	Benzoic	as in 73	4-Picoline as in 73	96–118
76	1,2-Epoxy-3-phenoxypropane	Benzoic	as in 73	Pyridine as in 73	96–118
77	1,2-Epoxy-3-phenoxypropane	Benzoic	as in 73	2,4-Lutidine as in 73	96–118
78	1,2-Epoxy-3-phenoxypropane	Benzoic	as in 73	2-Picoline as in 73	96–118
79	1,2-Epoxy-3-phenoxypropane	Benzoic	as in 73	4-methylquinoline as in 73	96–118
80	1,2-Epoxy-3-phenoxypropane	Benzoic	as in 73	Quinoline as in 73	96–118
81	1,2-Epoxy-3-phenoxypropane	Benzoic	as in 73	2-Methylquinoline as in 73	96–118
82	1,2-Epoxy-3-phenoxypropane	Benzoic	as in 73	2,6-Lutidine as in 73	96–118
83	1,2-Epoxy-3-phenoxypropane	Benzoic	Toluene/nitrobenzene (mol: 1/0) $[A]_0 = [E]_0 = 0.3$ mol l^{-1}	Pyridine 2.45×10^{-3}–2.42×10^{-2} mol l^{-1}	80–108
84	1,2-Epoxy-3-phenoxypropane	Benzoic	as in 83 (mol: 0.75/0.25)	as in 83	80–108
85	1,2-Epoxy-3-phenoxypropane	Benzoic	as in 83 (mol: 0.50/0.50)	as in 83	80–108
86	1,2-Epoxy-3-phenoxypropane	Benzoic	as in 83 (mol: 0.24/0.76)	as in 83	80–108
87	1,2-Epoxy-3-phenoxypropane	Benzoic	as in 83 (mol: 0/100)	as in 83	80–108

Contin. p. 188

Table 3. (continued)

No.	Epoxide	Acid	Solvent	Catalyst	Temperature Range (°C)
88	1,2-Epoxy-3-phenoxypropane	Benzoic	Toluene/dioxane (mol: 1/0) $[A]_0 = [E]_0 = 0.24\ mol\ l^{-1}$	as in 83	98
89	1,2-Epoxy-3-phenoxypropane	Benzoic	as in 88 (mol: 0.8/0.2)	as in 83	98
90	1,2-Epoxy-3-phenoxypropane	Benzoic	as in 88 (mol: 0.29/0.71)	as in 83	98
91	1,2-Epoxy-3-phenoxypropane	Benzoic	as in 88 (mol: 0/1)	as in 83	98
92	1,2-Epoxy-3-phenoxypropane	Benzoic	Nitrobenzene/dioxane (mol: 1/0) $[A]_0 = [E]_0 = 0.24\ mol\ l^{-1}$	as in 83	98
93	1,2-Epoxy-3-phenoxypropane	Benzoic	as in 92 (mol: 0.88/0.12)	as in 83	98
94	1,2-Epoxy-3-phenoxypropane	Benzoic	as in 92 (mol: 0.18/0.82)	as in 83	98
95	1,2-Epoxy-3-phenoxypropane	Benzoic	as in 92 (mol: 0/1)	as in 83	98
96	1,2-Epoxy-3-phenoxypropane	Benzoic	Tetrahydrofuran $[A]_0 = [E]_0 = 0.24\ mol\ l^{-1}$	as in 83	64.5
97	1,2-Epoxy-3-phenoxypropane	Benzoic	Dioxane as in 96	as in 83	64.5
98	1,2-Epoxy-3-phenoxypropane	Benzoic	Diisopropylether as in 96	as in 83	64.5
99	1,2-Epoxy-3-phenoxypropane	Benzoic	Di-*n*-butylether as in 96	as in 83	64.5
100	1,2-Epoxy-3-phenoxypropane	Benzoic	Anisole as in 96	as in 83	64.5
101	1,2-Epoxy-3-phenoxypropane	Benzoic	Benzene as in 96	as in 83	64.5

102	1,2-Epoxy-3-phenoxypropane	Benzoic	Cyclohexane as in 96	as in 83	64.5
103	1,2-Epoxy-3-phenoxypropane	Benzoic	Nitrobenzene $[A]_0 = [E]_0 = 0.2$ mol l^{-1}	Pyridine 0.0124 mol l^{-1}	98
104	1,2-Epoxy-3-phenoxypropane	Benzoic	Nitrobenzene $[A]_0 = [E]_0 = 0.25$ mol l^{-1}	Pyridine 0.0016 mol l^{-1}	98
105	1,2-Epoxy-3-phenoxypropane	Benzoic	Benzonitrile as in 103	as in 103	98
106	1,2-Epoxy-3-phenoxypropane	Benzoic	Benzonitrile as in 104	as in 104	98
107	1,2-Epoxy-3-phenoxypropane	Benzoic	Chlorobenzene as in 103	Pyridine 0.0248 mol l^{-1}	98
108	1,2-Epoxy-3-phenoxypropane	Benzoic	Chlorobenzene as in 104	Pyridine 0.0031 mol l^{-1}	98
109	1,2-Epoxy-phenoxypropane	Benzoic	Toluene as in 103	Pyridine as in 107	98
110	1,2-Epoxy-3-phenoxypropane	Benzoic	Toluene as in 104	Pyridine as in 108	98
111	1,2-Epoxy-3-phenoxypropane	Benzoic	Toluene/nitrobenzene (mol: 1/0) $[A]_0 = [E_0] = 0.24$ mol l^{-1}	Pyridine 0.0108 mol l^{-1}	80–108
112	1,2-Epoxy-3-phenoxypropane	Benzoic	as in 111 (mol: 0.75/0.25)	as in 111	80–108
113	1,2-Epoxy-3-phenoxypropane	Benzoic	as in 111 (mol: 0.50/0.50)	as in 111	80–108
114	1,2-Epoxy-3-phenoxypropane	Benzoic	as in 111 (mol: 0.25/0.75)	as in 111	80–108
115	1,2-Epoxy-3-phenoxypropane	Benzoic	as in 111 (mol: 0/1)	as in 111	80–108
116	1,2-Epoxy-3-phenoxypropane	Benzoic	*o*-Nitrotoluene $[A]_0 = [E]_0 = 0.24$ mol l^{-1}	as in 111	80–108
117	1,2-Epoxy-3-phenoxypropane	Benzoic	*o*-Dichlorobenzene as in 116	as in 111	80–108
118	1,2-Epoxy-3-phenoxypropane	Benzoic	Chlorobenzene as in 116	as in 111	80–108

Contin. p. 190

Table 3. (continued)

No.	Epoxide	Acid	Solvent	Catalyst	Temperature Range (°C)
119	1,2-Epoxy-3-phenoxypropane	Benzoic	Toluene as in 116	as in 111	80–108
120	1,2-Epoxy-3-phenoxypropane	Benzoic	Xylene as in 116	as in 111	80–108
121	1,2-Epoxy-3-phenoxypropane	Benzoic	Dioxane as in 116	as in 111	80–108
122	1,2-Epoxy-3-phenoxypropane	Benzoic	Dioxane/toluene (mol: 0.2/0.8) $[A]_0 = [E]_0 = 0.24$ mol l^{-1}	as in 111	80–108
123	1,2-Epoxy-3-phenoxypropane	Benzoic	as in 122 (mol: 0.5/0.5)	as in 111	80–108
124	1,2-Epoxy-3-phenoxypropane	Benzoic	as in 122 (mol: 0.9/0.1)	as in 111	80–108
125	1,2-Epoxy-3-phenoxypropane	Benzoic	Dioxane/nitrobenzene (mol: 0.23/0.77) $[A]_0 = [E]_0 = 0.24$ mol l^{-1}	as in 111	80–108
126	1,2-Epoxy-3-phenoxypropane	Benzoic	as in 125 (mol: 0.45/0.55)	as in 111	80–108
127	1,2-Epoxy-3-phenoxypropane	Benzoic	as in 125 (mol: 0.83/0.17)	as in 111	80–108
128	1,2-Epoxy-3-phenoxypropane	Benzoic	Tetrahydrofurane as in 111	as in 111	65
129	1,2-Epoxy-3-phenoxypropane	Benzoic	Dioxane as in 111	as in 111	65
130	1,2-Epoxy-3-phenoxypropane	Benzoic	Diisopropylether as in 111	as in 111	65
131	1,2-Epoxy-3-phenoxypropane	Benzoic	Di-*n*-propylether as in 111	as in 111	65
132	1,2-Epoxy-3-phenoxypropane	Benzoic	Anisole as in 111	as in 111	65

133	1,2-Epoxy-3-phenoxypropane	Benzoic	*n*-Amylphenylether as in 111	as in 111	65
134	1,2-Epoxy-3-phenoxypropane	Benzoic	Benzene as in 111	as in 111	65
135	1,2-Epoxy-3-phenoxypropane	Benzoic	Cyclohexane as in 111	as in 111	65
136	1,2-Epoxy-3-phenoxypropane	Benzoic	Nitrobenzene $[A]_0 = [E]_0 = 0.2\ mol\ l^{-1}$	Pyridine $0.00124\ mol\ l^{-1}$	81
137	1,2-Epoxy-3-phenoxypropane	Benzoic	Benzonitrile as in 136	as in 136	81
138	1,2-Epoxy-3-phenoxypropane	Benzoic	*o*-Dichlorobenzene as in 136	Pyridine $0.00248\ mol\ l^{-1}$	81
139	1,2-Epoxy-3-phenoxypropane	Benzoic	Chlorobenzene as in 136	as in 138	81
140	1,2-Epoxy-3-phenoxypropane	Benzoic	Toluene as in 136	as in 138	81
141	1,2-Epoxy-3-phenoxypropane	Benzoic	Xylene $[A]_0 = [E]_0 = 0.5\ mol\ l^{-1}$	Triethylamine $0.0025\ mol\ l^{-1}$	100
142	1,2-Epoxy-3-phenoxypropane	Benzoic	as in 141	as in 141	100
143	1,2-Epoxy-3-phenoxypropane	Benzoic	as in 141	Tetraethylammonium benzoate $0.0025\ mol\ l^{-1}$	100
144	1,2-Epoxy-3-phenoxypropane	Benzoic	as in 141	as in 143	100
145	1,2-Epoxy-3-phenoxypropane	Benzoic	Nitrobenzene as in 141	as in 143	100
146	1,2-Epoxy-3-phenoxypropane	Benzoic	as in 145	as in 145	100
147	1,2-Epoxy-3-phenoxypropane	Benzoic	Acetonitrile as in 141	Tetramethylammonium iodide $0.015\ mol\ l^{-1}$	100
148	1,2-Epoxy-3-phenoxypropane	Benzoic	Acetonitrile as in 141	Tetraethylammonium iodide as in 147	100
149	1,2-Epoxy-3-phenoxypropane	Benzoic	Acetonitrile as in 141	Tetrabutylammonium iodide as in 147	100
150	1,2-Epoxy-3-phenoxypropane	Benzoic	Acetonitrile as in 141	Trimethylbutylammonium iodide as in 147	100

Contin. p. 192

Table 3. (continued)

No.	Epoxide	Acid	Solvent	Catalyst	Temperature Range (°C)
151	1,2-Epoxy-3-phenoxypropane	Benzoic	as in 147	Triisobutylmethylammonium iodide as in 147	100
152	1,2-Epoxy-3-phenoxypropane	Benzoic	as in 147	Tetraethylammonium as in 147	100
153	1,2-Epoxy-3-phenoxypropane	Benzoic	as in 147	Triethylethanolammonium bromide as in 147	100
154	1,2-Epoxy-3-phenoxypropane	Benzoic	as in 147	Diethyldiethanolammonium bromide as in 147	100
155	1,2-Epoxy-3-phenoxypropane	Benzoic	as in 147	Ethyltriethanolammonium bromide as in 147	100
156	1,2-Epoxy-3-phenoxypropane	Benzoic	Xylene $[A]_0 = [E]_0 = 0.5$ mol kg^{-1}	Trimethylamine 0.005 mol kg^{-1}	95–125
157	1,2-Epoxy-3-phenoxypropane	*m*-Methylbenzoic	as in 156	as in 156	95–125
158	1,2-Epoxy-3-phenoxypropane	*p*-Methylbenzoic	as in 156	as in 156	95–125
159	1,2-Epoxy-3-phenoxypropane	*p*-Methoxybenzoic	as in 156	as in 156	95–125
160	1,2-Epoxy-3-phenoxypropane	*p*-Methoxybenzoic	Chlorobenzene as in 156	as in 156	95–125
161	1,2-Epoxy-3-phenoxypropane	*p*-Methoxybenzoic	Nitrobenzene as in 156	as in 156	95–125
162	1,2-Epoxy-3-phenoxypropane	*p*-Chlorobenzoic	Xylene as in 156	as in 156	95–125
163	1,2-Epoxy-3-phenoxypropane	*m*-Ethylesterbenzoic	as in 162	as in 156	95–125
164	1,2-Epoxy-3-phenoxypropane	*m*-Nitrobenzoic	as in 162	as in 156	95–125
165	1,2-Epoxy-3-phenoxypropane	*m*-Nitrobenzoic	Chlorobenzene as in 156	as in 156	95–125
166	1,2-Epoxy-3-phenoxypropane	*m*-Nitrobenzoic	Nitrobenzene as in 156	as in 156	95–125

167	1,2-Epoxy-3-phenoxypropane	Hexanoic	No $[A]_0 = 8.2$ mol kg^{-1} $[E]_0 = 0.295$ mol kg^{-1}	Trihexylamine 0.015 mol kg^{-1}	90
168	1,2-Epoxy-3-phenoxypropane	Hexanoic	No as in 167	NaOH 0.05 mol kg^{-1}	100
169	1,2-Epoxy-3-phenoxypropane	Hexanoic	No $[A]_0 = 0.504$ mol kg^{-1} $[E]_0 = 6.28$ mol kg^{-1}	NaOH 0.2 mol kg^{-1}	90
170	1,2-Epoxy-3-phenoxypropane	Hexanoic	No $[A]_0 = 8.21$ mol kg^{-1} $[E]_0 = 0.290$ mol kg^{-1}	No	110
171	1,2-Epoxy-3-phenoxypropane	Hexanoic	No $[A]_0 = 0.213$ mol kg^{-1} $[E]_0 = 6.52$ mol kg^{-1}	No	140
172	1,2-Epoxy-3-phenoxypropane	Hexanoic	Chlorobenzene 4.63 mol kg^{-1} $[A]_0 = 0.253$ mol kg^{-1} $[E]_0 = 3.000$ mol kg^{-1}	No	120
173	1,2-Epoxy-3-phenoxypropane	Hexanoic	No	Trihexylamine concentration non indicated	80
174	1,2-Epoxy-3-phenoxypropane	Hexanoic	No	KOH concentration non indicated	100–160
175	1,2-Epoxy-3-phenoxypropane	Hexanoic	No	$NaCOOCH_3$ concentration non indicated	100–160
176	1,2-Epoxy-3-phenoxypropane	Hexanoic	No	NaOH concentration non indicated	100–140
177	1,2-Epoxy-3-phenoxypropane	Hexanoic	No $r = 1$	No	90–150
178	1,2-Epoxy-3-phenoxypropane	Hexanoic	No $r = 1$	Monobutoxyaminotitanate 2.3×10^{-2} mol l^{-1}	90–150

Contin. p. 194

Table 3. (continued)

No.	Epoxide	Acid	Solvent	Catalyst	Temperature Range (°C)
179	1,2-Epoxy-3-(4-*t*-butyl)-phenoxypropane	Hexanoic	No r = 1	No	
180	1,2-Epoxy-3-(4-*t*-butyl)-phenoxypropane	Hexanoic	No r = 1	Triethylamine	90–110
181	1,2-Epoxy-3-(4-methyl)-phenoxypropane	Benzoic	Xylene $[A]_0 = [E]_0 = 0.27\ mol\ l^{-1}$	Dimethyldodecylamine $0.023\ mol\ l^{-1}$	80–130
182	1,2-Epoxy-3-(4-methyl)-phenoxypropane	Benzoic	Chlorobenzene as in 181	as in 181	80–130
183	1,2-Epoxy-3-(4-methyl)-phenoxypropane	Benzoic	*o*-Dichlorobenzene as in 181	as in 181	80–130
184	1,2-Epoxy-3-(4-methyl)-phenoxypropane	Benzoic	Nitrobenzene as in 181	as in 181	80–130
185	1,2-Epoxy-3-(4-methyl-phenoxypropane	Hexanoic	No r = 1	No	90
186	1,2-Epoxy-3-(4-methyl)-phenoxypropane	Hexanoic	No r = 1	Trihexylamine concentration non indicated	90
187	1,2-Epoxy-3-(4-methoxy)-	Benzoic	Xylene as in 181	as in 181	80–130
188	1,2-Epoxy-3-(4-methoxy)-phenoxypropane	Benzoic	Chlorobenzene as in 181	as in 181	80–130
189	1,2-Epoxy-3-(4-methoxy)-phenoxypropane	Benzoic	*o*-Dichlorobenzene as in 181	as in 181	80–130
190	1,2-Epoxy-3-(4-methoxy)-phenoxypropane	Benzoic	Nitrobenzene as in 181	as in 181	80–130
191	1,2-Epoxy-3-(4-chloro)-phenoxypropane	Benzoic	Xylene as in 181	as in 181	80–130
192	1,2-Epoxy-3-(4-chloro)-phenoxypropane	Benzoic	Chlorobenzene as in 181	as in 181	80–130

193	1,2-Epoxy-3-(4-chloro)-phenoxypropane	Benzoic	*o*-Dichlorobenzene as in 181	as in 181	80–130
194	1,2-Epoxy-3-(4-chloro)-phenoxypropane	Benzoic	Nitrobenzene as in 181	as in 181	80–130
195	1,2-Epoxy-3-(4-nitro)-phenoxypropane	Benzoic	Xylene as in 181	as in 181	80–130
196	1,2-Epoxy-3-(4-nitro)-phenoxypropane	Benzoic	Chlorobenzene as in 181	as in 181	80–130
197	1,2-Epoxy-3-(4-nitro)-phenoxypropane	Benzoic	*o*-Dichlorobenzene as in 181	as in 181	80–130
198	1,2-Epoxy-3-(4-nitro)-phenoxypropane	Benzoic	Nitrobenzene as in 181	as in 181	80–130
199	1,2-Epoxy-3-(4-ethyl-ester)phenoxypropane	Benzoic	Xylene as in 181	as in 181	80–130
200	1,2-Epoxy-3-(4-ethyl-ester)phenoxypropane	Benzoic	Chlorobenzene as in 181	as in 181	80–130
201	1,2-Epoxy-3-(4-ethyl-ester)phenoxypropane	Benzoic	*o*-Dichlorobenzene as in 181	as in 181	80–130
202	1,2-Epoxy-3-(4-ethyl-ester)phenoxypropane	Benzoic	Nitrobenzene as in 181	as in 181	80–130
203	1,2-Epoxy-3-allyloxypropane	Hexanoic	No	No	90–110
204	1,2-Epoxy-3-allyloxypropane	Hexanoic	No	Trihexylamine concentration non indicated	90–110
205	1,2-Epoxy-3-butoxypropane	Hexanoic	No	No	90–110
206	1,2-Epoxy-3-butoxypropane	Hexanoic	No	Trihexylamine concentration non indicated	90–110
207	1,2-Epoxy-3-benzoate-propane	Benzoic	Xylene $[A]_0 = [E]_0 = 1.5$ mol l^{-1}	Benzyldimethylamine 0.175 mol l^{-1}	98–124
208	1,2-Epoxy-3-benzoate-propane	Benzoic	Nitrobenzene $[A]_0 = [E]_0 = 0.48$ mol l^{-1}	Benzyldimethylamine 0.12 mol l^{-1}	110
209	1,2-Epoxy-3-benzoate-propane	Benzoic	*o*-Nitrotoluene as in 208	as in 208	110

Contin. p. 196

Table 3. (continued)

No.	Epoxide	Acid	Solvent	Catalyst	Temperature Range (°C)
210	1,2-Epoxy-3-benzoate-propane	Benzoic	Acetophenone as in 208	as in 208	110
211	1,2-Epoxy-3-benzoate-propane	Benzoic	Chlorobenzene as in 208	as in 208	110
212	1,2-Epoxy-3-benzoate-propane	Benzoic	Xylene as in 208	as in 208	110
213	1,2-Epoxy-3-dodecyloxypropane	Octadecanoic	No $[A]_0 = [E]_0 = 1.9$ eq kg^{-1}	No	121
214	1,2-Epoxy-3-dodecyloxypropane	Octadecanoic	No $[A]_0 = [E]_0 = 1.851$ eq kg^{-1}	Dimethyldodecylamine 1.17×10^{-1} eq kg^{-1}	86.6–115.5
215	1,2-Epoxy-3-dodecyloxypropane	Octadecanoic	No $[A]_0 = [E]_0 = 1.834$ eq kg^{-1}	as in 214 1.60×10^{-1} eq kg^{-1}	86.6–115.5
216	1,2-Epoxy-3-dodecyloxypropane	Octadecanoic	No $[A]_0 = [E]_0 = 1.851$ eq kg^{-1}	as in 214 1.17×10^{-1} eq kg^{-1}	86.6–115.5
217	1,2-Epoxy-3-dodecyloxypropane	Octadecanoic	No $[A]_0 = [E]_0 = 1.847$ eq kg^{-1}	as in 214 1.27×10^{-1} eq kg^{-1}	86.6–115.5
218	1,2-Epoxy-3-dodecyloxypropane	Octadecanoic	No $[A]_0 = [E]_0 = 1.874$ eq kg^{-1}	as in 214 0.60×10^{-1} eq kg^{-1}	86.6–115.5
219	1,2-Epoxy-3-dodecyloxypropane	Octadecanoic	No $[A]_0 = [E]_0 = 1.872$ eq kg^{-1}	as in 214 0.64×10^{-1} eq kg^{-1}	86.6–115.5
220	1,2-Epoxy-3-dodecyloxypropane	Octadecanoic	No $[A]_0 = [E]_0 = 1.889$ eq kg^{-1}	as in 214 0.20×10^{-1} eq kg^{-1}	86.6–115.5
221	1,2-Epoxy-3-dodecyloxypropane	Octadecanoic	No $[A]_0 = 2.41$ eq kg^{-1} $[E]_0 = 1.21$ eq kg^{-1}	as in 214 1.01×10^{-1} eq kg^{-1}	86.6–115.5
222	1,2-Epoxy-3-dodecyloxypropane	Octadecanoic	No $[A]_0 = 2.10$ eq kg^{-1} $[E]_0 = 1.40$ eq kg^{-1}	as in 214 2.44×10^{-1}	86.6–115.5

223	1,2-Epoxy-3-dodecyloxypropane	Octadecanoic	No $[A]_0 = 1.48$ eq kg^{-1} $[E]_0 = 2.22$ eq kg^{-1}	as in 214 1.88×10^{-1}	86.6–115.5
224	1,2-Epoxy-3-dodecyloxypropane	Octadecanoic	No $[A]_0 = 1.26$ eq kg^{-1} $[E]_0 = 2.51$ eq kg^{-1}	as in 214 1.59×10^{-1}	86.6–115.5
225	1,2-Epoxy-3-dodecyloxypropane	2,4-Hexadienoic	Dimethylsulfoxide $[A]_0 = [E]_0 = 2.0$ eq kg^{-1}	Dimethyldodecylamine 2×10^{-2} eq kg^{-1}	120
226	1,2-Epoxy-3-dodecyloxypropane	2,4-Hexadienoic	as in 225	as in 225 4×10^{-2} eq kg^{-1}	120
227	1,2-Epoxy-3-dodecyloxypropane	2,4-Hexadienoic	as in 225	as in 225 10^{-1} eq kg^{-1}	120
228	1,2-Epoxy-3-dodecyloxypropane	2,4-Hexadienoic	as in 225	as in 225 2×10^{-2} eq kg^{-1}	80
229	1,2-Epoxy-3-dodecyloxypropane	2,4-Hexadienoic	as in 225	as in 225	100
230	1,2-Epoxy-3-dodecyloxypropane	2,4-Hexadienoic	as in 225	as in 225	140
231	2-(oxiranylmethoxy)-1-hydroxyethane (ethyleneglycolmono-glycidylether)	Hexadecanoic	Xylene $[A]_0 = 0.2598$ mol l^{-1} $[E]_0 = 0.2857$ mol l^{-1}	Benzyltrimethylammonium chloride 0.01115 mol l^{-1}	91–107.9
232	2-[2-(oxiranylmethoxy)ethoxy]-1-hydroxyethane (diethyleneglycolmono-glycidylether)	Hexadecanoic	as in 231	as in 231 0.0223	91–107.9
233	α,ω-diglycidyl polyoxypropylene	Acrylic	No	Triethylbenzylammonium chloride	80–110
234	α,ω-diglycidyl poly(oxypropylene-*o*-phthalate)	Acrylic	No	as in 233	80–110
235	2,2-bis[4-(2,3-epoxypropoxy)-phenyl]propane (diglycidylether of bisphenol A)	Acrylic	No	Triethylbenzylammonium chloride 6.7×10^{-2} eq kg^{-1}	80–110
236	2,2-bis[4-(2,3-epoxypropoxy)-phenyl]propane	Poly(methylmethacrylate-co-methacrylic)	No r = 4	Dimethylbenzylamine 7×10^{-3} mol kg^{-1}	160

Contin. p. 198

Table 3. (continued)

No.	Epoxide	Acid	Solvent	Catalyst	Temperature Range (°C)
237	2,2-bis[4-(2,3-epoxypropoxy)-phenyl]propane	Poly(methylmethacrylate-co-methacrylic)	No $r = 1$	as in 236	120
238	2,2-bis[4-(2,3-epoxypropoxy)-phenyl]propane	Poly(methylmethacrylate-co-methacrylic)	No $r = 1$	Quinoline 3.75×10^{-3} mol kg^{-1}	120
239	2,2-bis[4-(2,3-epoxypropoxy)-phenyl]propane	Poly(methylmethacrylate-co-methacrylic)	No $r = 1$	Quinoline 3.75×10^{6} mol kg^{-1}	140
240	2,2-bis[4-(2,3-epoxypropoxy)-phenyl]propane	Poly(methylmethacrylate-co-methacrylic	No $r = 1$	Dimethylbenzylamine concentration non indicated	140
241	2,2-bis[4-(2,3-epoxypropoxy)-phenyl]propane	Poly(methylmethacrylate-co-methacrylic)	No $r = 1$	Pyridine concentration non indicated	140
242	2,2-bis[4-(2,3-epoxypropoxy)-phenyl]propane	Poly(methylmethacrylate-co-methacrylic)	No $r = 1$	2-Picoline concentration non indicated	140
243	2,2-bis[4-(2,3-epoxypropoxy)-phenyl]propane	Poly(methylmethacrylate-co-methacrylic)	No $r = 1$	Quinoline concentration non indicated	40–70

(Table 3. Part 2)

No.	Overall order	Order in epoxide	Order in acid	Order in catalyst	k	ΔH≠ kJ mol⁻¹	ΔS≠ J K mol⁻¹	Ref.
1			0		k_1 (non-catalyzed reaction) $k_1 = 8.8 \times 10^{-3}\ min^{-1}$			91)
2			0		$k_0 = k_1[RCOOH] + k_1'[R_3N]$ $k_0 = 4.2 \times 10^{-3}\ mol\ l^{-1}\ min^{-1}$			91)
3			0		k_0 as in 2 $k_0 = 7.7 \times 10^{-3}\ mol\ l^{-1}\ min^{-1}$			91)
4			0		k_1 as in 1 $k_1 = 8.0 \times 10^{-3}\ min^{-1}$			91)
5			0		k_0 as in 2 $k_0 = 3.8 \times 10^{-3}\ mol\ l^{-1}\ min^{-1}$			91)
6			0		$k_0 = 9.0 \times 10^{-3}\ mol\ l^{-1}\ min^{-1}$			91)
7			0		k_1 as in 1 $k_1 = 7.1 \times 10^{-2}\ min^{-1}$			91)
8			0		k_0 as in 2 $k_0 = 37.0 \times 10^{-3}\ mol\ l^{-1}\ min^{-1}$			91)
9	2	1	pseudo 0	1	$[A] = [A]_0 - k_0 t$ [A]: concentration of terephthalic acid groups contained in both terephthalic acid and 2-hydroxyethylterephthalate. $-\frac{d[E]}{dt} = k_2[E][catalyst]$ $k_0 = 0.228\ mol\ kg^{-1}\ h^{-1}$ $k_2 = 1.74\ kg\ mol^{-1}\ h^{-1}$			164)
10	2	1	pseudo 0	1	$k_0 = 0.231\ mol\ kg^{-1}\ h^{-1}$ $k_2 = 1.76\ kg\ mol^{-1}\ h^{-1}$			164)
11	2	1	pseudo 0	1	$k_0 = 0.242\ mol\ kg^{-1}\ h^{-1}$ $k_2 = 1.84\ kg\ mol^{-1}\ h^{-1}$			164)
12	2	1	pseudo	1	k_0 and k_2 as in 9; k_1: 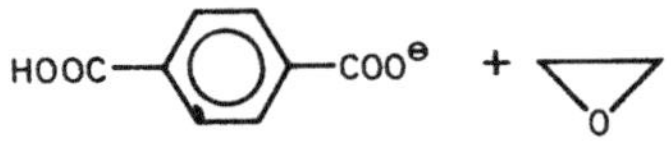			18)
			0		$k_0 = 3.125 \times 10^{-3}\ mol\ l^{-1}\ min^{-1}$ (100 °C, average value) $k_1 = 0.0771\ l\ mol^{-1}\ min^{-1}$ (100 °C, average value) $k_2 = 1.305 \times 10^{-3}\ mol\ l^{-1}\ min^{-1}$	80 98		

Contin. p. 200

Table 3. (continued)

No.	Overall order	Order in epoxide	Order in acid	Order in catalyst	k				ΔH* kJ mol^{-1}	ΔS* J K mol^{-1}	Ref.
13	3	1	1	1	k_3: $l^2\ mol^{-2}\ sec^{-1}$ (°C): 6.0 (145), 7.3 (150), 9.4 (155), 13.4 (165)				56		[4]
14					k: consumption of the mono-quaternary salt of the catalyst and of terephthalic acid; kg $mol^{-1}\ min^{-1}$		k: esterification kg $mol^{-1}\ min^{-1}$				[122]
						0.1	0.590				
15					as in 14	0.1	0.498				[122]
16					as in 14	0.00613	0.501				[122]
17					as in 14		0.639				[120]
18					as in 14		0.567				[120]
19					as in 14		0.529				[120]
20					as in 14	0.0324	0.447				[120]
21					as in 14	0.00453	0.543				[120]
22					as in 14	0.00168	0.648				[120]
23					as in 14	0.00162	0.518				[120]
24					as in 14	0.0154	0.533				[120]
25					as in 14	0.0152	0.463				[120]
26					as in 14	0.00464	0.467				[120]
27					as in 14	0.00410	0.413				[120]
28					as in 14	0.00387	0.432				[120]
29					$k_1(R_4N^{\oplus}X^{\ominus} + E + A \rightarrow XCH_2CH_2OH + A^{\ominus}R_4N^{\oplus})$ = 2.79 l $mol^{-1}\ min^{-1}$ $k_2(A + E \xrightarrow{A^{\ominus}R_4N^{\oplus}} ACOOCH_2CH_2OH)$ = 0.18 l · mol^{-1} · min^{-1} $k_3(A^{\ominus}R_4N^{\oplus} + XCH_2CH_2OH \rightarrow ACOOCH_2CH_2OH + R_4N^{\oplus}X^{\ominus})$ = 0.596 l mol min^{-1}						[231]
30					as in 29	k_1 = 8.09	k_2 = 0.18	k_3 = 0.851			[231]
31					as in 29	k_1 = 0.802	k_2 = 0.18	k_3 = 0.018			[231]
32					as in 29	k_1 = 8.67	k_2 = 0.79	k_3 = 0.934			[231]

33				as in 29 $k_1 = 0.25$ $k_2 = 0.01$ $k_3 = 0.003$			231)
34				k (formation of monoester of E) (107 °C) $= 5.21 \times 10^{-3}$ mol l^{-1} min^{-1}	78.2		277)
35			1	$k = 1.31\ l^2\ mol^{-2}\ sec^{-1}$ (120 °C)	45.1		218)
36				$k = 0.60 \times 10^{-4}\ sec^{-1}$			218)
37	1			$k_1 = 49 \times 10^{-3}\ min^{-1}$ (120 °C)	45.1		11)
38	1			$k_1 = 13 \times 10^{-3}\ min^{-1}$ (120 °C)			11)
39	1			$k_1 = 6.74 \times 10^{-3}\ min^{-1}$ (120 °C)			11)
40				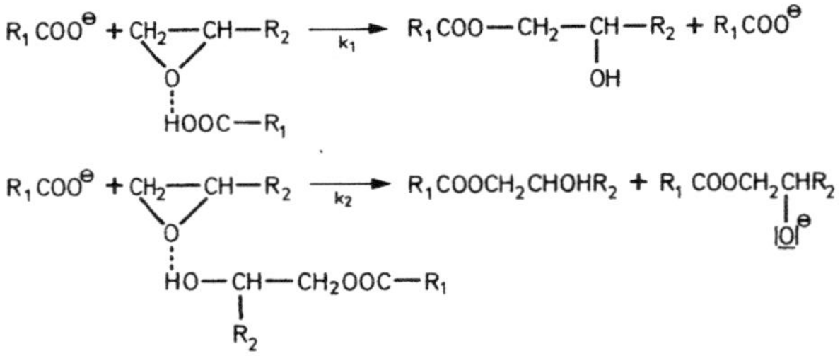 $k_1 \times 10^6$ ($l^2\ mol^{-2}\ sec^{-1}$) $k_2 \times 10^5$ ($l^2\ mol^{-2}\ sec^{-1}$) 4.14 (65.2 °C) 2.36 (65.2 °C)	75.4	—126	211)
41		0	1	$k = 1.24 \times 10^{-4}\ sec^{-1}$			228)
42		0	1	$k = 1.46 \times 10^{-4}\ sec^{-1}$			228)
43				k_1 and k_2 as in 40 5.63 2.08	68.8	—143	211)
44		1	0	$k = 1.50 \times 10^{-4}\ sec^{-1}$			228)
45	2			$k = 6.11 \times 10^{-3}\ kg\ mol^{-1}\ sec^{-1}$ (100 °C)	51.9	—187.6	15)
46	2			$k = 4.11 \times 10^{-3}\ kg\ mol^{-1}\ sec^{-1}$ (100 °C)	55.7	—177.9	15)
47	2			$k = 41.1 \times 10^{-3}\ kg\ mol^{-1}\ sec^{-1}$ (160 °C)			15)
48	2			$k = 40.7 \times 10^{-3}\ kg\ mol^{-1}\ sec^{-1}$ (160 °C)			15)
49	2			$k = 4.97 \times 10^{-3}\ kg\ mol^{-1}\ sec^{-1}$ (100 °C)			15)
50	2			$k = 6.73 \times 10^{-3}\ kg\ mol^{-1}\ sec^{-1}$ (120 °C)			15)
51	2			$k = 5.80 \times 10^{-3}\ kg\ mol^{-1}\ sec^{-1}$ (120 °C)			15)
52	2			$k = 25.5 \times 10^{-3}\ kg\ mol^{-1}\ sec^{-1}$ (160 °C)			15)
53	2			$k = 20.8 \times 10^{-3}\ kg\ mol^{-1}\ sec^{-1}$ (160 °C)			15)

Contin. p. 202

Table 3. (continued)

No.	Overall order	Order in epoxide	Order in acid	Order in catalyst	k	ΔH≠ kJ mol^{-1}	ΔS≠ J K mol^{-1}	Ref.
54	2				$k = 3.20 \times 10^{-3}$ kg mol^{-1} sec^{-1} (120 °C)			15)
55					$k_1(A^{\ominus} + E, HA \rightarrow AEH + A^{\ominus}) = 16.7 \times 10^{-7}$ l^2 mol^{-2} sec^{-1}			212)
					$k_2(A^{\ominus} + E, HEA \rightarrow AEH + AE^{\ominus}) = 45.59 \times 10^{-6}$ l^2 mol^{-2} sec^{-1}			
56	2	1	1		$k = 0.407 \times 10^{-6}$ kg mol^{-1} sec^{-1} (80 °C)	93	−113	236)
57	3	1	1	1	$k = 4.08 \times 10^{-6}$ kg mol^{-1} sec^{-1} (80 °C)	72	−115	236)
58					k_1, k_2 as in 40			211)
					$k_1 = 11.7 \times 10^{-6}$. l^2 mol^{-2} sec^{-1} $k_2 = 4.43 \times 10^{-5}$ l^2 mol^{-2} sec^{-1}	68.8	−143	
59	2				$k_2 = 3.48 \times 10^{-3}$ l mol^{-1} sec^{-1}	73.6	− 36.4	167)
60	2				$k_2 = 5.07 \times 10^{-3}$ l mol^{-1} sec^{-1}			167)
61	2	1	0	1	$k_2 = 19.11 \times 10^{-3}$ l mol^{-1} sec^{-1}			194)
62	3	1	1	1	$k = 22.1$ l^2 mol^{-2} h^{-1} (95 °C)			100)
63	3	1	1	1	$k = 71.7$ l^2 mol^{-2} h^{-1} (112 °C)			100)
64	3	1	1	1	$k = 222$ l^2 mol^{-2} h^{-1} (112 °C)			100)
65	3	1	1	1	$k = 1.54 \times 10^{-2}$ l^2 mol^{-2} sec^{-1} (80 °C)	74	− 87.4	116)
66	3	1	1	1	$k = 1.79 \times 10^{-2}$ l^2 mol^{-2} sec^{-1} (106.5 °C)			116)
67	3	1	1	1	$k = 2.18 \times 10^{-2}$ l^2 mol^{-2} sec^{-1} (106.5 °C)			116)
68	3	1	1	1	$k = 2.80 \times 10^{-2}$ l^2 mol^{-2} sec^{-1} (106.5 °C)			116)
69	2.5	1	0.5	1	$k_{2.5} = 8.6 \times 10^{-3}$ kg$^{1.5}$ eq$^{-1.5}$ s^{-1} (106.5 °C)	72	− 96.5	156, 158)
70	2.5	1	0.5	1	$k_{2.5} = 12.3 \times 10^{-3}$ kg$^{1.5}$ eq$^{-1.5}$ s^{-1} (106.5 °C)	74	− 90	156, 158)
71	2.5	1	0.5	1	$k_{2.5} = 16.0 \times 10^{-3}$ kg$^{1.5}$ eq$^{-1.5}$ s^{-1} (106.5 °C)	74	− 87	156, 158)
72	2	1	0	1	$k_2 = 6.1 \times 10^{-3}$ kg eq^{-1} s^{-1} (106.5 °C)	75	− 90.5	157, 158)
73	3	1	1	1	$k = 8.71 \times 10^{-2}$ l^2 mol^{-2} s^{-1} (96 °C)	69	−113	266)
74	3	1	1	1	as in 73 $k = 7.25 \times 10^{-2}$	71	−109	266)
75	3	1	1	1	as in 73 $k = 6.78 \times 10^{-2}$	71.5	−109	266)
76	3	1	1	1	as in 73 $k = 6.27 \times 10^{-2}$	71.5	−108	266)
77	3	1	1	1	as in 73 $k = 5.72 \times 10^{-2}$	72.3	−107	266)
78	3	1	1	1	as in 73 $k = 5.23 \times 10^{-2}$	74	−103	266)
79	3	1	1	1	as in 73 $k = 4.80 \times 10^{-2}$	74.5	−103	266)
80	3	1	1	1	as in 73 $k = 4.23 \times 10^{-2}$	76.6	−101	266)

81	3	1	1	1	as in 73	$k = 2.76 \times 10^{-2}$	82.8	— 84.4	266)
82	3	1	1	1	as in 73	$k = 1.87 \times 10^{-2}$	86.1	— 78.6	266)
83	3	1	1	1	$k_3 = 0.35 \times 10^{-4}$ l² mol⁻² s⁻¹ (80 °C)		71.9	— 69.0	117)
84	3	1	1	1	as in 83	$k_3 = 0.37 \times 10^{-4}$	68.1	— 71.9	117)
85	3	1	1	1	as in 83	$k_3 = 0.41 \times 10^{-4}$	64.8	— 74.4	117)
86	3	1	1	1	as in 83	$k_3 = 0.40 \times 10^{-4}$	62.3	— 77.3	117)
87	3	1	1	1	as in 83	$k_3 = 0.55 \times 10^{-4}$	60.6	— 79.4	117)
88	3	1	1	1	$k_3 = 2.41 \times 10^{-2}$ l² mol⁻² s⁻¹ (98 °C)		71.9	— 69	117)
89	3	1	1	1	as in 88	$k_3 = 1.41 \times 10^{-2}$	75.7	— 59	117)
90	3	1	1	1	as in 88	$k_3 = 0.80 \times 10^{-2}$	84.9	— 38.9	117)
91	3	1	1	1	as in 88	$k_3 = 0.73 \times 10^{-2}$	90.7	— 24.7	117)
92	3	1	1	1	$k_3 = 9.46 \times 10^{-2}$ l² mol⁻² s⁻¹ (64.5 °C)		60.6	— 79.4	117)
93	3	1	1	1	as in 92	$k_3 = 8.15 \times 10^{-2}$	64.4	— 71.9	117)
94	3	1	1	1	as in 92	$k_3 = 2.67 \times 10^{-2}$	86.9	— 25.1	117)
95	3	1	1	1	as in 92	$k_3 = 0.50 \times 10^{-2}$	97.4	— 7.5	117)
96	3	1	1	1	as in 92	$k_3 = 3.6 \times 10^{-2}$			117)
97	3	1	1	1	as in 92	$k_3 = 4.2 \times 10^{-2}$			117)
98	3	1	1	1	as in 92	$k_3 = 5.4 \times 10^{-2}$			117)
99	3	1	1	1	as in 92	$k_3 = 7.1 \times 10^{-2}$			117)
100	3	1	1	1	as in 92	$k_3 = 11 \times 10^{-2}$			117)
101	3	1	1	1	as in 92	$k_3 = 11.5 \times 10^{-2}$			117)
102	3	1	1	1	as in 92	$k_3 = 46.8 \times 10^{-2}$			117)
103	3	1	1	1	$k_3 = 0.8 \times 10^{-3}$ l² mol⁻² min⁻¹				117)
104	3	1	1	1	as in 103	$k_3 = 1.65 \times 10^{-3}$			117)
105	3	1	1	1	$k_3 = 6.73 \times 10^{-2}$ l² mol⁻² min⁻¹				117)
106	3	1	1	1	as in 105	$k_3 = 14.7 \times 10^{-2}$			117)
107	3	1	1	1	as in 105	$k_3 = 2.73 \times 10^{-2}$			117)
108	3	1	1	1	as in 105	$k_3 = 4.10 \times 10^{-2}$			117)
109	3	1	1	1	as in 105	$k_3 = 2.65 \times 10^{-2}$			117)
110	3	1	1	1	as in 105	$k_3 = 2.94 \times 10^{-2}$			117)
111	3	1	1	1	$k_3 = 2.34 \times 10^{-3}$ l² mol⁻² s⁻¹ (80 °C)		67.7	—106.2	267)
112	3	1	1	1	as in 111	$k_3 = 3.04 \times 10^{-3}$	63.5	—115.8	267)
113	3	1	1	1	as in 111	$k_3 = 4.13 \times 10^{-3}$	69.2	—122.5	267)
114	3	1	1	1	as in 111	$k_3 = 5.20 \times 10^{-3}$	57.7	—127.9	267)

Contin. p. 204

Table 3. (continued)

No.	Overall order	Order in epoxide	Order in acid	Order in catalyst	k	ΔH≠ kJ mol⁻¹	ΔS≠ J K mol⁻¹	Ref.
115	3	1	1	1	as in 111 $k_3 = 6.55 \times 10^{-3}$	56.0	—130.4	267)
116	3	1	1	1	as in 111 $k_3 = 4.86 \times 10^{-3}$	58.9	—124.6	267)
117	3	1	1	1	as in 111 $k_3 = 3.59 \times 10^{-3}$	61.8	—118.7	267)
118	3	1	1	1	as in 111 $k_3 = 2.73 \times 10^{-3}$	65.2	—111.6	267)
119	3	1	1	1	as in 111 $k_3 = 2.34 \times 10^{-3}$	67.7	—106.2	267)
120	3	1	1	1	as in 111 $k_3 = 2.36 \times 10^{-3}$	67.7	—105.8	267)
121	3	1	1	1	as in 111 $k_3 = 0.162 \times 10^{-3}$	94	— 54.8	267)
122	3	1	1	1	as in 111 $k_3 = 0.709 \times 10^{-3}$	76.5	— 91.1	267)
123	3	1	1	1	as in 111 $k_3 = 0.390 \times 10^{-3}$	81.5	— 81.9	267)
124	3	1	1	1	as in 111 $k_3 = 0.226 \times 10^{-3}$	89	— 66	267)
125	3	1	1	1	as in 111 $k_3 = 2.85 \times 10^{-3}$	64.4	—113.7	267)
126	3	1	1	1	as in 111 $k_3 = 1.77 \times 10^{-3}$	70.6	— 99.9	267)
127	3	1	1	1	as in 111 $k_3 = 0.813 \times 10^{-3}$	78.6	— 84.4	267)
128	3	1	1	1	$k_3 = 2.58 \times 10^{-5}$ l² mol⁻² s⁻¹			267)
129	3	1	1	1	as in 128 $k_3 = 3.39 \times 10^{-5}$			267)
130	3	1	1	1	as in 128 $k_3 = 4.63 \times 10^{-5}$			267)
131	3	1	1	1	as in 128 $k_3 = 5.76 \times 10^{-5}$			267)
132	3	1	1	1	as in 128 $k_3 = 8.40 \times 10^{-5}$			267)
133	3	1	1	1	as in 128 $k_3 = 9.80 \times 10^{-5}$			267)
134	3	1	1	1	as in 128 $k_3 = 74.6 \times 10^{-5}$			267)
135	3	1	1	1	as in 128 $k_3 = 109 \times 10^{-5}$			267)
136	3	1	1	1	$k_3 = 6.81 \times 10^{-3}$ l² mol⁻² s⁻¹			268)
137	3	1	1	1	as in 136 $k_3 = 5.50 \times 10^{-3}$			268)
138	3	1	1	1	as in 136 $k_3 = 3.69 \times 10^{-3}$			268)
139	3	1	1	1	as in 136 $k_3 = 2.87 \times 10^{-3}$			268)
140	3	1	1	1	as in 136 $k_3 = 2.34 \times 10^{-3}$			268)
141					$k_2 = 1.98 \times 10^{-3}$ l mol⁻¹ min⁻¹; $k_1 = 0.98 \times 10^{-3}$ min⁻¹ } t = 60 min			276)

142					$k_2 = 7.17 \times 10^{-3}$ l mol^{-1} min^{-1} $k_1 = 2.23 \times 10^{-3}$ min^{-1}	t = 360 min	276)
143	2	1	0	1	$k_2 = 8.69 \times 10^{-3}$ l mol^{-1} min^{-1} $k_1 = 4.04 \times 10^{-3}$ min^{-1}	t = 45 min	276)
144	2	1	0	1	$k_2 = 18.53 \times 10^{-3}$ l mol^{-1} min^{-1} $k_1 = 3.70 \times 10^{-3}$ min^{-1}	t = 270 min	276)
145	2	1	0	1	$k_2 = 14.83 \times 10^{-3}$ l mol^{-1} min^{-1} $k_1 = 6.66 \times 10^{-3}$ min^{-1}	t = 35 min	276)
146	2	1	0	1	$k_2 = 22.6 \times 10^{-3}$ l mol^{-1} min^{-1} $k_1 = 6.28 \times 10^{-3}$ min^{-1}	t = 175 min	276)
147	2	1	0	1	$k_1 = 3.64 \times 10^{-3}$ min^{-1}		276)
148	2	1	0	1	$k_1 = 3.74 \times 10^{-3}$ min^{-1}		276)
149	2	1	0	1	$k_1 = 3.71 \times 10^{-3}$ min^{-1}		276)
150	2	1	0	1	$k_1 = 3.69 \times 10^{-3}$ min^{-1}		276)
151	2	1	0	1	$k_1 = 3.75 \times 10^{-3}$ min^{-1}		276)
152	2	1	0	1	$k_1 = 3.76 \times 10^{-3}$ min^{-1}		276)
153	2	1	0	1	$k_1 = 2.72 \times 10^{-3}$ min^{-1}		276)
154	2	1	0	1	$k_1 = 2.14 \times 10^{-3}$ min^{-1}		276)
155	2	1	0	1	$k_1 = 1.90 \times 10^{-3}$ min^{-1}		276)
156	3	1	1	1	$k_3 = 22.1$ l^2 mol^{-2} h^{-1} (95 °C)		100)
157	3	1	1	1	as in 156 $k_3 = 19.1$		100)
158	3	1	1	1	as in 156 $k_3 = 15.8$		100)
159	3	1	1	1	as in 156 $k_3 = 14.0$		100)
160	3	1	1	1	$k_3 = 49.9$ l^2 mol^{-2} h^{-1} (105 °C)		100)
161	3	1	1	1	as in 160 $k_3 = 161$		100.
162	3	1	1	1	as in 160 $k_3 = 71.2$		100)
163	3	1	1	1	as in 156 $k_3 = 50.1$		100)
164	3	1	1	1	as in 156 $k_3 = 66.2$		100)
165	3	1	1	1	as in 160 $k_3 = 160$		100)
166	3	1	1	1	as in 160 $k_3 = 345$		100)
167		1.011			$k_1 = 2.40 \times 10^{-5}$ s^{-1} (90 °C)		236)
168		0.91			$k_1 = 6.38 \times 10^{-5}$ s^{-1} (100 °C)		236)
169			0.900		$k_1 = 4.85 \times 10^{-5}$ s^{-1} (90 °C)		236)
170		0.83			$k_1 = 1.75 \times 10^{-5}$ s^{-1} (110 °C)		236)

Contin. p. 206

Table 3. (continued)

No.	Overall order	Order in epoxide	Order in acid	Order in catalyst	k	ΔH≠ kJ mol⁻¹	ΔS≠ J K mol⁻¹	Ref.
171			1.100		$k_1 = 3.26 \times 10^{-5}\ s^{-1}$ (140 °C)			236)
172			0.932		$k_1 = 0.553 \times 10^{-5}\ s^{-1}$ (120 °C)			236)
173					$k = 5.23 \times 10^{-6}$ kg $mol^{-1}\ s^{-1}$ (110 °C)	75.3	—106.6	236)
174					$k = 3.42 \times 10^{-4}\ kg^2\ mol^{-2}\ s^{-1}$ (100 °C)	76	—118	236)
175					as in 174 $k = 2.09$	64.4	—118	236)
176					as in 174 $k = 2.00$	64.4	—152.6	236)
177						70.6	—174	133)
178						67.5	—159	133)
179					$k = 1.085 \times 10^{-6}$ kg $mol^{-1}\ s^{-1}$ (90 °C)	76.3	—158	236)
180					as in 179 $k = 7.24$	69.3	—123	236)
181	3	1	1	1	$k_3 = 0.211 \times 10^{-2}\ l^2\ mol^{-2}\ s^{-1}$ (80 °C)	74	— 87	116)
182	3	1	1	1	$k_3 = 1.71 \times 10^{-2}\ l^2\ mol^{-2}\ s^{-1}$ (106.5 °C)			116)
183	3	1	1	1	as in 182 $k_3 = 2.04$			116)
184	3	1	1	1	as in 182 $k_3 = 2.52$			116)
185					$k = 1.05 \times 10^{-6}$ kg $mol^{-1}\ s^{-1}$ (90 °C)	76.3	—157	236)
186					as in 185 $k = 7.66$	70.2	—121	236)
187	3	1	1	1	as in 181 $k_3 = 0.175$	76.5	— 82	236)
188	3	1	1	1	as in 182 $k_3 = 1.58$			236)
189	3	1	1	1	as in 182 $k_3 = 1.88$			236)
190	3	1	1	1	as in 182 $k_3 = 2.40$			236)
191	3	1	1	1	as in 181 $k_3 = 0.246$	73	— 90	236)
192	3	1	1	1	as in 182 $k_3 = 1.92$			236)
193	3	1	1	1	as in 182 $k_3 = 2.41$			236)
194	3	1	1	1	as in 182 $k_3 = 3.14$			236)
195	3	1	1	1	as in 181 $k_3 = 0.356$	67	—102	236)
196	3	1	1	1	as in 182 $k_3 = 2.52$			236)
197	1	1	1	1	as in 182 $k_3 = 3.20$			236)
198	3	1	1	1	as in 182 $k_3 = 4.31$			236)
199	3	1	1	1	as in 181 $k_3 = 0.323$	69	— 99	236)

200	3	1	1	1	as in 182	$k_3 = 2.32$			236)
201	3	1	1	1	as in 182	$k_3 = 2.86$			236)
202	3	1	1	1	as in 182	$k_3 = 4.03$			236)
203	2	1	1		$k = 1.24 \times 10^{-6}$ kg mol^{-1} s^{-1} (90 °C)		83.6	—137	236)
204	2	1	0	1	$k = 5.00 \times 10^{-4}$ kg mol^{-1} s^{-1} (90 °C)		67	—134	236)
205	2	1	1		$k = 1.24 \times 10^{-6}$ kg mol^{-1} s^{-1} (90 °C)		84	—137	236)
206	2	1	0	1	$k = 5.00 \times 10^{-4}$ kg mol^{-1} s^{-1} (90 °C)		65	—138	236)
207	2	0.52	1.06	0.60	$k = 0.505 \times 10^{-2}$ l$^{0.5}$ mol$^{-0.5}$ s^{-1} (98 °C)		70.6		64)
208	2	0.52	1.06	0.60	$k = 0.027$ l$^{0.5}$ mol$^{-0.5}$ s^{-1} (110 °C)				64)
209	2	0.52	1.06	0.60	as in 208	$k = 0.024$			64)
210	2	1.06	0.52	0.60	as in 208	$k = 0.017$			64)
211	2	1.06	0.52	0.60	as in 208	$k = 0.013$			64)
212	2	1.06	0.52	0.60	as in 208	$k = 0.994$			64)
213	2	1	1		$k_2 = 4.2 \times 10^{-6}$ kg eq^{-1} s^{-1}				159)
214	2.5	1	0.5	1	$k^A_{2.5} = 1.05 \times 10^{-3}$ kg$^{1.5}$ eq$^{-1.5}$ s^{-1} $k^E_{2.5} = 1.25 \times 10^{-3}$ kg$^{1.5}$ eq$^{-1.5}$ s^{-1} } (86.6 °C)		70.5	—108.0	159)
215	2.5	1	0.5	1	as in 214	$k^A_{2.5} = 1.25 \times 10^{-3}$ $k^E_{2.5} = 1.55 \times 10^{-3}$ } (90 °C)			159)
216	2.5	1	0.5	1	as in 214	$k^A_{2.5} = 1.75 \times 10^{-3}$ $k^E_{2.5} = 1.90 \times 10^{-3}$ } (96 °C)			159)
217	2.5	1	0.5	1	as in 214	$k^A_{2.5} = 2.35 \times 10^{-3}$ $k^E_{2.5} = 3.05 \times 10^{-3}$ } (102.4 °C)			159)
218	2.5	1	0.5	1	as in 214	$k^A_{2.5} = 3.85 \times 10^{-3}$ $k^E_{2.5} = 4.25 \times 10^{-3}$ } (106.5 °C)			159)
219	2.5	1	0.5	1	as in 214	$k^A_{2.5} = 4.2 \times 10^{-3}$ $k^E_{2.5} = 4.75 \times 10^{-3}$ } (110.4 °C)			159)
220	2.5	1	0.5	1	as in 214	$k^A_{2.5} = 5.75 \times 10^{-3}$ $k^E_{2.5} = 6.65 \times 10^{-2}$ } (115.5 °C)			159)
221	2.5	1	0.5	1	$k_{2.5} = 1.2 \times 10^{-3}$ kg$^{1.5}$ eq$^{-1.5}$ s^{-1} (86 °C)				159)
222	2.5	1	0.5	1	as in 221	$k_{2.5} = 1.1 \times 10^{-3}$			159)
223	2.5	1	0.5	1	as in 221	$k_{2.5} = 1.2 \times 10^{-3}$			159)
224	2.5	1	0.5	1	as in 221	$k_{2.5} = 1.1 \times 10^{-3}$			159)

Contin. p. 208

Table 3. (continued)

No.	Overall order	Order in epoxide	Order in acid	Order in catalyst	k	ΔH* kJ mol⁻¹	ΔS* J K mol⁻¹	Ref.
225	2.5	1	0.5	1	$k_{2.5} = 6.6 \times 10^{-3}$ $kg^{1.5}$ $eq^{-1.5}$ s^{-1}	68	—116	[203]
226	2.5	1	0.5	1	as in 225 $k_{2.5} = 7.8 \times 10^{-3}$			[203]
227	2.5	1	0.5	1	as in 225 $k_{2.5} = 7.2 \times 10^{-3}$			[203]
228	2.5	1	0.5	1	$k_{2.5} = 0.63 \times 10^{-3}$ $kg^{1.5}$ $eq^{-1.5}$ s^{-1} (80 °C)			[203]
229	2.5	1	0.5	1	as in 228 $k_{2.5} = 2.8 \times 10^{-3}$ (100 °C)			[203]
230	2.5	1	0.5	1	as in 228 $k_{2.5} = 19.7 \times 10^{-3}$ (140 °C)			[203]
231	2	1	0	1	$k_2 = 4.2 \times 10^{-3}$ $l\ mol^{-1}$ s^{-1} (91 °C)	67.3		[194]
232	2	1	0	1	$k_2 = 2.02 \times 10^{-3}$ $l\ mol^{-1}$ s^{-1} (91.8 °C)	75.2		[194]
233	1				$k_1 = 2.66 \times 10^{-3}$ min^{-1} (80 °C)	70		[188]
234	1				$k_1 = 4.53 \times 10^{-3}$ min^{-1} (80 °C)			[188]
235	1				$k_1 = 3.65 \times 10^{-3}$ min^{-1} (80 °C)	69		[188]
236	1.5	1	0.25	0.25	$k = 4.8$% conv. min^{-1}			[63]
237	1.5	1	0.25	0.25	as in 236 $k = 1.4$			[63]
238	1.5	1	0.25	0.25	$k = 16.1 \times 10^{-3}$ $kg^{0.25}$ $mol^{-0.25}$ min^{-1}			[63]
239	1.5	1	0.25	0.25	$k = 17.2 \times 10^{-3}$ $kg^{0.25}$ $mol^{-0.25}$ min^{-1}			[63]
240	2	1	0.5	0.5		41.8		[63]
241	2	1	0.5	0.5		63		[63]
242	2	1	0.5	0.5		54		[63]
243	2	1	0.5	0.5		67		[63]

Kakiuchi and Tanaka [116] found that the reaction is first order with respect to acid, epoxide, and catalyst, respectively:

$$-\frac{d[\text{epoxide}]}{dt} = -\frac{d[\text{acid}]}{dt} = k_3[\text{Amine}][\text{Acid}][\text{Epoxide}] \tag{34}$$

k_3 was determined for several substituents (see Table 3).

As already stated, Kakiuchi and Tanaka [116] criticized the hypothesis of a pure ionic mechanism that they did not find consistent with the dependence of the system on the temperature and solvent. Their kinetic study provides supplementary clues to the critics: kinetic relations resulting from the hypothesis of a pure ionic mechanism do not fit experimental values. On the other hand, if the reaction proceeds as shown in (44), (45), and (46), the rate of formation of IV is given by:

$$\frac{d[\text{IV}]}{dt} = k_1k_3/k_2[\text{I}][\text{III}][\text{Base}] \tag{35}$$

with the reasonable assumption that $k_2 \gg k_3[\text{III}]$, which fits experimental results. Equilibrium (44) is sustained by the formation of complexes such as $R_3N(RCOOH)_n$ ($n > 2$) [3]. Moreover, the catalytic effect of such complexes has already been shown in the formation of chlorhydrin esters from 1,2-epoxy-3-chloropropane [54,65,160,226,256].

Tanaka [266] interestingly discussed the factors which influence the stereochemistry of the transition-state species IV for the reaction of benzoic acid with 1,2-epoxy-3-phenoxypropane catalyzed by substituted pyridines in xylene. As in Ref. [116], it was shown that reaction (47) to (50) do not take place and that the kinetics is second order with respect to acid and epoxide. The plot of the rate constant k against [Cat] is a straight line which does not pass through the origin:

$$k = k''[\text{Amine}] + k' \tag{36}$$

The plot of the activation energy or of the logarithm of the rate constant vs. the basicity of these pyridines in water does not show any one relationship covering all the catalysts studied in this work. According to Tanaka [266], the deviations from a linear relationship, which have been observed with other reactions, should not be due to the fact that the pK_A and the kinetic values were measured in different solvents, but to the difference in the steric requirements of these reactions.

The formation of complex IV shows steric hindrance, determined kinetically, which exceeds that in the corresponding ionizations. In other words, the steric constraint in IV should be greater than in the corresponding pyridinium ion. The steric constraints in IV, which are not very great, may arise from a steric interaction of the bulky groups in position 2 and 6 of the pyridine ring with the other parts of IV. They can be estimated from the assumption used by Brown and Horowitz [27] evaluating the steric hindrance in the reaction of 2-alkyl pyridines with BF_3, I Me, and CH_3SO_3H.

The importance of the steric hindrance in complex IV is estimated by comparing the activation enthalpies relative to the compounds substituted in *ortho* and *para*, respectively:

0.6 for 2-methylpyridine: from E_A(2-methylpyridine)-E_A(4-methylpyridine)
3.1 for 2,6-dimethylpyridine: from E_A(2,6-dimethylpyridine)-E_A(2,4-dimethylpyridine)
0.6 for 2,6-dimethylquinoline: from E_A(quinoline)-E_A(isoquinoline)
3.6 for 2-methylquinoline: from E_A(2-methylquinoline)-E_A(4-methylquinoline)

Comparison of these results with those reported in Ref. [27] shows that the steric constraint in the formation of IV, when catalyzed by substituted pyridines, is greater than those obtained for the reaction of 2-methylpyridines with Me I, but smaller than those corresponding to the reaction of 2-methylpyridines with BF_3. Another clue is that the least active amines are those with a substituent in *ortho* positions of the nitrogen atom in pyridine. Each step of the reactional schemes (44) to (46) is discussed by Tanaka, who gives very convincing evidence to support the mechanism.

The structure which is proposed for IV fits the classical characteristics of a hydrogen bond.

Using infra-red spectra analysis, Pirozhnaya and Gromov [198] claimed that the formation of complex IV is not sufficient to explain the kinetics of the tertiary amine-catalyzed reaction of butanoic acid with 1,2-epoxy-3-phenoxypropane. According to them, aromatic and aliphatic tertiary amines have similar kinetic activity, but differ greatly in their ability to form complexes with butanoic acid.

In the same way, they dismiss the hypothesis of a quaternary ammonium salt resulting from the reaction of the base with the oxide, since aromatic amines do not react with cyclic ethers. On the other hand, they observed the formation of an ammonium salt accumulating in the system:

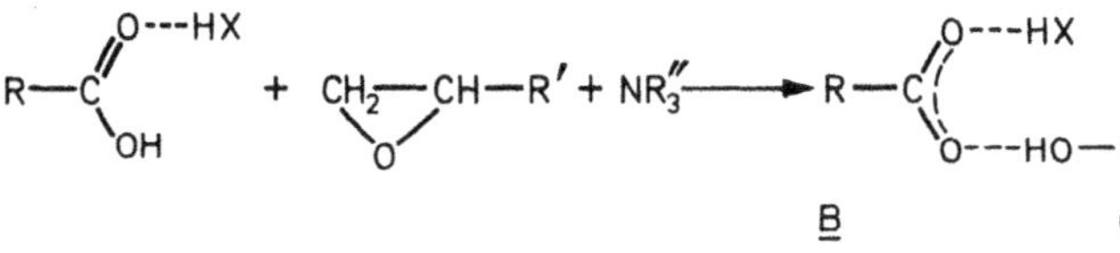

(51)

HX is either the acid or the alcohol resulting from the esterification. When added to a mixture of acid and glycidyl ether, complex *B* accelerates the reaction but is not consumed.

Studying the pyridine-catalyzed reaction of 1,2-epoxy-3-phenoxypropane with benzoic acid, Kakiuchi and Endo [117] found first order in acid, epoxide, and amine, respectively. They proposed a reaction scheme involving the simultaneous contribution of reactions (44) to (46) and of reactions (52) and (53):

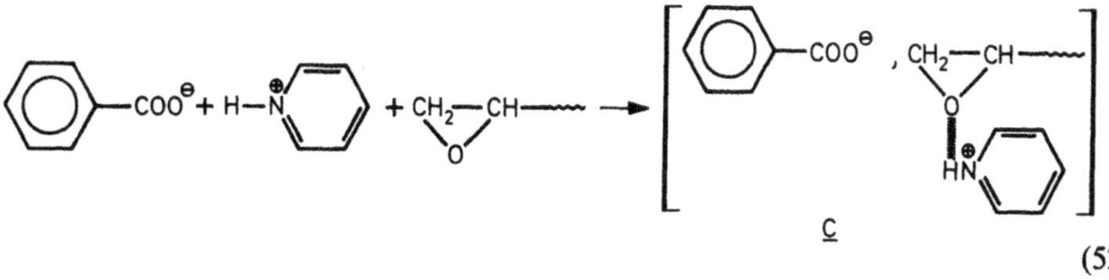

(52)

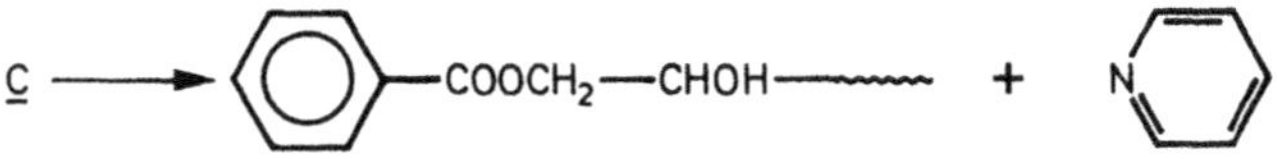

(53)

In fact, to obtain a kinetic relation fitting experimental orders, they were obliged to neglect reactions (52) and (53). They supported the occurrence of reactions (44) and (45) by UV and IR analysis.

Madec and Maréchal [156-158] studied the system N,N-dimethyldodecylamine/benzoic acid/1,2-epoxy-3-phenoxypropane in different solvents. They confirmed the formation of a cyclic complex and, at the same time, they showed that a side reaction contributes to the general kinetic pattern. This will be developed in Sect. 4.2.4.

4.2.3 Influence of Substituents and of Solvents

As already stated, several studies report on the dependence of kinetic parameters on the nature and the position of substituents [100,116,215]. Ishii et al. [100] compared the influence of substituents on the reactivity of 1,2-epoxy-3-phenoxypropane with phenols and with benzoic acids. Hammett's plots are linear but indicate that m-NO_2 behaves abnormally. In the case of phenols, the sign of ϱ depends on the nature of the solvent (positive for xylene, negative for nitrobenzene). On the other hand, when the reactant is a substituted benzoic acid, ϱ is positive in both solvents. This fact indicates that the mechanism is affected by the nature of the active hydrogen compound.

Studying the reaction of substituted 1,2-epoxy-3-phenoxypropane with benzoic acid in the presence of a tertiary amine, Kakiuchi and Tanaka [116] reported Hammett's plots obtained in solvents such as xylene, mono- and dichlorobenzene, and nitrobenzene. They obtained the following values of ϱ:

in xylene: 0.26 (80 °C); 0.17 (106.5 °C); 0.13 (120 °C), and 0.097 (130 °C) at 106 °C: 0.26 (nitrobenzene); 0.22 (*o*-dichlorobenzene); 0.19 (chlorobenzene).

Positive values indicate that electron-withdrawing substituents enhance the rate and that the reaction is of SN_2 type. 1,2-epoxy-3-phenoxypropane substituted by electron-releasing groups give lower constants and higher activation parameters than those with electron-withdrawing substituents.

These authors [116] established the following relations between ϱ and reaction temperature:

$$\varrho = -1.054(1 - 440.5/T) \quad (37)$$

and between ϱ and the dielectric constant D of the reaction medium:

$$\varrho = a \log D + b \quad \text{with} \quad a = 0.0784 \quad \text{and} \quad b = 0.149 \quad (38)$$

They showed that an isokinetic relationship holds for this reaction (Fig. 1).

$$\Delta H^{\neq} = \Delta H_0^{\neq} + \beta \Delta S^{\neq} \quad (39)$$

where $\Delta H_0^{\neq} = 27.2$ and $\beta = 450$. Relations (37) and (39) show that ϱ tends towards zero when the temperature approaches the isokinetic temperature. This cannot be explained by the classical Hammett's relation between ϱ, T, and D, but fits Leffer and Brunwald's relation [149]:

$$\varrho_T = \varrho_\infty(1 - \beta/T) \quad (40)$$

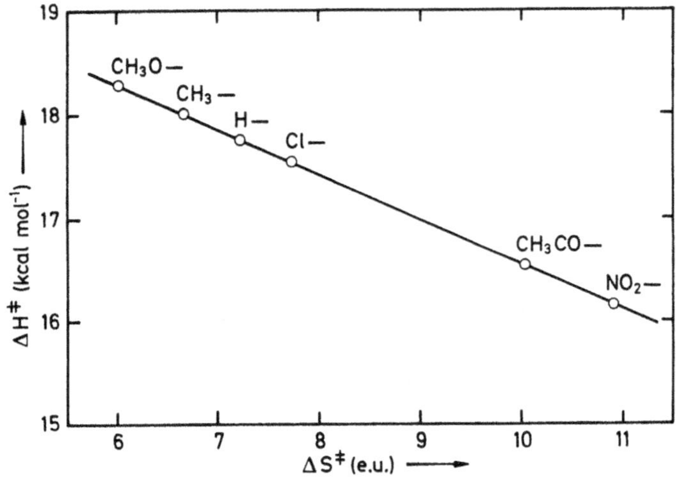

Fig. 1. Isokinetic relationship in the reaction of 1,2-epoxy-3-phenoxypropane and benzoic acid with dimethyldodecylamine in xylene. Symbols are R from $RC_6H_4OCH_2CHCH_2$ (epoxide O bridging CH and CH_2) [116]

where ϱ_T and ϱ_∞ are the reaction constants at temperature T and ∞. Parameter β has the dimension of an absolute temperature and can be identified as an effective or virtual temperature at which all the differences in the rate or the equilibrium constant vanish: $\delta_R\Delta G^{\neq} = 0$ when $T = \beta$.

In addtion, the plot of ϱ against σ remains linear when the temperature changes; this fits relation (40), but not that proposed by Hammett [86] for describing the dependence of ϱ on T and D.

Kakiuchi and Endo [117] studied the influence of the nature of the solvent on the pyridine-catalyzed reaction of benzoic acid with 1,2-epoxy-3-phenoxypropane. The reaction is first order for acid, epoxide, and catalyst; it was carried out in toluene-nitrobenzene and in toluene-dioxane mixtures as well as in different pure solvents. The dielectric constant was measured before the reaction started and after the reaction stopped it was noted that this constant hardly changes during the course of the reaction. A linear relationship between log ϱ and D was observed in toluene-nitrobenzene, but not in dioxane-nitrobenzene; according to the authors, this could be due to a specific solvating effect of dioxane.

As stated previously, the mechanism is described by reactions (44) to (46); the plots of $\Delta H^{\neq}$ against $\Delta S^{\neq}$ are linear in toluene-nitrobenzene, toluene-dioxane, and nitrobenzene-dioxane systems, showing that the mechanism hardly varies with the nature of the reaction medium. Tentative explanations of the dependence of k, $\Delta H^{\neq}$, and $\Delta S^{\neq}$ on the composition of the mixture are given in the discussion of the experimental results [100].

Tanaka and Takeuchi [267] studied the same system in various pure and mixed aromatic hydrocarbons, such as toluene, monochlorobenzene, *o*-dichlorobenzene, nitrotoluene, nitrobenzene, and toluene-nitrobenzene mixtures, at temperatures ranging from 80 to 108 °C. The mechanism is described by reactions (44) to (46) and it involves two catalytic effects: by acid and by amine as shown in relation (36).

In pure or mixed aromatic hydrocarbon solvents, the reaction rate constant and parameters increase with increasing dielectric constant D of the solvent: $\Delta H^{\neq}$ and log k do not follow a linear relationship with D, (D − 1)/(2D + 1) or 1/D; on the other hand, there is a satisfactory linear relationship between log k″ or $\Delta H^{\neq}_{298}$ and log D. According to the authors, complex IV [Reaction (45)], although non-ionic, should have a large dipole moment and therefore has general properties close to a salt. Thus, in highly polar solvents, such as nitrobenzene and a-nitrotoluene, this complex is more soluble and more ionized, resulting in a high rate of reaction. Using Kirkwood relation [128] the authors [267] give possible explanations of the various solvent effects.

Uejima and Munakata [276] obtained somewhat different kinetic results. Thus, studying the 1,2-epoxy-3-phenoxypropane/benzoic acid system, they showed that, when the catalyst is a quaternary ammonium salt, the reaction rate is:

$$-\frac{d[\text{epoxide}]}{dt} = -\frac{d[\text{acid}]}{dt} = k[\text{epoxide}][\text{catalyst}] \tag{41}$$

This kinetical behavior was not observed when the catalyst was an amine. In this case, a mechanism was suggested, which includes the formation of an intermediate quaternary ammonium carboxylate according to the following scheme:

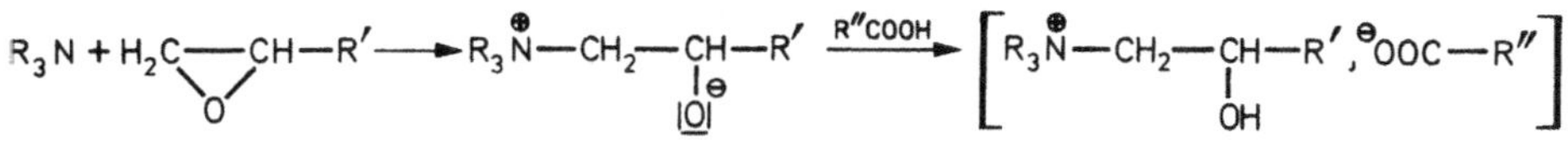

(54)

In the same way, Pinazzi et al. [26], studying the reaction of epoxidized polyalcadiene with benzoic acid, observed that the order was nearly always 2 but depending more or less on conversion.

Madec and Maréchal [156–158] discussed the influence of the nature of the solvent on the reaction rate of the main and of one of the side reactions in the N,N-dimethyl-dodecylamine-catalyzed reaction of benzoic acid with 1,2-epoxy-3-phenoxypropane; their results are discussed in Sect. 4.2.4.

4.2.4 Side Reactions

We have already discussed the contribution of some side reactions in Sect. 1 and 4.2. In Sect. 4.2.4. we limit our discussion to the system: benzoic acid/1,2-epoxy-3-phenoxy-propane/dodecyl dimethylamine. The kinetics obeys a law which depends on the nature of the reaction solvent.

In xylene, chlorobenzene, or o-dichlorobenzene, which are solvents with low or medium dielectric constants, results fit the following kinetic law:

$$-\frac{d[A]}{dt} = k[A]^{0.5}[E][Cat] \tag{42}$$

The fact that the order in acid is not an integer, shows that the global phenomenon is complex and that at least two reactions contribute to the kinetics:

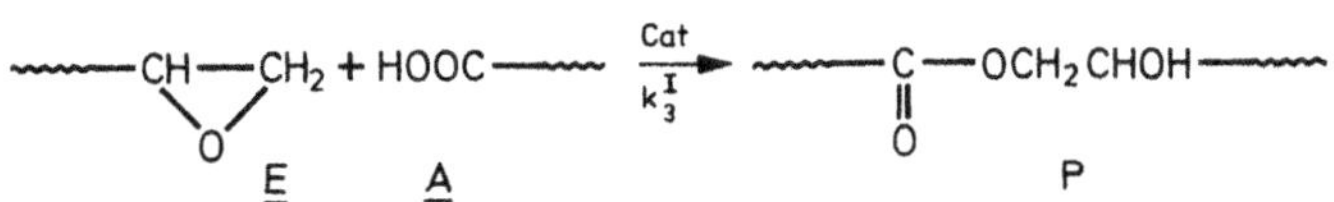

(55)

Here, the amine behaves as a catalyst and is reformed. This is the main reaction (I).

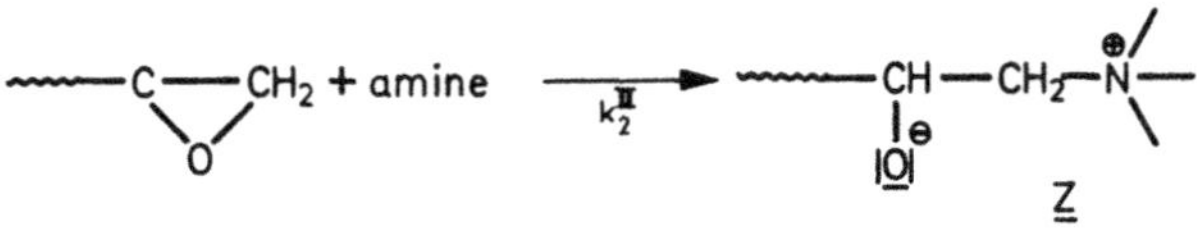

(56)

Here, the amine behaves as a reactant and is consumed. This is a side reaction (II). Thus:

$$-\frac{d[A]}{dt} = k_3^I[E][A][Amine] + k_2^{II}[E][Amine] \quad (43)$$

Determination of k, k_3^I, and k_2^{II} in a very wide range of experimental conditions showed that the overall rate constant k depends on the stoichiometry of reactants; on the other hand, k_2^{II} and k_3^I depend little on this parameter. This shows that the main reaction is complex but that reactions I and II are not.

Amine consumption in reaction II could make it difficult to use the epoxy-carboxy reaction to prepare esters or polyesters. In fact, most of the zwitterion Z, which is formed from E and the amine, is transformed into the ester *P* and, at least in solution, only a small quantity of side product is formed. Moreover, titrations carried out during the reaction show that epoxy and acid groups are consumed at roughly the same rate, demonstrating that even if the contribution of reaction II to overall kinetics and mechanism is important it does not forbid the use of this method for preparation of polyesters.

In nitrobenzene, which is a polar solvent, the results are different to some extent as shown in Fig. 2. When conversion is below 70%, the orders in E, A, and Cat. are, respectively, 0, 1, and 1. When conversion is above 70%, the orders in E and Cat.

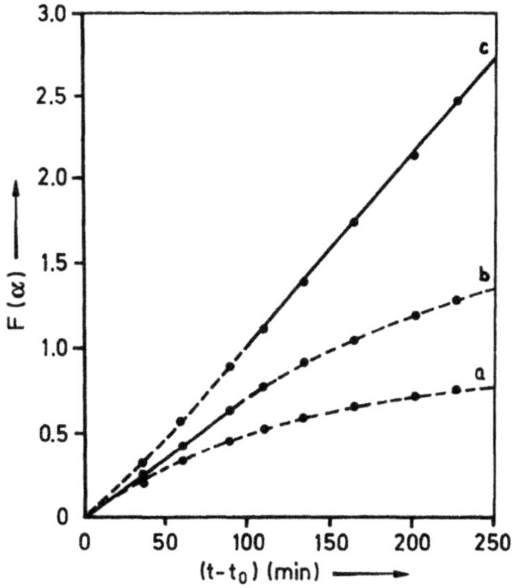

Fig. 2. F(α) as a function of $(t - t_0)$. Catalyzed reaction of benzoic acid and 1,2-epoxy-3-phenoxypropane in nitrobenzene. $[A]_0 = [E]_0 = 0.397$ eq · kg^{-1}; [DMDA] $= 2.95 \times 10^{-2}$ eq $\times$ kg^{-1}. Temperature: 106.5 °C. Global orders in reactants: curve. **a** (0.5), **b** (1), **c** (1.5)

do not change, but the order in acid tends toward 0.5. The fact that, when conversion is below 70%, the order in acid is zero shows that the rate-determining step is only reaction II. Thus:

$$-\frac{d[A]}{dt} = k_2^{II}[E][Cat] \quad (44)$$

and $k = k_2^{II}$.

When conversion is above 70%, both reactions I and II take place and the overall rate is the same as in nonpolar solvents:

$$\text{Overall rate} = k_{2.5}[E][A]^{0.5}[Cat] = k_2^{II}[E][A] + k_3^{I}[E][A][Cat] \quad (45)$$

Relations (44) and (45), respectively, allow a direct and an indirect determination of k_2^{II}. The values obtained by both these methods are very close. Another proof of the existence of these two reactions and of the validity of their rate constant determinations is given in Fig. 3, where log k_3^{I} and log k_2^{II} are plotted against the logarithm of the dielectric constant of the solvent. These plots are linear, which is not the case when log k is plotted against log D.

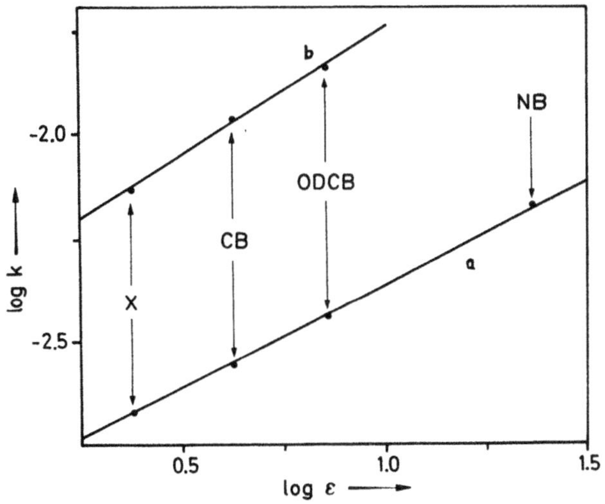

Fig. 3. Variations of log k_2^{II} (**a**) and log k_3^{I} (**b**) with respect to log ε. Temperature 106.5 °C. Solvents: xylene (X), chlorobenzene (CB), o-dichlorobenzene (ODCB) and nitrobenzene (NB)

When the reaction is carried out in the melt, as in most industrial applications, the situation can be very different because, due to the strong interaction of amine and epoxy groups, a non-negligible polymerization of oxirane can take place. This was observed [159] with the system octadecanoic acid/1,2-epoxy-3-dodecyloxypropane which was studied in the melt. In spite of the fact that E is consumed more rapidly than A, it was shown that both reactions I and II take place, and k_2^{II} and k_3^{I} were determined showing that the contribution of II to the kinetics is more important in the melt than in solution.

A global interpretation of these kinetic results has been provided [158].

Scheme I (Main reaction)

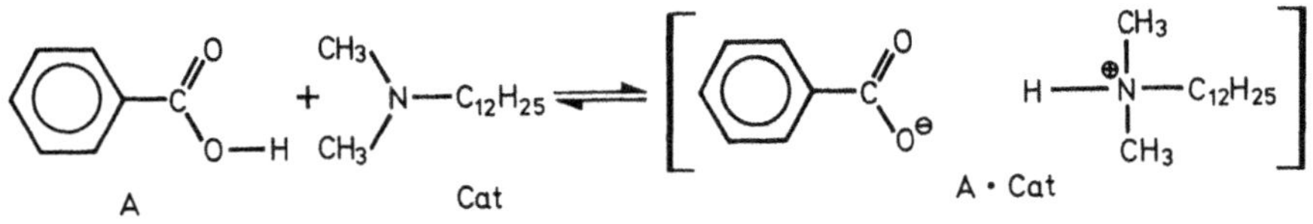

(57)

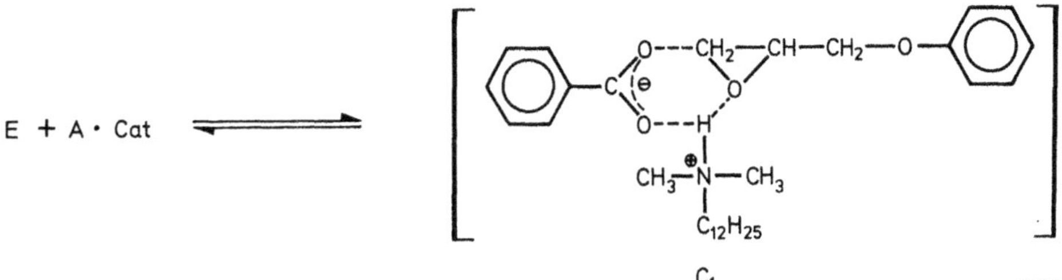

(58)

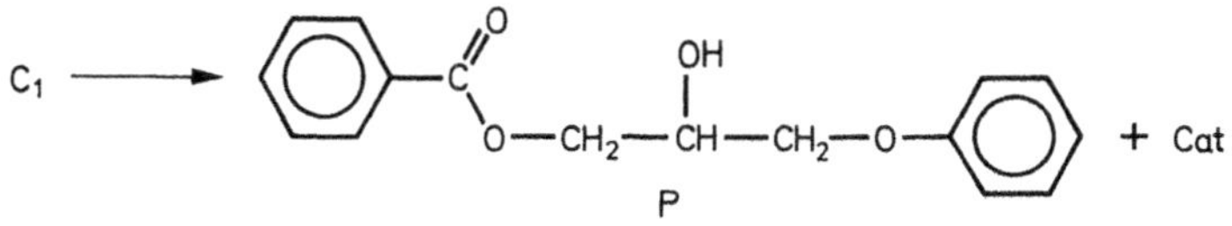

(59)

Scheme II (Side reaction)

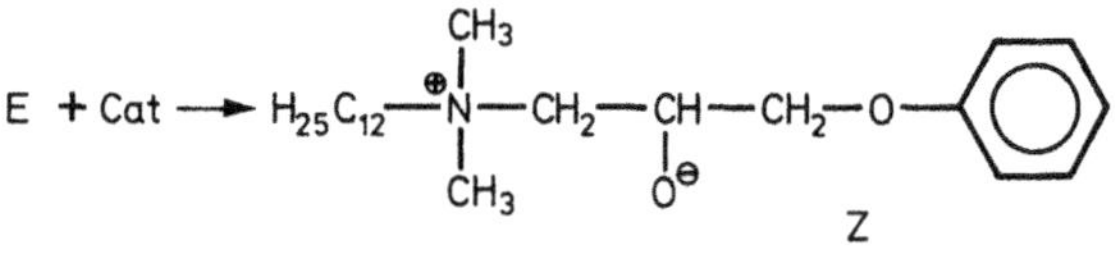

(60)

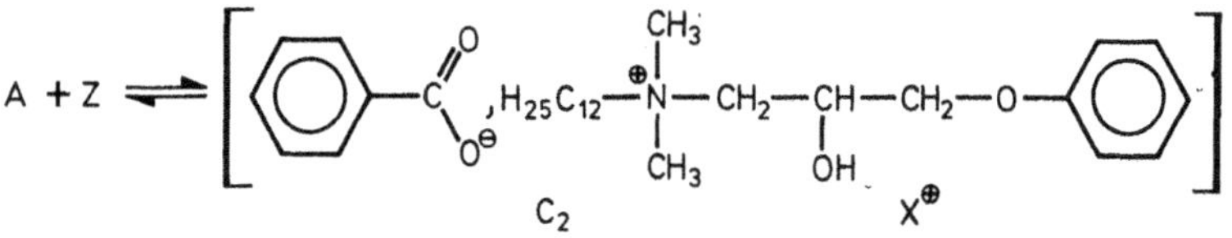

(61)

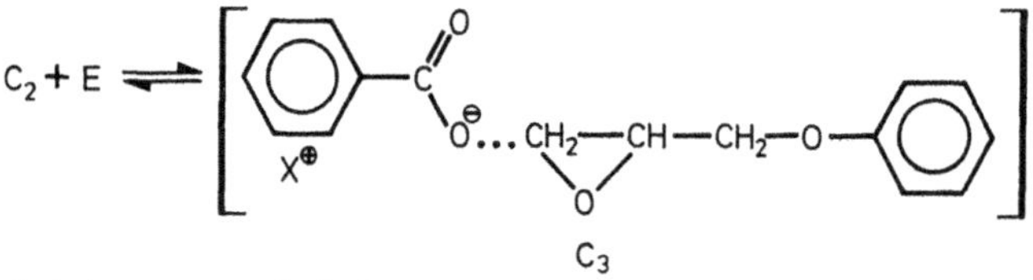

(62)

$C_3 + A \longrightarrow P + C_2$ (63)

Scheme I corresponds to reaction I which is the main reaction. It starts by the reaction of A and Cat to form an ion pair A · Cat which, through a quadripolar interaction, associates with the oxirane group of E to form complex C_1. Reaction (58) is the rate-determining step. Complex C_1 then dissociates rapidly, forming the expected product P and regenerating the catalyst. Scheme I consumes most A and E. However, as already stated, another mechanism takes place; this is described by scheme II and corresponds to the side reaction. Its kinetics is ruled by reaction (60) which forms zwitterion Z; Z reacts with A to form complex C_2, and C_2 reacts with the epoxide to form complex C_3. Counterion $X^{\oplus}$ is, of course, the same in reactions (61) and (62).

It is important to underline that, in constrast to A · Cat, complex C_2 is not able to convert to a complex similar to C_1. This is probably because of the steric hindrance due to the $-CH_2CHOHCH_2OC_6H_5$ group which forbids the formation of the six-membered ring. On the other hand, due to the large size of the cation, the ion pair very probably dissociates which makes the corresponding carboxylate very reactive towards the α site of the oxirane ring. The anion of C_3 is rapidly protonated by A, giving the product and regenerating C_2. In scheme II, C_2 behaves as a catalytic species; in fact, the reaction medium contains two catalysts: B and C_2.

The whole process can be represented by the following very simplified scheme:

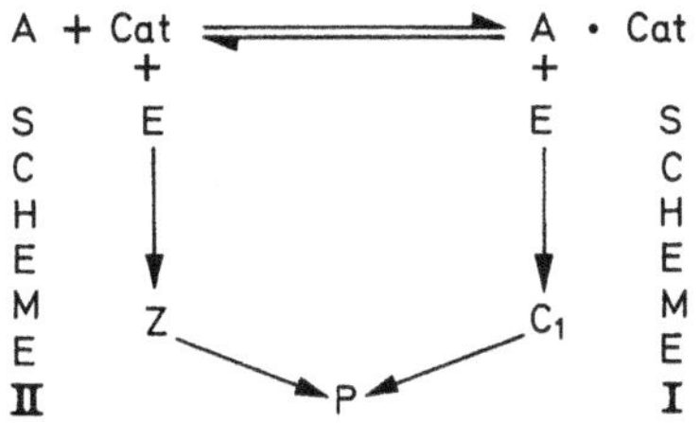

Use of a highly basic amine such as dimethyldodecylamine favors mechanism I and, in consequence, the free amine concentration in the medium is low. However, its nucleophilic character favors the attack of the oxirane ring, i.e., mechanism II. Thus, the basic strength of the amine acts on both mechanisms I and II.

The influence of the solvent is noteworthy. In less polar solvents, most of the acid and amine is used to form complex A · Cat and, to a smaller extent, complexes such as $A_2 \cdot$Cat and $A_3 \cdot$Cat. Low or medium polar solvents favor mechanism I; however, a small fraction of amine remains and is able to attack the α site of oxirane. Acid-base equilibrium [reaction (57)] controls the free amine concentration, i.e., the contribution of mechanism I. A similar contribution of the two kinetics, which have close activation energies, was observed.

In a more polar solvent, such as nitrobenzene, dissociation of complex A · Cat (and of complexes $A_2 \cdot$Cat and $A_3 \cdot$Cat) is favored and greatly decreases the relative contribution of mechanism I. For this reason, during most of the process (conversion below 70–80%), only mechanism II is observed. During the last part of the reaction, the contribution of mechanism I is higher since the increase of the ratio [Amine]/[Acid] enhances the relative contribution of the A · Cat complex.

The decrease of the reaction temperature favors complexation, i.e., mechanism I. Thus, in an experiment carried out in xylene at 84 °C, practically only reaction I took place.

Madec's pattern [158] was applied to the chemical modification of a poly(butyl-acrylate-Co-glycidylacrylate) by 4-hydroxybenzoic acid [37,38]. The reaction was studied both on the polymer and on models:

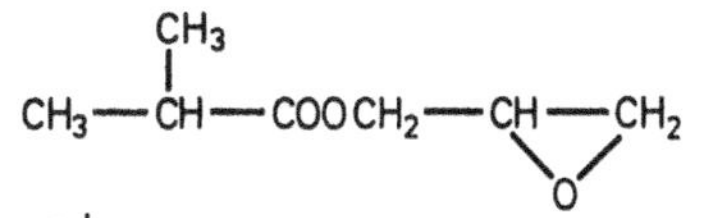

and

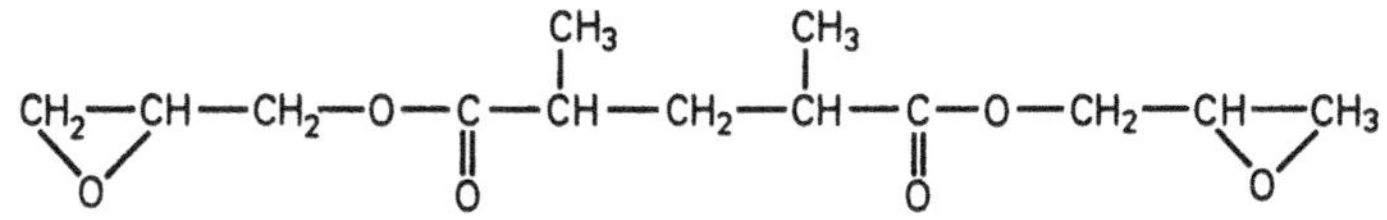

Results fit Madec's and Maréchal's pattern [158]. However, Crépeau and Maréchal [37,38] showed that contribution of scheme II is higher than for the benzoic/1,2-epoxy-3-phenoxypropane/dimethyldodecylamine system and provided explanation thereof.

Another kinetic study of reactions I and II was carried out on the 1,2-epoxy-3-dodecyloxypropane/2,4-hexadienoic acid system [203]. As in the above example, results fit the schemes reported in [158].

4.3 Miscellaneous Catalysts

4.3.1 Basic Catalysts other than Amines

Barès et al. [11] studied the system octadecanoic acid/ethylene oxide/potassium octadecanoate, in the bulk, mostly at 120 °C.

The process involves three stages:

— Formation of (hydroxy-2 ethyl)octadecanoate:

$$C_{11}H_{23}COOH + \underbrace{CH_2-CH_2}_{O} \rightarrow \underset{\text{I}}{C_{11}H_{23}COOCH_2CH_2OH} \qquad (64)$$

— Formation of ethylene bis-octadecanoate:

$$C_{11}H_{23}COOCH_2CH_2OH + C_{11}H_{23}COOH \rightarrow H_2O + \underset{\text{II}}{C_{11}H_{23}COOCH_2CH_2OOCC_{11}H_{23}} \qquad (65)$$

— Bis(hydroxy-2 ethyl)octadecanoate transesterification:

$$2\,\text{I} \rightleftharpoons HOCH_2CH_2OH + C_{11}H_{23}COOCH_2CH_2OOCC_{11}H_{23} \qquad (66)$$

Reaction (64) is unidirectional and is not accompanied by side or consecutive reactions. On the other hand, (66) starts at the same time as (65). The maximum content in I is attained at the maximum of conversion of octadecanoic acid. The kinetic was studied only during the first stage. The overall order is 1.

The equilibrium constant:

$$K = \frac{[\text{Monoester}]^2}{[\text{Diester}][\text{Glycol}]}$$

was determined from reaction (66) and found equal to 3.3.

The same reaction was studied by Shvets et al. [229] who prepared the catalyst (K octadecanoate) in situ by introducing potassium hydroxide into molten octadecanoic acid. They not only considered the main reaction:

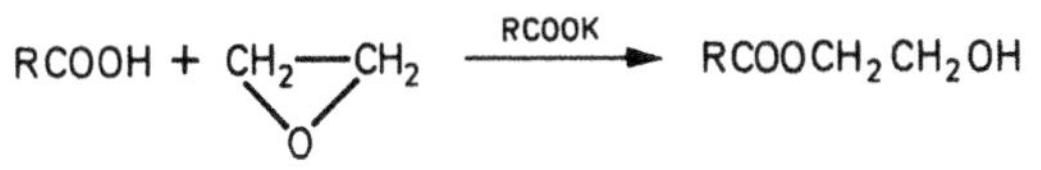

(67)

but also the formation of mono- and diesters of polyglycols: $RCOO(CH_2CH_2O)_{n+1}H$ and $RCOO(CH_2CH_2O)_nOCR$. They showed that the formation of polyglycols and their esters takes place only after the addition of one mole of ethylene oxide per mole of acid. Theoretical calculation are reported, but we found no rate constant value.

Sorokin and Gershanova [236] analyzed the system hexanoic acid/1,2-epoxy-3-phenoxypropane/MeOH, where Me is an alkaline metal (K or Na). The true catalyst is obviously RCOOMe. They found the following kinetic relation:

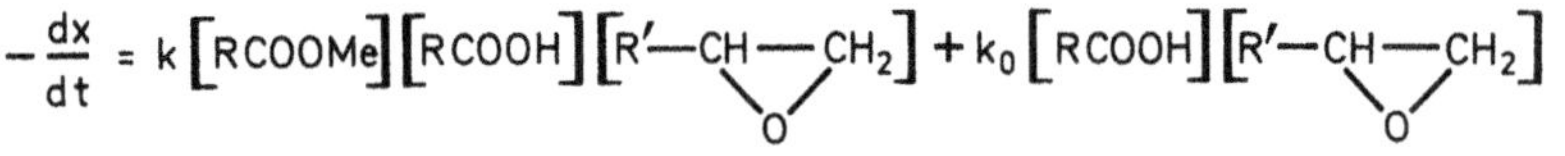

(46)

k_0 and k relate to the non-catalyzed and to the catalyzed reaction, x refers to the acid or the oxide concentration, which were both determined. $k_0 \times 10^5$ is in the range of 0.6 to 1.9 (kg mol^{-1} sec^{-1}) and $k \times 10^4$ is equal to 2 (kg^2 mol^{-2} sec^{-1}) at 100 °C when the catalyst is RCOONa.

On the basis of the kinetic data, the authors proposed the following mechanism:

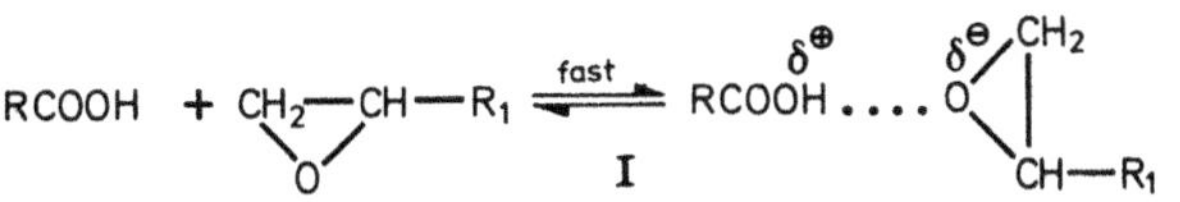

(68)

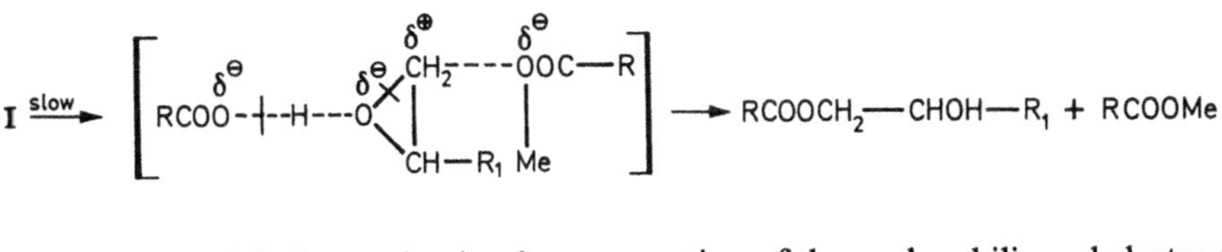

(69)

Reaction (69) is due to the simultaneous action of the nucleophilic and electrophilic reagents which attack the oxide molecule from different sides.

This mechanism is different from the one proposed for the amine-catalyzed reaction [236].

4.3.2 Metal Derivatives

As shown in Table 1, many metal derivatives can be used for catalyzing the epoxy-carboxy reaction. Unfortunately, in most cases, there is no study of the kinetics and of the mechanisms of the reaction.

The behavior of an acid-epoxy system largely depends on the nature of the catalyst. Thus, Pishnamazzade and Mamishov [199], studying the reaction of 1,2-epoxy-3-allyloxypropane with various carboxylic acids, found that the direction of the epoxide ring opening depends on the Lewis acid or on the ion exchanges they used as catalyst. When they saturated an epoxy-acid mixture with hydrogen chloride in the presence of the catalyst, two reactions were observed:

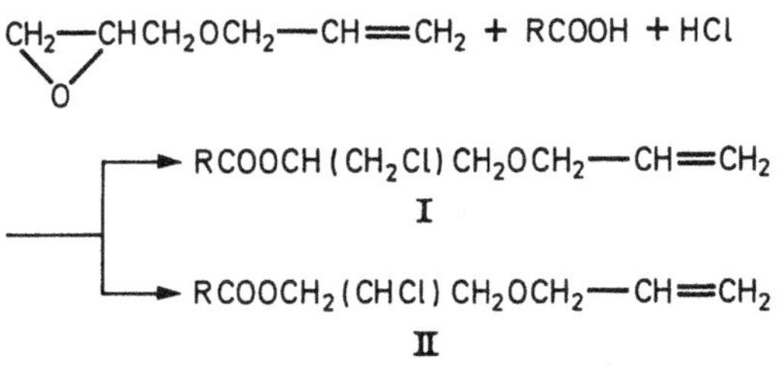

(70)

(71)

When ion exchanger or zinc chloride were used as catalysts, a mixture of I and II was obtained, I predominating. On the other hand, in the presence of aluminium chloride, only ester I was obtained with traces of II.

Stannic chloride was used by Shechter et al. [223] for catalyzing the reaction of styrene oxide with octanoic acid. They obtained a mixture of isomers:

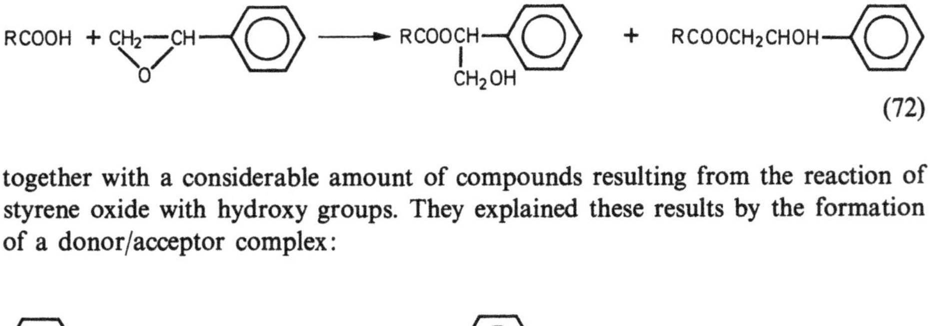

(72)

together with a considerable amount of compounds resulting from the reaction of styrene oxide with hydroxy groups. They explained these results by the formation of a donor/acceptor complex:

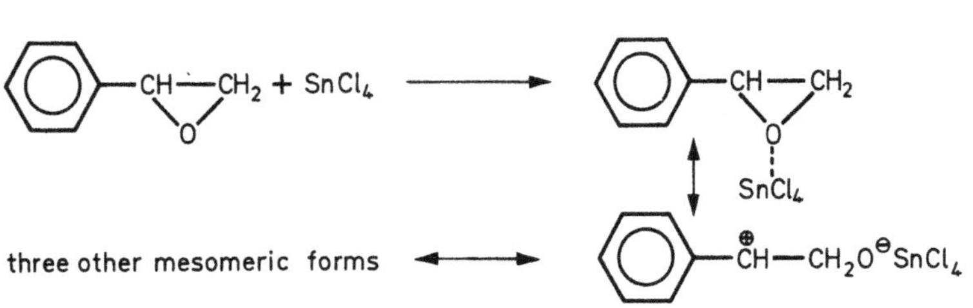

(73)

However, this mechanism must be considered very cautiously, since in the presence of traces of water (hardly avoidable in such a system), stannic chloride gives protons which are also catalysts, according to:

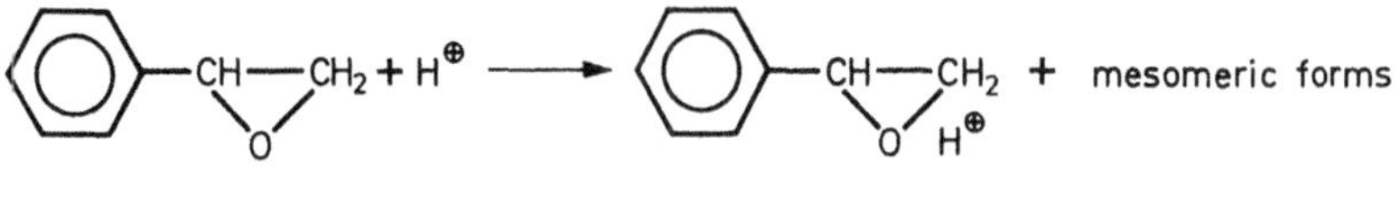
(74)

Rzhevskaya et al. [214] studied the ferric chloride-catalyzed reaction of methacrylic acid with propylene oxide. They showed that the reaction is first order with respect to the catalyst and to the propylene oxide, but zero order with respect to acid. From a spectroscopic study (mainly UV and a little NMR), they concluded that the mechanism involves the following stages:

— Formation of a dimeric complex which is the true catalyst of the reaction:

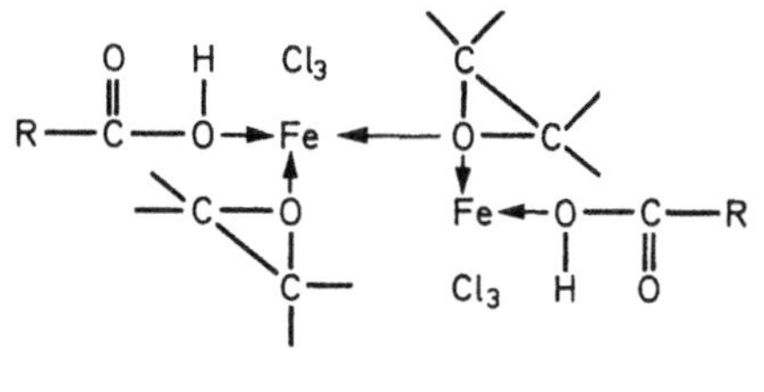

the zero order with respect to the acid can be explained by the strong polarizing influence of $Fe^{3\oplus}$ ions in the ternary-complex catalyst, on the O—H bond of the coordinated acid.

— Interaction between the coordinatively bound acid and a labile epoxide molecule in the second sphere of the iron ion, which leads to the formation of the reaction product. When a propylene oxide molecule forms an oxygen bridge, it does not take part in the reaction.

— Replacement of the resulting hydroxyester by a stronger ligand, i.e., a molecule of acid.

Chromium III derivatives have been used for catalyzing epoxy-carboxy reactions (see Table 1). Unfortunately, most publications are patents which describe the process without any information on the mechanisms and the kinetics. Let Cr(OOC R^a) be an "activated chromium tricarboxylate salt", i.e., a chromium salt having readily available coordination sites. Steele et al. [255] propose the following mechanism:

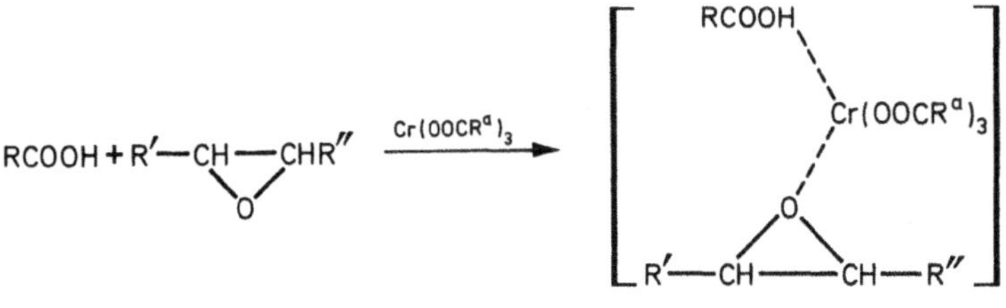
(75)

I

$$\text{I} \longrightarrow RCOOCHR'\text{—}CHOHR'' + Cr(OOCR^a)_3 \qquad (76)$$

5 Concluding Remarks

Epoxy-carboxy esterification has many applications, mainly in polymer chemistry. It is not only used to build polymer chains but also for chemical modification. Its most important advantage is that it can be carried out at moderate temperature with the formation of β-hydroxyester linkages and without elimination of volatiles. However, it can be spoiled by side reactions, and it is important to define the conditions which decrease the contribution of these latter. Most of the works that we have reviewed describe the reactional systems under different experimental conditions (catalyzed or non-catalyzed reaction, in solution or in the bulk) and, sometimes, give the values of kinetic parameters (partial or global orders, activation parameters).

First of all, we collected the main techniques used in the kinetic studies of the reaction (reaction procedures, analytical methods, treatment of experimental data). In fact, unambiguous results can be obtained only if some basic conditions are observed. The survey of the literature shows a great diversity of the catalysts used in epoxy-carboxy esterifications or polyesterifications; however, at least in the case of kinetic studies, tertiary amines and ammonium salts are largely predominant. From the fundamental studies carried out with organic models, and mostly in solution, several mechanisms were proposed involving the formation of a complex which can be cyclic or not.

In most articles relative to the study of the base-catalyzed reaction, a general mechanism is put forward; unfortunately the contribution of the side reactions is very often disregarded. If the contribution of some of them (esterification or etherification of the hydroxy side groups, hydrolysis, etc.) can be avoided or at least reduced to a low extent, that of some others (especially those resulting from the amine/epoxide interaction) is more difficult to eliminate. Some studies provided information on the kinetic contribution of these side reaction.

The knowledge of the side reactions should lead to a more efficient use of this esterification process. If their contribution cannot be eliminated or, at least, decreased to a very low value, there will be a limit in the optimization of the reaction and, therefore, in its use as a powerful tool in macromolecular synthesis. In that way, the discovery of new catalytic systems or, at least, a better knowledge of those which have been scarcely studied (e.g., chromium derivatives) would be very valuable.

6 References

1. Alvey, F. B.: J. Appl. Polym. Sci. *13*, 1473 (1969)
2. Alvey, F. B.: J. Polym. Sci., Part A-1, *7*, 2117 (1969)
3. Andrews, L. J., Keefer, R. M.: J. Amer. Chem. Soc. *84*, 2886 (1962)
4. Anfinogentov, A. A., Mikhailov, G. D., Samsonova, T. I., Malykh, V. A., Chegolya, A. S.: Zh. Prikl. Khim (Leningrad) *51*, 710 (1978)
5. Aoshima, A., Tanaka, K., Yoshioka, Y., Doi, T.: Jap. Kokai 76133227 (1976)
6. Arndt, P. J., Gruber, W., Loeffer, G., Lowitz, J., Partheil, G.: Ger. Offen. 2446944 (1976)
7. Bagrov, B. M., Mikhailov, G. D., Malykh, V. A.: Khim. Volokna *6*, 10 (1979)
8. Bagrov, B. M.: Vses. Konf. Mat. Modelir. i Apparaturn. Oformlenie Polimerizatsion. Protsessov, Vladimir, 122 (1979)
9. Bamford, C. H., Tipper, C. F. H.: "Comprehensive Chemical Kinetics", Elsevier, Amsterdam (1969)

10. Bares, M., Coupek, J., Pokorny, S., Hanzalova, J., Zajic, J.: Tenside Deterg. *12*, 155 (1975)
11. Bares, M., Bleha, M., Jeneralova, B., Zajic, J., Coupek, J.: Tenside Deterg. *12*, 162 (1975)
12. Bares, M., Bleha, M., Votovova, E., Zajic, J.: Tr. Mezhdunar. Kongr. Poverkhn. — Akt. Veschestvan, 7th, *1*, 238 (1976)
13. Bares, M., Bleha, M., Navratil, M., Votavova, E., Zajic, J.: Tenside Deterg. *16*, 74 (1979)
14. Bares, M., Bleha, M., Votavova, E., Marek, J., Vackova, E., Zajic, J.: Tenside Deterg. *16*, 308 (1979)
15. Batog, A. E., Stepko, O. P., Nikonova, L. P.: Zh. Org. Khim. *16*, 1126 (1980)
16. Batten, G. L. Jr., Miller, H. B.: J. Org. Chem. *39*, 3058 (1974)
17. Betts, A. T., Uri, N.: Ger. Offen. 2030704 (1971)
18. Bhatia, S., Rao, M. G., Rao, M. S.: J. Appl. Chem. Biotechnol. *26*, 295 (1976)
19. Bhatia, S., Rao, M. G., Rao, M. S.: Chem. Eng. Sci. *31*, 427 (1976)
20. Bleha, M., Bares, M., Votavova, E., Zajic, J., Coupek, J.: Tenside Deterg. *14*, 123 (1977)
21. Bodnaryuk, F. N., Korshunov, M. A., Melekhov, V. M., Britneva, T. P., Nemtsev, A. G., Baranov, E. L.: U.S.S.R. 357196 (1972)
22. Bowen, R. L.: J. Dent. Res. *58*, 1101 (1979)
23. Boyd, D. R., Marle, E. R.: J. Chem. Soc. *105*, 2117 (1914)
24. Boyd, D. R., Marle, E. R.: J. Chem. Soc. *114*, 1239 (1919)
25. Briegleb, G., Bieber, A.: Z. Elektrochem. *55*, 250 (1971); *53*, 350 (1949)
26. Brosse, J. C., Soutif, J. C., Pinazzi, C.: Makromol. Chem. *180*, 2109 (1979)
27. Brown, H. C., Horowitz, R. H.: J. Amer. Chem. Soc. *77*, 1733 (1955)
28. Browne, A. A. B., Mc Intyre, J. E.: Ger. Offen. 2600980 (1976)
29. Brzozowski, Z. K., Reda, Z.: Polimery (Warsaw) *19*, 617 (1974)
30. Burnett, J. F.: "Investigation of Rates and Mechanisms of Reactions", in "Technics of Chemistry", Ed. E. S. Lewis, 3rd ed., p. 101, New-York, Wiley Interscience (1974)
31. Capellos, C., Bielski, H. J.: "Kinetic Systems — Mathematical Description of Chemical Kinetics in Solution", Wiley-Interscience, New-York (1972)
32. Chegolya, A. S., Samsonova, T. I., Bagrov, B. M., Mikhailov, G. D., Malykh, V. A.: Ref. Dokl. Soobshch — Mendeleevsk. S'ezd Obshch. Prikl. Khim., 11th, *2*, 225 (1975)
33. Chegolya, A. S., Mikhailov, G. D., Malykh, V. A.: Lenzinger Ber. 43112 (1977)
34. Chimura, K., Takashima, S., Kawashima, M., Nishibara, H.: Japan Kodai 7424933 (1974)
35. Ciba-Geigy, A. G.: Fr. Demande 2225498 (1974)
36. Clark, C., Dalton, R. J., Harrison, A. C.: Brit. 1208277 (1970)
37. Crépeau, C., Maréchal, E.: Europ. Polym. J. *19*, 1099 (1983)
38. Crépeau, C., Maréchal, E.: Europ. Polym. J. *19*, 1107 (1983)
39. De Benedictus, A., Sobolev, I.: Ger. 1793834 (1978)
40. Distler, H., Widder, R., Schneider, K.: Ger. Offen. 2157455 (1973)
41. Distler, H., Schneider, K.: Ger. Offen. 2361907 (1975)
42. Dobinson, B., Green, G. E.: S. African 7302573 (1974)
43. Dobinson, B., Green, G. E.: S. African 7303227 (1974)
44. Domide, C. T., Mihaiu, T., Popescu, L.: Rom. 67494 (1979)
45. Doorakian, G. A., Schmidt, D. L., Cornell, M. C.: Ger. Offen. 2609475 (1977)
46. Dow Chemical Co: Neth. Appl. 7800475 (1979)
47. Doyle, T. E., Fekete, F., Keenan, P. J., Plant, W. J.: U.S. 3317465 (1967)
48. Dukes, C. F., Welch, R. W.: U.S. 3859314 (1975)
49. Dumitriu, E., Oprea, S., Dima, M., Avramescu, S.: Rev. Chim. (Bucharest) *32*, 225 (1981)
50. Durbetaki, A. J.: Anal. Chem. *28*, 2000 (1956)
51. Ebich, Y. R., Blokh, G. A., Grigor'yants, I. K., Stonoga, O. A., Rudaya, I. S.: Vses. Nauch. — Tekh. Konf., Novye Mater. Protsessy Rezin. Prom. 1, 38 (1973)
52. Ebich, Y. R., Blokh, G. A., Rudaya, I. S.: Tezisy Dokl. — Resp. Konf. Vysokomol. Soedin., 3rd., 54 (1973)
53. Ebich, Y. R., Blokh, G. A., Grigor'yants, I. K., Stonoga, O. A., Rudaya, I. S.: Evdokimenko, N. M., Vopr. Khim. Khim. Teknol. *40*, 118 (1975)
54. Edwards, P.: U.S. Patent 2537981 (1951)
55. Enikolopiyan, N. S.: Pure and Appl. Chem. *48*, 317 (1976)
56. Enoki, Y., Adachi, S., Sakata, S., Kumaki, M.: Japan 2418967 (1964)
57. Enoki, Y., Takahashi, S., Adachi, S., Ikari, E.: Japan 6800638 (1968)

58. Enoki, Y., Takahashi, S., Takakura, H., Takashima, Y.: Japan 6813236 (1968)
59. Enoki, Y., Takahashi, S., Adachi, S., Takashima, Y.: Japan 6813237 (1968)
60. Enoki, Y., Takahashi, S., Adachi, S.: Japan 6906837 (1969)
61. Enoki, Y., Takahashi, S., Sakata, S., Ikari, E.: Japan 6929052 (1969)
62. Fedoseev, M. S., Batyr, D. G., Gavrilenko, G. Y., Nirka, V. E.: Koord. Khim. *3*, 648 (1977)
63. Fellers, J.: Macromolecules *3*, 202 (1970)
64. Fiala, V., Lidarik, M.: Angew. Makromol. Chem. *12*, 157 (1970)
65. Fugukawa, N., Kato, F., Shimonura, K. S.: Yohilara, Report 17th Annual Meeting of the Chemical Society of Japan, Tokyo (1964)
66. Fujita, T., Watanabe, S., Suga, K.: Austr. J. Chem. *27*, 2205 (1974)
67. Fujita, T., Suga, K., Watanabe, S., Taguchi, Y., Sakurai, K.: Yakagaku *25*, 480 (1976)
68. Fujita, T., Suga, K., Watanabe, S., Nakayama, H., Hokyo, M.: Yakagaku *26*, 720 (1977)
69. Fujita, T., Watanabe, S., Suga, K., Hokyo, M.: J. Appl. Chem. Biotechnol. *27*, 539 (1977)
70. Fujita, Y.: Japan 6926653 (1969)
71. Fujita, Y., Morimoto, T.: Japan 7215943 (1972)
72. Fujita, Y., Nakamura, I., Yoshioka, T.: Japan 7221980 (1972)
73. Furukawa, Y., Yamashita, G., Yamanaka, Y.: Japan 7201612 (1972)
74. Gander, R. J.: Braz. Pedido P I 7406844 (1976)
75. Girvan, I. J. M.: Brit. 1096035 (1967)
76. Gololobova, A. A., Sineokov, A. P., Salakhetdinova, Z. S., Shtefan, V. N., Zubkova, V. T., Neumoin, Y. V.: U.S.S.R. 505627 (1976)
77. Green, G. E.: S. African 7302101 (1973)
78. Groff, G. L.: U.S. 3576903 (1971)
79. Grossmann, H.: Tenside Deterg. *12*, 16 (1975)
80. Gruber, W., Walter, H.: Ger. Offen. 2345394 (1975)
81. Gruzinov, E. V., Trifonova, V. P.: Polim. Stroit. Mater. *45*, 57 (1977)
82. Gurgiolo, A. E.: U.S. 4069242 (1978)
83. Gurgiolo, A. E.: Ger. Offen. 2801422 (1979)
84. Gurgiolo, A. E.: Braz. Pedido P I 7800250 (1979)
85. Hajek, K., Svoboda, B., Kaska, J., Cerny, J., Makes, J.: Czech. 166330 (1976)
86. Hammett, L. P.: J. Amer. Chem. Soc. *59*, 96 (1937)
87. Hashi, N., Takase, Y.: Japan 7247366(1972)
88. Hayami, H.: Japan Kokai 7739618 (1977)
89. Hayashi, Y., Sasaguri, K., Nakamura, K., Matsumoto, Y., Matsuo, S., Sato, M., Uda, B.: Japn. Kokai 7656834 (1976)
90. Hess, L. Ġ.: U.S. 4223160 (1980)
91. Hetflejs, J., Mares, F., Bazant, V., Collect. Czech. Chem. Commun. *34*, 3098 (1969)
92. Horsley, L. H., U.S. 3433824 (1969)
93. Ichikawa, Y., Yamashita, G., Yamanaka, Y.: Japan 7137578 (1971)
94. Ichikawa, Y., Yamashita, G., Yamanaka, Y.: Japan 7137579 (1971)
95. Ichikawa, Y., Yamashita, G., Yamanaka, Y.: Japan 7141295 (1971)
96. Ichikawa, Y., Yamashita, G., Yamanaka, Y.: Japan 7141296 (1971)
97. Ichikawa, Y., Yamashita, G., Yamanaka, .: Japan 7137578 (1971)
98. Ichikawa, Y., Kobayashi, O.: Japan Kokai 7396537 (1973)
99. Imperial Chemical Industries Ltd.: Neth. Appl. 6611627 (1967)
100. Ishii, Y., Sakai, S., Sugiyama, T.: Bull. Japan Petrol. Inst. *5*, 44 (1963)
101. Ito, H., Kimura, K.: Japan 7241332 (1972)
102. Ito, H., Kimura, K.: Japn. Kokai 7300416 (1973)
103. Ito, H., Kimura, K.: Japn. Kokai 7322414 (1973)
104. Ito, H., Kimura, K.: Japan 7311083 (1973)
105. Ito, H., Kimura, K.: Japan 7400815 (1974)
106. Ivkina, A. V., Trofimov, V. A., Belov, P. S.: Zh. Prikl. Khim. (Leningrad) *47*, 2135 (1974)
107. Izawa, N., Iizuka, Y., Kubota, Y., Obata, K.: Japan 7246581 (1972)
108. Izawa, N., Iizuka, Y., Kubota, Y., Obata, K.: Japan 7246582 (1972)
109. Jardnickova, Z., Krupicka, I.: Chem. Pruem *18*, 554 (1968)
110. Jay, R. R.: Anal. Chem. *36*, 667 (1964)
111. Jeffrey, G. C., Embrey, W. E.: U.S. 3804884 (1974)

112. Johnson & Johnson: Neth. Appl. 7411068 (1976)
113. Jung, V. G., Kleeberg, W.: Kunstoffe *51*, 714 (1961)
114. Kadoma, Y., Warashina, N., Noguchi, M., Sakai, M.: Japan Kokai Tokkyo Koho 8011525 (1980)
115. Kakiuchi, H., Tanaka, Y.: Kobunshi Kagaku *20*, 619 (1963)
116. Kakiuchi, H., Tanaka, Y.: J. Org. Chem. *31*, 1559 (1966)
117. Kakiuchi, H., Endo, T.: Bull. Chem. Soc. Jap. *40*, 892 (1967)
118. Kalal, J., Svec, F., Marousek, V.: J. Polym. Sci., Polym. Symp. 47 (1974)
119. Kamatani, H.: Japan 7202623 (1972)
120. Kamatani, H.: Nippon Kagaku Kaishi *10*, 1505 (1977)
121. Kamatani, H.: Nippon Kagaku Kaishi *9*, 1271 (1978)
122. Kamatani, H.: Kobunshi Ronbunshu *36*, 311 (1979)
123. Katsurada, H., Hikita, M., Shioyama, Y., Kawahara, K.: Japn. Kokai 7595216 (1975)
124. Katzakian, A. Jr., Steele, R. B., Scigliano, J. J.: Ger. Offen. 2402992 (1974)
125. Kimura, K., Yamada, A., Suzuki, K.: Japan Kokai 75101312 (1975)
126. Kimura, K., Suzuki, K.: Japan Kokai 76133227 (1976)
127. Kimura, T., Kobayashi, S., Nakamoto, H.: Japan 7525461 (1975)
128. Kirkwood, J. G.: J. Chem. Phys. *2*, 351 (1934)
129. Kleeman, A., Kolb, H., Rindfuss, A., Schreyer, G., Schwoerzer, L. K., Schuster, G.: Ger. Offen. 2449847 (1976)
130. Kondo, K., Matsumoto, M., Dobashi, S.: Japan Kokai 7584514 (1975)
131. Kovalenko, O. N.: Zh. Prikl. Khim. (Leningrad) *52*, 1128 (1979)
132. Krasnova, L. M., Kozlova, T. S., Sumin, I. G.: U.S.S.R. 327169 (1972)
133. Kuznetsov, B. V., Marchenko, G. N.: Kinet. Katal. *13*, 1351 (1972)
134. Labana, S. S., Golovoy, A.: U.S. 3787521 (1971)
135. Labana, S. S., Chang, Y. G.: Ger. Offen. 2240312 (1973)
136. Labana, S. S., Golovoy, A.: Ger. Offen. 2307748 (1973)
137. Labana, S. S.: Can. 985825 (1976)
138. Labsky, J., Exner, J.: Ger. Offen. 2408501 (1974)
139. Larkin, D. R.: Ger. Offen. 1919782 (1969)
140. Lebedev, N. N., Gus'kov, K. A.: Kinet. Katal. *4*, 116 (1963)
141. Lebedev, N. N., Gus'kov, K. A.: Kinet. Katal. *4*, 581 (1963)
142. Lebedev, N. N., Gus'kov, K. A.: Kinet. Katal. *5*, 446 (1964)
143. Lebedev, N. N., Gus'kov, K. A.: Kinet. Katal. *5*, 787 (1964)
144. Lebedev, N. N., Kozlov, V. M.: Zh. Organ. Khim. *2*, 261 (1966)
145. Lebedev, N. N., Shvets, V. F.: Kinet. Katal. *9*, 504 (1968)
146. Lebedev, N. N., Shvets, V. F., Tyokova, O. A.: Zh. Organ. Khim. *7*, 1851 (1971)
147. Lebedev, N. N., Shvets, V. F., Romashkin, A. V.: U.S.S.R. 372206 (1973)
148. Lee, H., Neville, K.: "Handbook of Epoxy-Resins", Mc Graw Hill, N.Y. (1967)
149. Leffer, J. E., Grundwald, E.: "Rates and Equilibria of Organic Reactions", John Wiley and Sons.
150. Liao, R., Xie, Q., Wei, D., Wang, J.: Cuihua Xuebao *2*, 92 (1981)
151. Lipanov, A. M., Brio, B. S., Loktionova, L. F.: Fiz. Goreniya Vzry va *9*, 813 (1973)
152. Lok, C. M.: Ger. Offen. 2937516 (1980)
153. Maatschappij, B. V.: Neth. Appl. 7612362 (1977)
154. Mc Connel, R. L., Coover, H. W. Jr.: U.S. 3298998 (1967)
155. Madec, P. J., Maréchal, E.: J. Polym. Sci., Pol. Chem. Ed. *16*, 3165 (1978)
156. Madec, P. J., Maréchal, E.: Makromol. Chem. *184*, 323 (1983)
157. Madec, P. J., Maréchal, E.: Makromol. Chem. *184*, 335 (1983)
158. Madec, P. J., Maréchal, E.: Makromol. Chem. *184*, 343 (1983)
159. Madec, P. J., Maréchal, E.: Makromol. Chem. *184*, 357 (1983)
160. Maerker, G., Carmichael, J. F., Port, W. S.: J. Org. Chem. *26*, 2681 (1961)
161. Makimura, O., Yoshida, S.: Japan 7538675 (1975)
162. Malek, J., Silhavy, P.: Collect. Czech. Commun. *41*, 84 (1976)
163. Malkemus, J. D., Swan, J. D.: J. Amer. Oil Chemists' Soc. *34*, 342 (1957)
164. Mares, F., Hetflejs, J., Bazant, V.: Collect. Czech. Chem. Commun. *34*, 3086 (1969)
165. Mares, F., Silhavy, P., Malek, J., Bazant, V.: Czech. 130531 (1969)

166. Martinez Utrilla, R., Olano Villen, A.: Grasas Aceites (Seville) *24*, 13 (1973)
167. Martinez Utrilla, R., Olano Villen, A.: Grasas Aceites (Seville) *24*, 71 (1973)
168. Matejka, L., Bouchal, K., Pokorny, S., Ryska, M., Dusek, K.: Makrotest, Sb. Prednasek,
169. Matejka, L., Pokorny, S., Dusek, K.: Polymer Bull. *7*, 123 (1982)
170. Matsuda, T., Motohashi, T., Dandan, H.: Japan Kokai Tokkyo Koho 78132518 (1978)
171. Matsuzawa, K., Suzuki, Y., Murao, Y.: Japan Kokai 7308742 (1973)
172. May, C. A., Tanaka, Y.: "Epoxy Resins — Chemistry and Technology", Marcel Dekker, N.Y. (1973)
173. Mitch, E. L., Kaplan, S. L.: Soc. Plast. Eng., Tech. Pap. *21*, 219 (1975)
174. Mitsui, T.: Japan Kokai Tokyyo Koho 8187537 (1981)
175. Mizuno, K.: Ger. Offen. 2405330 (1974)
176. Mleziva, J., Jarusek, J., Cermak, V., Vecera, M.: FATIPEC Congr. 14, 451 (1978)
177. Morozova, T. V., Selyakova, V. A., Sineokov, A. P.: Osnovn. Org. Sint. Neftekhim *8*, 45 (1977)
178. Mravec, D., Kalamar, J., Rattay, V., Repasova, I., Foltinova, M., Mikula, F.: Czech 186589 (1978)
179. Munakata, H., Ukai, T., Kamijima, A.: Japan 7028388 (1970)
180. Munakata, H., Uejima, T., Ukai, T.: Japan 7135254 (1971)
181. Munakata, H., Kamaya, H., Kamijima, T., Ukai, T., Mizumoto, T., Yasukawa, T.: Japan 7238872 (1972)
182. Munakata, H., Kamishima, A., Ukai, T.: Japan 7307268 (1973)
183. Munakata, H., Kamatani, H., Mizumoto, T.: Japan 7539711 (1975)
184. Murayama, M.: Japan 7008970 (1970)
185. Nagase, K., Sakaguchi, K.: Kogyo Kagaku Zashi *64*, 1035 (1961)
186. Nakahara, S., Yashikawa, H.: Japan Kokai 74109313 (1974)
187. N'Guyen, C. H., Mleziva, J.: Angew. Makromol. Chem. *31*, 83 (1973)
188. Nishikubo, T., Imaura, M., Mizuko, T., Takaoka, T.: J. Appl. Polym. Sci. *18*, 3445 (1974)
189. Nishikubo, T., Ibuki, S., Mizuko, T., Takaoka, T.: Kobunshi Ronbunshu *32*, 604 (1975)
190. Nishikubo, T.: Kobunshi Ronbunshu *35*, 673 (1978)
191. Nishinohara, M.: Japan Kokai 7643708 (1976)
192. Noma, K., Nakamura, C., Niwa, M.: Doshisha Daigaku Rikogaku Kenkyu Hokoku *15*, 60 (1974)
193. Ogrel, A. M., Medvedev, V. P., Lyapichev, V. E., Tuzhikov, O. I.: Kauch Rezina *3*, 15 (1981)
194. Olano Villen, A.: Grasas Aceites (Seville) *26*, 136 (1975)
195. Ootani, H., Morita, K., Takemuchi, O.: Japan Kokai 7374534 (1973)
196. Oprea, S., Dumitriu, E., Stefanel, L.: Rev. Roum. Chim. *25*, 1039 (1980)
197. Percy, E. J., Wickings, J. A.: Brit. 1120301 (1968)
198. Pirozhnaya, L. N., Gromov, V. V.: Zh. Org. Khim. *9*, 210 (1973)
199. Pishnamazzade, B. F., Mamishov, A. K.: Zh. Org. Khim. *9*, 1365 (1973)
200. Podgornova, V. A., Oreklova, N. F., Ustavshchikova, B. F., Kopylova, V. D., Kargman, V. B., Galitskaya, N. B.: Osnovn. Org. Sint. Neftekhim. *2*, 89 (1975)
201. Podgornova, V. A., Shpekht, V. A., Vokhmyanina, E. V., Kargman, V. B., Smurygina, N. N.: Osnovn. Org. Sint. Neftkhim. *8*, 41 (1977)
202. Podgornova, V. A., Shpekht, V. A., Vokhmyanina, E. V., Ustavshchikov, B. F., Kargman, V. B., Galitskaya, N. B.: Zh. Prikl. Khim. (Leningrad) *50*, 878 (1977)
203. Pourdjavadi, A., Madec, P. J., Maréchal, E.: Europ. Polym. J. *20*, 311 (1984)
204. Pyun, H. C., Park, W. B., Kim, K. Y., Sung, K. Y.: Report 1980, KAERI — 413/RR-146/80. Avail. INIS.
205. Randall, G. C. W.: Ger. Offen. 1911447 (1970)
206. Randall, G. C. W.: Eur. Pat. Appl. 1904(1979)
207. Rattay, V., Repasova, I., Kalamar, J., Mravec, D., Mikula, F., Molendova, B.: Czech. 181443 (1980)
208. Rattay, V., Repasova, I., Kalamar, J., Mravec, D., Mikula, F., Molendova, B.: Czech. 186368 (1980)
209. Rehfuss, J. W.: S. African 6801333 (1968)
210. Reid, J. C. Jr.: U.S. 3746744 (1973)
211. Rokaszewski, E.: Pol. J. Chem. *52*, 1487 (1978)
212. Rokaszewski, E.: Pol. J. Chem. *53*, 1303 (1979)

213. Rowton, R. L.: U.S. 3441598 (1969)
214. Rzhevskaya, N. N., Stepanov, E. G., Svitych, R. B., Podgornova, V. A., Ustavshchikov, B. F.: Zh. Obshch. Khim. *48*, 2083 (1978)
215. Sakai, S., Ueda, K., Ishii, Y.: Kogyo Kagaku Zasshi *64*, 2159 (1961)
216. Sallay, P., Morgos, J., Farkas, L., Rusznak, I., Bartha, B.: Tenside Deterg. *16*, 17 (1979)
217. Samsonova, T. I., Mikhailov, G. D., Malykh, V. A., Chegolya, A. S., Soboleva, N. I.: Zh. Prikl. Khim. (Leningrad) *48*, 2267 (1975)
218. Samsonova, T. I., Mikhailov, G. D., Malykh, V. A., Chegolya, A. S.: Khim. Volokna 20 (1977)
219. Sanders, H. L., Braunwarth, J. B., Connel, R. B., Swenson, R. A.: J. Amer. Oil Chemist's Soc. *46*, 167 (1969)
220. Schmid, R., Fisch, W.: S. African 6800481 (1968)
221. Schmid, R., Lohse, F., Batzer, H. S.: African 6805738 (1969)
222. Shechter, L., Wynstra, J.: Ind. Eng. Chem. *48*, 86 (1956)
223. Shechter, L., Wynstra, J., Kurkjy, R. P.: Ind. ENg. Chem. *49*, 1107 (1957)
224. Shibata, M., Iwasawa, N., Watanabe, T., Yoshihara, I.: Japan Kokai 7616338 (1976)
225. Shimbo, M., Ochi, M., Yamada, M.: Kobunshi Ronbunshu *37*, 57 (1980)
226. Shokal, E. C., Creek, W., Mueller, A. G.: U.S. 2895047 (1959)
227. Shvets, V. F., Tyukova, O. A.: Zh. Org. Khim. *7*, 1847 (1971)
228. Shvets, V. F., Lebedev, N. N., Tyukova, O. A.: Zh. Org. Khim. *7*, 1851 (1971)
229. Shvets, V. F., Trushin, A. M., Lebedev, N. N.: Zh. Prikl. Khim. (Leningrad) *45*, 829 (1972)
230. Shvets, V. F., Romashkin, A. V.: Kinet. Katal. *13*, 885 (1972)
231. Shvets, V. F., Romashkin, A. V., Yudina, V. V.: Kinet. Katal. *14*, 928 (1973)
232. Siggia, S., Starke, A. C., Garis, J. J., Stahl, C. R.: Anal. Chem. *30*, 115 (1958)
233. Smith, O. W., Koleske, J. V., Brezinski, J. J.: Ger. Offen. 2419331 (1974)
234. Sorokin, M. F.. Khinchina, E. L.: Tr. Mosk. Khim.-Technol. Inst. *48*, 85 (1965)
235. Sorokin, M. F., Khinchina, E. L.: Tr. Mosk. Khim.-Teknol. Inst. *48*, 94 (1965)
236. Sorokin, M. F., Gershanova, E. L.: Kinet. Katal. *8*, 512 (1967)
237. Sorokin, M. F., Gershanova, E. L., Lipkina, S. I.: Lakokrasoch. Mater. Ikh. Primen. *1*, 1 (1968)
238. Sorokin, M. F., Chan Win Suu: Tr. Mosk. Khim.-Teknol. Inst. *61*, 81 (1969)
239. Sorokin, M. F., Chan Win Suu: Tr. Mosk. Khim.-Teknol. Inst. *61*, 85 (1969)
240. Sorokin, M. F., Shode, L. G.: Shteinpress, A. B., Stokozenko, V. N., Plast. Massy 20 (1973)
241. Sorokin, M. F., Shode, L. G., Stokozenko, V. N., Sinitsyna, G. N., Marukhina, S. M.: Tr. Mosk. Khim.-Teknol. Inst. *74*, 23 (1973)
242. Sorokin, M. F., Denisova, R. I., Tishkova, G. S.: Tr. Mosk. Khim.-Teknol. Inst. *80*, 42 (1974)
243. Sorokin, M. F., Shode, L. G., Veslov, V. V., Nogteva, S. I.: Deposited Doc. 1975 VINITI 3671
244. Sorokin, M. F., Shode, L. G., Veslov, V. V., Petrova, L. P.: Deposited Doc. 1975, VINITI 3673
245. Sorokin, M. F., Shode, L. G., Veslov, V. V.: Deposited Doc. 1976, VINITI 4249
246. Sorokin, M. F., Shode, L. G., Veslov, V. V.: Lakokras. Mater. Ikh. Primen. *5*, 9 (1976)
247. Sorokin, M. F., Gershanova, E. L., Mikhitarova, Z. A., Trubnikova, Z. A., Kuz'hina, L. I.: Deposited Doc. 1978, VINITI 2291
248. Sorokin, M. F., Kochnova, Z. A., Nikolaev, P. V., Finyakin, L. N.: Izv. Vyssh. Uchebn. Zaved. Khim. Khim. Teknol. *22*, 947 (1979)
249. Sorokin, M. F., Shode, L. G., Veslov, V. V.: Sintez i Issled. Plenkoobrazuyushch. Veshchetv i Pigmentov (Yaroslavl) *4*, 46 (1979)
250. Sorokin, M. F., Shode, L. G., Pavelyukov, S. A., Onosova, L. A.: Deposited Doc. 1979, VINITI 403
251. Sorokin, M. F., Gershanova, E. L., Trubnikova, N. A., Meshalkina, M. G.: Deposited Doc. 1980, VINITI 461
252. Sorokin, M. F., Gershanova, E. L., Mikhitarova, Z. A., Trubnikova, N. A., Kuz'kina, L. I.: Tr. Mosk. Khim.-Teknol. Inst. im. D. I. Mendeleeva *110*, 91 (1980)
253. Steele, R. B., Katzakian, A. Jr.: Ger. Offen. 2439352 (1975)
254. Steele, R. B., Katzakian, A. Jr.: U.S. 3968135 (1976)
255. Steele, R. B., Katzakian, A. Jr.: U.S. 4017429 (1977)
256. Stein, G.: U.S. 2224026 (1940); German Patent 708463 (1941)

257. Sueling, C., Wilms, H., Roos, E.: Ger. 1568537 (1974)
258. Suzuki, K., Kimura, K., Ito, H.: Japan Kokai 7401510 (1974)
259. Swanson, R. G., Walus, A. N.: U.S. 3816557 (1974)
260. Szakacs, S., Gobolos, S., Nagy, F.: Magy. Kem. Foly. *86*, 276 (1980)
261. Takamatsu, M., Sakakibara, T.: Japan Kokai 7367238 (1973)
262. Takamatsu, M., Sakakibara, T.: Japan Kokai 7376829 (1973)
263. Takayama, Y., Saito, S.: Japan 6902685 (1969)
264. Takeshi, A., Kunio, I., Tsumoto, F.: J. Chem. Soc. Japan, Ind. Chem. Soc. 66941 (1963)
265. Takiyama, E., Hokamura, S.: Brit. 1293971 (1972)
266. Tanaka, Y.: J. Org. Chem. *32*, 2405 (1967)
267. Tanaka, Y., Takeuchi, H.: Tetrahedron *24*, 6433 (1968)
268. Tanaka, Y., Okada, A., Suzuki, M.: Can. J. Chem. *48*, 3258 (1970)
269. Taniyama, O., Kamatani, H.: Japan 7223291 (1972)
270. Tanizaki, Y., Kubo, Y.: Japan 7251328 (1972)
271. Tardenaka, A.: Issled. Obl. Khim. Drev., Tezisy Dokl., Konf. Molodykh Uch., 1st, 20 (1974)
272. Taylor, A. E., Speight, J. H.: Brit. 1308250 (1973)
273. Tsubuko, K., Tasaka, M., Matsubayashi, K., Hashimoto, J.: Japan Kokai Tokkyo Koho 79132517 (1979)
274. Tsiji, I., Kato, H., Tatemichi, H.: Japan Kokai 7388126 (1973)
275. Tsukumi, H., Yamazaki, Y., Fujiki, S.: Japan Kokai 7505316 (1975)
276. Uejima, A., Munakata, H.: Nippon Kagaku Kaishi *8*, 1496 (1973)
277. Uejima, A.: Nippon Kagaku Kaishi *6*, 974 (1981)
278. Umbach, W., Stein, W.: Tenside *7*, 132 (1970)
279. Umbach, W., Stein, W.: J. Amer. Oil Chem. Soc. 48394 (1971)
280. United Kindom Ministry of Technology, Fr. 2097494 (1972)
281. Vanlerberghe, G., Sebag, H.: Ger. Offen. 2535802 (1976)
282. Virtanen, P. O. I.: Suom. Kem. *38*, 231 (1965)
283. Voronina, T. A., Fomina, N. V.: Lakokras. Mater. Ikh. Primen. *4*, 8 (1981)
284. Watanabe, I., Tanaka, K.: Japan Kokai 74135913 (1974)
285. Wrighley, A. N., Smith, F. D., Shirton, A. J.: J. Amer. Oil Chemist's Soc. *34*, 39 (1957)
286. Wrighley, A. N., Smith, F. D., Shirton, A. J.: J. Amer. Oil Chemist's Soc. *36*, 34 (1959)
287. Wright, H. J., Bremmer, J. F.: Ger. 1443633 (1970)
288. Yamahara, T., Fujioka, S.: Japan 6819537 (1968)
289. Yamauchi, T., Suzuki, T., Hasegawa, T.: Japan Kokai 7338310 (1973)
290. Yoshida, S., Shimizu, N.: Japan Kokai 74124015 (1974)
291. Yoshida, S., Kobayashi, D., Shimizu, N.: Japan Kokai 7626810 (1976)
292. Yoshikawa, H.: Japan Kokai 7582009 (1975)
293. Yukhnovskii, G. L., Kuznetsova, V. M., Postoeva, M. E.: Plast. Massy 60 (1970)
294. Zaripov, I. N., Nasybullin, S. A., Lazareva, L. M., Faizullin, I. N.: Vysokomol. Soed., Ser. A, *22*, 640 (1980)
295. Zimmerschied, W. J., Meyer, D. H.: Ger. Offen. 2247116 (1974)

Prof. H.-J. Cantow (Editor)
Received November 19, 1984

Swelling Equilibrium Studies of Elastomeric Network Structures

J. P. Queslel and J. E. Mark
Department of Chemistry and the Polymer Research Center,
The University of Cincinnati, Cincinnati, Ohio 45221, USA

Until recently, interpretation of swelling equilibrium experiments rested on the Flory-Rehner equation developed in 1943 for networks which deform affinely. The relationship made possible at least the ranking of series of networks of the same polymer, according to their cross-link densities. However, the theory does not explain the important fact (first observed by Gee in 1965) that there is a maximum in the dependence of $\lambda \ln (a_1^c/a_1^u)$ on λ (where λ is the isotropic deformation $v_2^{-1/3}$, v_2 the volume fraction of polymer in the polymer-solvent system, and a_1^u and a_1^c the solvent activities respectively in uncross-linked and cross-linked polymers). Decisive progress in this field has been achieved by the formulation of a molecular theory of real networks by Flory and Erman in 1979. The present paper reviews the main theoretical advances concerning swelling and swelling equilibrium, and describes useful methods for characterizing both model networks (of controlled structure) and networks in which the cross linking is highly random and essentially uncontrolled.

1 Introduction

Swelling by an organic solvent is one of the simplest methods for characterizing elastomeric networks, rivaling stress-strain measurements in its simplicity. Swelling of an elastomeric network is the result of two thermodynamic phenomena: an increase of entropy of the network-solvent system by introduction of small molecules of diluent, and a decrease of entropy of the polymer chains by the isotropic deformation. For the first effect, the mixing of polymer and solvent was successfully accounted for by the liquid lattice theory of Flory and Huggins [1-3]. Use of their relationship and a statistical mechanical expression for the Gibbs elastic free energy change with dilation lead Flory and Rehner to propose a relation between swelling and degree of cross-linking [4-6] for an affine network (one in which the cross-links move linearly with the dimensions of the elastomeric sample). Their equation has been widely used to characterize a variety of networks [7]. Recently, the molecular treatment of rubberlike elasticity has been improved by Flory and Erman [8,9], including detailed swelling-structure relationships [10-14]. Both these relationships and the results of phenomenological treatments are reviewed in this paper, and are applied to the characterization of model and randomly cross-linked networks.

2 General Theory of Swelling

The change ΔA of the Gibbs free energy due to mixing a solvent 1 with a polymer 2 is given by [3]

$$\Delta A = A_M - A_1 - A_2 \tag{1}$$

where A_M, A_1 and A_2 are respectively the free energies of the system polymer-solvent, pure solvent, and pure polymer. Chemical potentials of solvent in the uncrosslinked polymer, μ_1^u, and of pure solvent, μ_1^0, are obtained by differentiation of ΔA with respect to the number of moles of solvent n_1 at constant n_2 of polymer:

$$\left(\frac{\partial \Delta_A}{\partial n_1}\right)_{n_2,T,p} = \left(\frac{\partial A_M}{\partial n_1}\right)_{n_2,T,p} - \left(\frac{\partial A_2}{\partial n_1}\right)_{n_2,T,p} = \mu_1^u - \mu_1^0 \tag{2}$$

The difference $\mu_1^u - \mu_1^0$ is related to the vapor pressures of solvent p_1^u over the solution and p_1^0 of pure solvent by

$$\mu_1^u - \mu_1^0 = RT \ln (p_1^u/p_1^0) \tag{3}$$

Flory and Huggins [1,2] used a liquid lattice theory to express $\mu_1^u - \mu_1^0$ as a function of the volume fraction v_2 of polymer in the mixture, the polymer-solvent interaction parameter χ [15-18], and the ratio x of molar volumes of polymer V_2 and solvent V_1. Specifically,

$$\mu_1^u - \mu_1^0 = RT(\ln (1 - v_2) + \chi v_2^2 + v_2(1 - 1/x)) \tag{4}$$

with

$$x = V_2/V_1 = (M_2/M_1)(\varrho_1/\varrho_2) \quad (5)$$

where M_1 and M_2 are the number-average molecular weights of solvent and polymer, and ϱ_1 and ϱ_2 their densities. The term $1/x$ is small and equal to zero for a network (which is a molecule of infinite molecular weight).

The difference between the chemical potential of solvent in cross-linked polymer, μ_1^c, and of pure solvent, μ_1^0, is related to the vapor pressure p_1^c of solvent in the swollen network, and over pure solvent, p_1^0:

$$\mu_1^c - \mu_1^0 = RT \ln (p_1^c/p_1^0) \quad (6)$$

Combination of Eq. (3) and (6) at constant volume fraction of solvent leads to

$$\mu_1^c - \mu_1^u = RT \ln (p_1^c/p_1^u) \quad (7)$$

The difference $\mu_1^c - \mu_1^0$ arises from two contributions: a mixing component $(\mu_1^c - \mu_1^0)_{mix}$ approximately equal to $\mu_1^u - \mu_1^0$ for large x, and an elastic component $(\mu_1^c - \mu_1^0)_{el}$. It is generally assumed that these two contributions are separable. Introducing μ_1^0 on the left side of Eq. (7) leads to

$$(\mu_1^c - \mu_1^0)_{el} = RT \ln (p_1^c/p_1^u) \quad (8)$$

The quantity $(\mu_1^c - \mu_1^0)_{el}$ results from the elastic deformation (dilation) of the network and is related to the elastic free energy change ΔA_{el} by

$$(\mu_1^c - \mu_1^0)_{el} = \left(\frac{\partial \Delta A_{el}}{\partial n_1}\right)_{T,p} \quad (9)$$

Combining equations 8 and 9 with the expressions for the activities of solvent over the uncross-linked polymer ($a_1^u = p_1^u/p_1^0$), and over the network ($a_1^c = p_1^c/p_1^0$) at constant volume fraction v_2 of polymer gives

$$RT \ln(a_1^c/a_1^u) = \left(\frac{\partial \Delta A_{el}}{\partial n_1}\right)_{T,p} \quad (10)$$

If solvent is present in excess, absorption equilibrium is reached when the chemical potential of solvent in the swollen network is equal to that of pure solvent outside the network, i.e. $\mu_1^c = \mu_1^0$. Hence,

$$\mu_1^c - \mu_1^0 = (\mu_1^c - \mu_1^0)_{mix} + (\mu_1^c - \mu_1^0)_{el} = 0 \quad (11)$$

where $(\mu_1^c - \mu_1^0)_{mix}$ is given by equation 4 with $1/x = 0$. The volume fraction of polymer at equilibrium (maximum) swelling is designated v_{2m}. Then,

$$RT(\ln (1 - v_{2m}) + \chi v_{2m}^2 + v_{2m}) + \left(\frac{\partial \Delta A_{el}}{\partial n_1}\right)_{T,p} = 0 \quad (12)$$

3 Elastic Free Energy of Deformation and Swelling Equations

3.1 Statistical Mechanical Approach

3.1.1 Elastic Free Energy

A molecular model for the deformation is required. The Flory-Rehner Eqs. (4) and (5) were developed for a network deforming affinely, i.e. a network without junction fluctuations. A more general treatment by Flory and Erman [9] which includes such fluctuations is described in this Section.

In the phantom network model, chains may move freely through one another [19], and junctions fluctuate around their mean positions due to Brownian motion and are independent of deformation. The instantaneous distribution of chain vectors r is not affine in the strain because it is the convolution of the distribution of mean vectors $\bar{r}$, which is affine, with the distribution of fluctuations Δr, which is independent of the strain. The elastic free energy of such a network is [20]

$$\Delta A_{ph} = \left(\frac{1}{2}\right) \xi kT(I_1 - 3) \tag{13}$$

where I_1 is the first invariant of the tensor of deformation:

$$I_1 = \lambda_x^2 + \lambda_y^2 + \lambda_z^2 \tag{14}$$

The quantities λ_x, λ_y, λ_z are the principal extension ratios, which specify the strain relative to an isotropic state of Reference Chapt. 5. The cycle rank ξ was first introduced by Flory [20] and is the number of independent circuits in the network or the number of chains which have to be cut to reduce the network to an acyclic structure or tree. Subsidiary quantities called the number of effective chains and junctions noted, respectively, ν_e and μ_e can be defined by the relationship

$$\xi = \nu_e - \mu_e + 1 \simeq \nu_e - \mu_e \tag{15}$$

In reality, diffusion of junctions about their mean positions may be severely restricted by neighboring chains sharing the same region of space. The extreme case is the affine network, where fluctuations are completely suppressed. The instantaneous distribution of chain vectors is then affine in the strain, and the elastic free energy of deformation is given by [20]

$$\Delta A_{aff} = \left(\frac{1}{2}\right) \nu_e kT(I_1 - 3) - (\nu_e - \xi)\, kT \ln (V/V_0) \tag{16}$$

where V is the volume of the deformed sample.

Stress-strain measurements in uniaxial extension have revealed that real networks have a behavior closest to the affine limit at small deformations and approach the phantom limit at large deformations. The recent molecular theory developed by Flory and Erman accounts for this transition. In this model, the restrictions on junction fluctuations due to neighboring chains are represented by domains of constraints. At small deformations, the stress is enhanced by restrictions on the junction fluctuations. At large strains, the effects of these restrictions vanish and the relationship of stress to strain converges to that for a phantom network. In a later theoretical refinement, the nature of real networks is characterized by two parameters, the most important one being $\varkappa$, which measures the severity of entanglement constraints relative to those imposed by a phantom network. Another parameter ζ takes into account the non-affine transformation of the domains of constraints with strain. The elastic free energy change is written as the sum of the elastic free energy ΔA_{ph} for a phantom network, and a term ΔA_c which accounts for entanglement constraints:

$$\Delta A_{el} = \Delta A_{ph} + \Delta A_c \tag{17}$$

The second term can be expressed [9)]

$$\Delta A_c = \left(\frac{1}{2}\right) \mu kT \sum_{t=x,y,z} \{(1 + g_t) B_t - \ln((B_t + 1)(g_t B_t + 1))\} \tag{18}$$

with

$$B_t = (\lambda_t - 1)(1 + \lambda_t - \zeta\lambda_t^2)(1 + g_t)^{-2} \tag{18a}$$

$$g_t = \lambda_t^2(\varkappa^{-1} + \zeta(\lambda_t - 1)) \tag{18b}$$

The Flory-Erman theory has given a good account of experimental results in elongation and compression [21–26)].

3.1.2 Structural Characteristics of a Network

As already mentioned, a network is characterized by its cycle rank density ξ/V_0. Scanlan and Case [27)] have defined an active junction as one joined by at least three paths to the gel network and an active chain as one terminated by active junctions at both its ends. Pearson and Graessley [28a)] have shown that for a randomly interconnected network whose junctions are of even functionality

$$\xi = \nu_a - \mu_a + 1 \simeq \nu_a - \mu_a \tag{19}$$

where ν_a and μ_a are respectively the numbers of active chains and junctions. This relationship was generalized to networks of any kind by Flory [28b)].

3.1.2.1 Model Networks

Model networks are networks obtained in a controlled manner, for example by cross-linking end-reactive chains (of number-average molecular weight M_n) with multifunctional agents [29)]. The molecular weight between cross-linking points is then $M_c = M_n$ and the functionality of the network is the functionality of the cross-linking agent. The following relationships then hold for a perfect end-linked network (no chain ends or loops):

$$\nu_e/V_0 = \nu_a/V_0 = \varrho/M_c \tag{20a}$$

where ϱ is the polymer density,

$$\mu_e/\nu_e = \mu_a/\nu_a = 2/\varphi \tag{20b}$$

and

$$\xi/V_0 = (\nu_a - \mu_a)/V_0 = (1 - 2/\varphi)\,\nu_a/V_0 = (1 - 2/\varphi)\,\varrho/M_c \tag{21}$$

It is also possible to obtain expressions for a model network with a known number of dangling ends of known length. As a first step, a perfect network is visualized as being formed with junctions of functionality φ and a weight fraction Φ_1 of chains of number-average molecular weight M_{n1} and a weight fraction Φ_2 of chains of number-average molecular weight $M_{n2}(\Phi_1 + \Phi_2 = 1)$. Since the network is perfect, Eq. (20b) holds. The number density of effective chains ν_1/V_0 is given by

$$\nu_1/V_0 = \varrho\Phi_1/M_{n1} + \varrho\Phi_2/M_{n2} \tag{22}$$

so that

$$\mu_1/V_0 = (2/\varphi)\,(\varrho\Phi_1/M_{n1} + \varrho\Phi_2/M_{n2}) \tag{23}$$

In the second step, chains of weight M_{n2} are cut at their midpoints. Experimentally, the resulting network can be formed by mixing φ-functional agents, di end-reactive chains of weight M_{n1} and mono end-reactive chains of molecular weight $M_0 = M_{n2}/2$. Cutting one chain is equivalent to decreasing the cycle rank by one. Therefore at the second step

$$\xi_2/V_0 = \xi_1/V_0 - \varrho\Phi_2/M_{n2} = \tag{24}$$

$$(1 - 2/\varphi)\,\varrho\Phi_1/M_{n1} - (2/\varphi)\,\varrho\varphi_2/M_{n2} \tag{25}$$

The gel threshold corresponds to $\xi_2 = 0$. (To calculate Φ_2 such that the network is reduced to an acyclic structure, it is necessary to use the exact formula $\xi = \nu_e - \mu_e + 1$ and $\Phi_1 = 1 - \Phi_2$. The number density of effective chains of networks of functionality φ is

$$\nu_2/V_0 = \varrho\Phi_1/M_{n1} - 2\varrho\Phi_2/(M_{n2}(\varphi - 2)) \tag{26}$$

since the number of effective chains is $\nu_e = \xi(1 - 2/\varphi)^{-1}$. If two chains of weight M_{n2} are connected at the same junction (supposed to be tetrafunctional), cutting these chains leads to a remaining difunctional junction which is not active. An exact treatment would involve a probability study of the different possible topologies of the network junctions. Therefore ν_2 must not be used as the number of active chains. The knowledge of M_{n1}, M_0, Φ_1 and φ enables one to determine ξ_2, which can then be used to calculate the phantom modulus. This phantom modulus is predicted by Flory and Erman to be the modulus exhibited by real networks at infinite deformation. Queslel and Mark have calculated the phantom modulus of perfect poly(dimethylsiloxane) networks and by comparison with stress-strain data deduced the parameters $\varkappa$ and ζ [21]. This could also be done for networks with dangling ends.

Another way to obtain networks with irregularities is to mix chains end and cross-linking agent in a non-stoichiometric ratio of reactive groups [30]. The cycle rank and the topology of the resulting networks can be predicted by Monte-Carlo simulation [31].

3.1.2.2 Randomly Cross-linked Networks

The topology of a regular randomly cross-linked network having no irregularities other than dangling chains can be studied. In the treatment presented below, junctions to which are connected two or three dangling chains are ignored. However the number of these junctions can be considered to be relatively small and our relationships will be valid in a first approximation. Another consequence of this assumption is that all the junctions are active.

The network is formed by cross-linking primary chains of number-average molecular weight M_n. The functionality of cross-links at this stage is four. If the chain ends have a molecular weight M_0, there are $(M_n - 2M_0)/M_c$ active segments per chain, where M_c is the average molecular weight between cross-links. Since there are ϱ/M_n primary chains per unit volume, the total number density of active segments is

$$\nu_a/V_0 = (\varrho/M_c)(1 - 2M_0/M_n) \tag{27}$$

The Flory expression for the chain-end correction [6] is recovered for $M_0 = M_c$.

If the tetrafunctional cross-links to which are connected four chains leading to the network are suppressed, the network (first-stage) is formed only with the cross-links having three chains connected to the mesh. Each chain end is connected to this type of cross-link, whose number density is therefore

$$\mu(\varphi = 3)/V_0 = 2\varrho/M_n \tag{28}$$

each primary chain having two chain ends. To form this first-stage network, it is necessary to distribute the $2\varrho/M_n$ junctions over ϱ/M_n chains, each one already having two tetrafunctional junctions. This results in each primary chain having four junctions and therefore three active segments. This is illustrated in Fig. 1. Hence, the number density of active segments in the first-stage network is

$$\nu_1/V_0 = 3\varrho/M_n \tag{29}$$

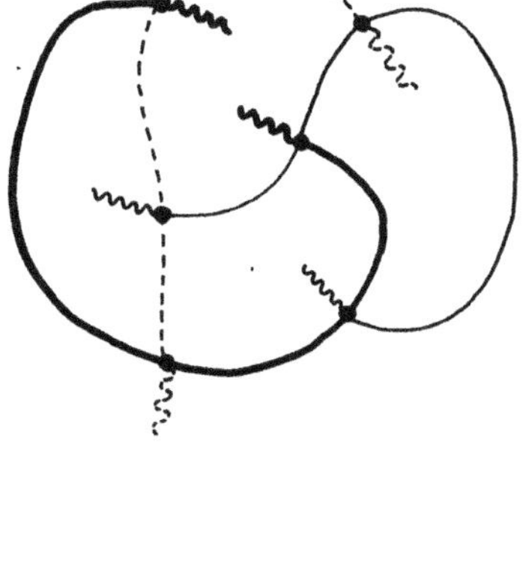

Fig. 1. First-stage network formed with three chains; $\mu(\varphi = 3) = 6$ and $\nu_a = 9$

A number $\mu(\varphi = 4)$ of tetrafunctional junctions are then added per unit volume, each cross-link creating two active segments. The resulting number density of active chains is then

$$v_a/V_0 = 3\varrho/M_n + 2\mu(\varphi = 4) \tag{30}$$

The total number density of junctions is the sum

$$\mu_a/V_0 = (\mu(\varphi = 3) + \mu(\varphi = 4))/V_0 \tag{31}$$

This leads to

$$v_a/V_0 = 2\mu_a/V_0 - \varrho/M_n \tag{32}$$

where μ_a can be deduced by making use of Eq. (27):

$$\mu_a/V_0 = \varrho/2M_c + \varrho/2M_n - \varrho M_0/M_cM_n \tag{33}$$

Hence

$$\xi/V_0 = \varrho/2M_c - \varrho/2M_n - \varrho M_0/M_cM_n \tag{34}$$

In a perfect tetrafunctional network, $\mu_a/V_0 = 2\varrho/M_c$ and thus equation 33 leads to $M_0 = M_c/2$, which means that chain ends have been connected two by two to obtain this perfect network.

Equations (33) and (34) can be used to interpret stress-strain measurements with the Flory-Erman theory [31]. The general Flory-Erman relationship for uniaxial extension of a swollen network is

$$f_S v_2^{1/3}/A_d(\alpha - \alpha^{-2}) = (\xi/V_0)\,RT\{1 + (\mu/\xi)\,(\alpha K(\alpha^2 v_2^{-2/3}) - \alpha^{-2}K(\alpha^{-1}v_2^{-2/3}))\,(\alpha - \alpha^{-2})^{-1}\} \tag{35}$$

where f_S is the measured force, α the extension ratio along the stretching direction (relative to the undeformed swollen state), A_d the cross-sectional area of the undeformed dry sample, and v_2 the volume fraction of polymer. The quantity μ is the total number of junctions that are subject to entanglement constraints. The function K is defined by

$$K(\lambda_t^2) = B_t(\dot{B}_t(B_t + 1)^{-1} + g_t(\dot{B}_t g_t + \dot{g}_t B_t)\,(g_t B_t + 1)^{-1}) \tag{36}$$

with

$$\dot{B}_t = B_t\{(2\lambda_t(\lambda_t - 1))^{-1} + (1 - 2\zeta\lambda_t)\,(2\lambda_t(1 + \lambda_t - \zeta\lambda_t^2))^{-1} - 2\dot{g}_t(1 + g_t)^{-1}\} \tag{36a}$$

$$\dot{g}_t = \varkappa^{-1} - \zeta(1 - 3\lambda_t/2) \tag{36b}$$

The quantities B_t and g_t are given by Eq. (18a) and (18b). Fitting stress-strain data at different degrees of swelling by combination of Eq. (33), (34) and (35) enables one to determine $\varkappa$, ζ, and M_c (presumed to be equal to M_0). Flory and Erman [32] have performed the fitting by assuming $\mu/\xi = 1$ and have determined $\varkappa$, ζ and the phantom modulus for randomly cross-linked poly(dimethylsiloxane) (PDMS) networks. These results were recently further exploited by Queslel and Mark.

3.1.3 Swelling Equations

An expression for $\ln(a_1^c/a_1^u)$ is obtained through Eq. (10) by differentiating ΔA_{el} with respect to n_1.

3.1.3.1 Isotropic Swelling

The resulting isotropic extension ratios are

$$\lambda = \lambda_x = \lambda_y = \lambda_z = ((n_1 V_1 + V_0)/V_0)^{1/3} = v_2^{-1/3} \tag{37}$$

The molar volume V_1 of typical solvents is 89.39, 108.74, 116.11 cm^3/mol for benzene, cyclohexane, and *n*-pentane, respectively. Similarly, V_0 is the volume of the dry network. The differentiation is thus

$$\left(\frac{\partial \Delta A_{el}}{\partial n_1}\right)_{T,p} = \left(\frac{\partial \Delta A_{el}}{\partial \lambda^2}\right)_{T,p} \left(\frac{\partial \lambda^2}{\partial n_1}\right)_{T,p} \tag{38}$$

with

$$\left(\frac{\partial \lambda^2}{\partial n_1}\right)_{T,p} = 2V_1/3\lambda V_0 \tag{39}$$

The derivative of ΔA_{el} as a function of λ^2 is the sum of the corresponding derivatives of ΔA_{ph} and ΔA_c:

$$\left(\frac{\partial \Delta A_{ph}}{\partial \lambda^2}\right)_{T,p} = (3/2)\,\xi kT \tag{40}$$

$$\left(\frac{\partial \Delta A_c}{\partial \lambda^2}\right)_{T,p} = 3\left(\frac{\partial \Delta A_c}{\partial \lambda_t^2}\right)_{T,p,\lambda_{t'} \neq t} = (3/2)\,\mu kTK(\lambda^2) \tag{41}$$

The final result is

$$\lambda RT \ln(a_1^c/a_1^u) = (\xi/V_0)\,kTV_1\{1 + (\mu/\xi)\,K(\lambda^2)\} \tag{42}$$

At swelling equilibrium, Eq. (12) gives the relationship

$$\ln(1 - v_{2m}) + \chi v_{2m}^2 + v_{2m} = -(\xi/V_0)\,V_1 v_{2m}^{1/3}\{1 + (\mu/\xi)\,K(v_{2m}^{-2/3})\} \tag{43}$$

where ξ is expressed in mols. Swelling equations for phantom networks are recovered through $\varkappa = 0$ and $K(\lambda^2) = 0$:

$$\lambda \ln (a_1^c/a_1^u) = (\xi/V_0)\, V_1 \tag{44}$$

$$\ln (1 - v_{2m}) + \chi v_{2m}^2 + v_{2m} = -(\xi/V_0)\, V_1 v_{2m}^{1/3} \tag{45}$$

For a perfect network, ξ is given by Eq. (21). For a randomly cross-linked network with $M_0 = M_c$, M_c is given by

$$M_c = (3/M_n - 2(V_1 v_{2m}^{1/3} \varrho)^{-1} (\ln (1 - v_{2m}) + \chi v_{2m}^2 + v_{2m}))^{-1} \tag{46}$$

For affine networks, $\varkappa = \infty$ and $K(\lambda^2) = 1 - \lambda^{-2}$,

$$\lambda \ln (a_1^c/a_1^u) = (\xi/V_0)\, V_1 \{1 + (\mu/\xi)(1 - \lambda^{-2})\} \tag{47}$$

$$\ln (1 - v_{2m}) + \chi v_{2m}^2 + v_{2m} = -(\xi/V_0)\, V_1 v_{2m}^{1/3} (1 + (\mu/\xi)(1 - v_{2m}^{2/3})) \tag{48}$$

For a perfect network, ξ and μ are deduced from Eq. (20a), (20b) and (21). Again for a randomly cross-linked affine network with $M_0 = M_c$,

$$M_c = (3/M_n - 2(V_1 \varrho (2 v_{2m}^{1/3} - v_{2m}))^{-1} (\ln (1 - v_{2m}) + \chi v_{2m}^2 + v_{2m}))^{-1} \tag{49}$$

The simplest swelling measurement is the equilibrium determination of v_{2m}. Without knowledge of the network structure (i.e. $\varkappa$ and ζ), it enables one to determine the range where M_c lies (between M_{cph} obtained by making use of Eq. (45) and M_{caff} by Eq. (48)). The calculated M_{cph} is lower than M_{caff}. In a phantom network, junction fluctuations decrease the impact of chain entropy changes. It is therefore necessary to have a phantom network with a higher density of crosslinks or a smaller M_c to counteract this effect and to give the same elastic contribution as in an affine network.

3.1.3.2 Swelling of Deformed Networks

For stress-strain measurements on swollen networks, it is advantageous to immerse the samples completely in excess solvent. Thus errors due to solvent vaporization during the stretching of a swollen network exposed to the air are thereby suppressed. However, the swelling capacity of an elastomer is not constant with deformation [33, 34, 35]; fortunately, a swelling equilibrium equation can also be obtained for this case making use of the Flory-Erman theory [32].

An unswollen sample is deformed uniaxially. The principal extension ratio along the stretching direction is denoted $\lambda_{\|}$. The sample is then immersed in solvent, and swelling occurs in directions normal to the stretching axis; this is accompanied by a lateral change in dimensions. The quantities $\lambda_{\perp}$ are the two principal extension ratios in the lateral directions with respect to the unswollen sample of volume V_0. The product of the three principal extension ratios is the inverse of the volume fraction of polymer, v_2', in the swollen network

$$\lambda_{\|} \lambda_{\perp}^2 = (n_1 V_1 + V_0)/V_0 = v_2'^{-1} \tag{50}$$

The elastic free energy ΔA_{el} can now be differentiated with respect to n_1:

$$\left(\frac{\partial \Delta A_{el}}{\partial n_1}\right)_{T,p,\lambda_{\|}} = \left(\frac{\partial \Delta A_{el}}{\partial \lambda_{\perp}^2}\right)_{T,p,\lambda_{\|}} \left(\frac{\partial \lambda_{\perp}^2}{\partial n_1}\right)_{T,p,\lambda_{\|}} \tag{51}$$

where $\lambda_{\perp}$ is introduced since it is the extension ratio which varies with swelling at constant $\lambda_{\|}$.

By making use of Eq. (50),

$$\left(\frac{\partial \lambda_{\perp}^2}{\partial n_1}\right)_{T,p,\lambda_{\|}} = V_1/V_0\lambda_{\|} = \lambda_{\perp}^2 v_2' V_1/V_0 \tag{52}$$

The phantom free energy change may be expressed by

$$\Delta A_{ph} = (1/2)\,\xi k T(\lambda_{\|}^2 + 2\lambda_{\perp}^2 - 3) \tag{53}$$

Hence,

$$\left(\frac{\partial \Delta A_{ph}}{\partial \lambda^2}\right)_{T,p,\lambda_{\|}} = \xi k T \tag{54}$$

$$\left(\frac{\partial \Delta A_c}{\partial \lambda_{\perp}^2}\right)_{T,p,\lambda_{\|}} = 2\left(\frac{\partial \Delta A_c}{\partial \lambda_t^2}\right)_{T,p,\lambda_{\|}} = \mu k T K(\lambda_{\perp}^2) \tag{55}$$

The swelling equilibrium relationship (Eq. (12)) becomes

$$\ln(1 - v_2') + \chi v_2'^2 + v_2' = -(\xi/V_0)\,V_1\lambda_{\perp}^2 v_2'(1 + (\mu/\xi)\,K(\lambda_{\perp}^2)) \tag{56}$$

Thus, for a phantom network

$$\ln(1 - v_2') + \chi v_2'^2 + v_2' = -(\xi/V_0)\,V_1\lambda_{\|}^{-1} \tag{57}$$

while for an affine network

$$\ln(1 - v_2') + \chi v_2'^2 + v_2' = -(\xi/V_0)\,V_1\lambda_{\|}^{-1}(1 + (\mu/\xi)\,(1 - \lambda_{\|}v_2')) \tag{58}$$

It is possible to predict whether an already-stretched network swells more or less than an undeformed network at equilibrium. The threshold is obtained by comparing Eqs. (45) and (57), (48) and (58), with $v_2' = v_{2m}$. The solution is $\lambda_{\|} = v_{2m}^{-1/3}$. The physical explanation was given by Treloar [35]. If the dry sample is uniaxially deformed with an extension ratio $\lambda_{\|}$ lower than the isotropic deformation $v_{2m}^{-1/3}$, then, when swelling occurs, the clamps act as restrictions to swelling deformation and the net result is a compression. The trend in swelling when $\lambda_{\|} \neq v_{2m}^{-1/3}$ can be deduced from a numerical application of Eqs. (45), (48), (57) and (58).

Poly(dimethylsiloxane) has a density $\varrho = 0.97\ g/cm^3$ and an interaction parameter with benzene of $\chi = 0.484 + 0.330 v_2$ in the range of interest [17]. The molar volume of benzene is $V_1 = 89.39\ cm^3/mol$. The network is presumed to be perfect and tetrafunctional: $\varphi = 4$, $\mu/\xi = 1$, $M_c = 5000$ g/mol. Hence, $\xi/V_0 = (1 - 2/\varphi)\, \varrho/M_c = 9.7 \cdot 10^{-5}$ mol/cm³. Equations (45), (48), (57), and (58) are transformed numerically to Eqs. (59), (60), (61), and (62), respectively:

$$-\{\ln(1 - v_{2m}) + \chi v_{2m}^2 + v_{2m}\} = 8.7 \times 10^{-3} v_{2m}^{1/3} \tag{59}$$

$$-\{\ln(1 - v_{2m}) + \chi v_{2m}^2 + v_{2m}\} = 8.7 \times 10^{-3}(2v_{2m}^{1/3} - v_{2m}) \tag{60}$$

$$-\{\ln(1 - v_2') + \chi v_2'^2 + v_2'\} = 8.7 \times 10^{-3} \lambda_{\|}^{-1} \tag{61}$$

$$-\{\ln(1 - v_2') + \chi v_2'^2 + v_2'\} = 8.7 \times 10^{-3}(2\lambda_{\|}^{-1} - v_2') \tag{62}$$

Graphical solutions can be obtained for Eq. (59) (right-hand side function of v_{2m} represented by curve P_i in Fig. 2 and Eq. (60) (right-hand side represented by curve A_i). For networks considered to be phantom and affine, respectively, this leads to values of v_{2m} of 0.332 and 0.372. The swelling threshold value is calculated to be $\lambda_{\|} = 1.444$ and 1.390, respectively. Curves P_d and A_d in Fig. 2 represent the right-hand sides of Eqs. (61) and (62) respectively for $\lambda_{\|} = 2$, and thus are higher than the threshold value. Graphical solution gives $v_2' = 0.302$ and 0.330. Thus a deformed network swells more than an isotropic network if $\lambda_{\|} > v_{2m}^{-1/3}$ and less in the case of the reverse inequality.

If a samples is stretched when immersed completely in excess solvent, a new swelling equilibrium occurs at each extent of deformation. Generally, the force f_s

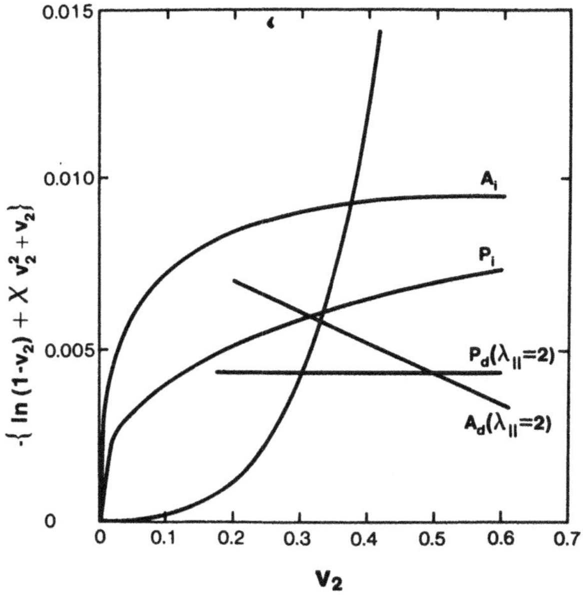

Fig. 2. Graphical solution of swelling equilibrium for an isotropic and a deformed perfect PDMS network in benzene ($M_c = 5000$ g/mol, $\varphi = 4$). Curves P_i, A_i, P_d and A_d represent the right-hand sides of Eq. (59), (60), (61) and (62), respectively

is monitored as a function of α, the ratio of the length L_V of the deformed sample of volume V, to the initial length L_V^i at the same volume:

$$\alpha = L_V/L_V^i \tag{63}$$

It is customary to plot the reduced force

$$[f^*] = f_S v_2'^{1/3}/A_d(\alpha - \alpha^{-2}) \tag{64}$$

against α. To calculate $[f^*]$ and α, it is necessary to know v_2' and L_V^i at each extent of deformation. The extension ratios $\lambda_{||}$ and $\lambda_{\perp}$ are defined with respect to the unswollen sample. For example,

$$\lambda_{||} = L_V/L_{V_0}^i \tag{65}$$

where $L_{V_0}^i$ is the length of the dry isotropic sample. By drawing on the sample two ink marks parallel to the stretching axis and two along the lateral direction, it is possible to measure $\lambda_{||}$ and $\lambda_{\perp}$ with a cathetometer, and then to calculate v_2' from

$$v_2' = (\lambda_{||}\lambda_{\perp}^2)^{-1}$$

and

$$L_V^i = v_2'^{-1/3} L_{V_0}^i$$

3.2 Phenomenological Approach

3.2.1 General Comments

Continuum mechanics is also used to account for the observed stress-strain behavior exhibited by elastomers [36, 37]. The most general form of the strain energy function (which vanishes at zero strain) is the power series

$$W/V_0 = \sum_{i,j,k=0}^{\infty} C_{ijk}(I_1 - 3)^i (I_2 - 3)^j (I_3 - 1)^k \tag{66}$$

where the C_{ijk} coefficients have units of N/m^2. The quantities I_1, I_2, and I_3 are the three strain invariants, I_1 being given by Eq. (14), and I_2, I_3 by

$$I_2 = \lambda_x^2\lambda_y^2 + \lambda_y^2\lambda_z^2 + \lambda_z^2\lambda_x^2 \tag{67a}$$

$$I_3 = \lambda_x^2\lambda_y^2\lambda_z^2 \tag{67b}$$

It was found that a simple form of Eq. (66) can match experimental results [17, 38–40], specifically,

$$W/V_0 = C_1(I_1 - 3) + C_2(I_2 I_3^{-1+m/2} - 3) - C_3 \ln I_3 \tag{68}$$

where m is a parameter which can be adjusted to provide the best fit of the data. The last term $C_3 \ln I_3$ is introduced to yield a form similar to Eq. (16), i.e., I_3 is $(V/V_0)^2$. For a tetrafunctional network, C_3/C_1 is then 1/2 by analogy with Eq. (16), making use of Eq. (21).

3.2.2 Stress-Strain Data on Swollen Networks

During uniaxial stress-strain measurements on a swollen network, the reference state is the undeformed swollen sample of volume V. The force f_S exhibited by this network is recorded as a function of the extension ratio $\alpha = \alpha_x$ along the stretching direction, relative to the reference state of volume V. It can be obtained by differentiation of the elastic energy of the swollen network W_s (per unit volume of swollen network) with respect to α:

$$W_s = W/V = v_2 W/V_0 \tag{69}$$

where v_2 is the volume fraction of polymer in the swollen state, i.e. V_0/V.

By combination of Eq. (68) and (69) with $I_3 = \lambda_x^2\lambda_y^2\lambda_z^2 = (V/V_0)^2 = v_2^{-2}$, one obtains

$$W_s = v_2 C_1(I_1 - 3) + v_2 C_2(v_2^{2-m} I_2 - 3) + 2C_3 \ln v_2 \tag{70}$$

It is now necessary to express W_s as a function of α. The quantities L_V, L_V^i, and L_{V0}^i are the lengths of the deformed swollen sample, isotropic swollen sample and isotropic unswollen sample, respectively. The following relations hold in uniaxial extension.

$$\alpha_x\alpha_y\alpha_z = 1$$

$$\lambda_x\lambda_y\lambda_z = v_2^{-1}$$

$$\alpha = \alpha_x = L_V/L_V^i$$

$$\lambda = \lambda_x = L_V/L_{V_0}^i$$

$$L_V^i = L_{V_0}^i v_2^{-1/3} \tag{71}$$

Hence,

$$\lambda = \lambda_x = \alpha v_2^{-1/3} \tag{72}$$

$$\lambda_y = \lambda_z = \alpha^{-1/2} v_2^{-1/3} \tag{73}$$

By making use of Eq. (72) and (73), Eq. (70) is transformed to

$$W_s = W/V = C_1(\alpha^2 v_2^{1/3} + 2\alpha^{-1} v_2^{1/3} - 3) + C_2(2\alpha v_2^{5/3-m} + \alpha^{-2} v_2^{5/3-m} - 3) + 2C_3 \ln v_2 \tag{74}$$

The elementary work of deformation of the swollen network is

$$dW = f_s dL_V \tag{75}$$

Introducing the cross-sectional area of the undeformed unswollen sample $A_d = A_s v_2^{2/3}$, where A_s is the cross-sectional area of the undeformed swollen sample equal to V/L_V^i, Eq. (75) becomes

$$(1/V)\frac{dW}{d\alpha} = v_2^{2/3} f_s/A_d \tag{76}$$

Differentiating W (given by Eq. (74)) with respect to α and rearranging gives for the reduced force

$$[f^*] = f_s v_2^{1/3} A_d^{-1}(\alpha - \alpha^{-2})^{-1} = 2C_1 + 2C_2 v_2^{4/3-m}\alpha^{-1} \tag{77}$$

Mullins, studying natural rubber swollen with *n*-decane, found m = 0 [39]. However, Flory and Tatara have reported m = 1/2 for poly(dimethylsiloxane) in benzene [17] and m between 0 and 1/2 for the results on natural rubber in *n*-decane obtained by Allen and co-workers [41]. In summary, the decrease of the slope of [f*] versus α^{-1} with swelling shows a dependence on either $v_2^{4/3}$ (m = 0) or $v_2^{5/6}$ (m = 1/2).

An important point concerns the influence of the sol fraction on stress-strain measurements. If at stress equilibrium the sol fraction is considered to be only an inert diluent, then the slope of [f*] versus α^{-1} recorded for unextracted networks is predicted to be lower by $v_2^{4/3}$ or $v_2^{5/6}$ than that of extracted networks, v_2 being the ratio of the volume of extracted sample to that of the unextracted one. For example, an unextracted network with a 10 percent sol fraction ($v_2 = 0.90$) will exhibit a slope equal to 0.87 times that of the extracted network (if m is assumed to be zero). This complication may be avoided of course by conducting stress-strain experiments only on already-extracted samples.

3.2.3 Isotropic Swelling Equations

By combination of Eqs. (14), (37), (67) and (68) with $C_3/C_1 = 1/2$, the stored elastic energy can be expressed as

$$W = 2C_1 V_0(\lambda^2 - 1 - (1/2)\ln\lambda^2) + 3C_2 V_0((\lambda^2)^{3m/2-1} - 1) \tag{78}$$

Equations (10), (38), (39) and (78) then lead to the isotropic swelling equation

$$\lambda \ln(a_1^c/a_1^u) = 2(C_1 V_1/RT)\{1 - 2/\lambda^2 - (C_2/2C_1)(2 - 3m)\lambda^{3m-4}\} \tag{79}$$

Use of the Mooney-Rivlin free energy expression, i.e. Eq. (78) with m = 0 and without the logarithmic term, leads to the expression

$$\lambda \ln(a_1^c/a_1^u) = 2(C_1 V_1/RT)(1 - C_2/C_1\lambda^4) \tag{80}$$

4 Apparatus and Procedures

4.1 Swelling Equilibrium

This simple experiment consists of immersing completely a sample in solvent and waiting until swelling equilibrium occurs. Then v_{2m} is determined after weighing the sample. An accurate value is obtained only with already-extracted samples.

4.2 Determination of Interaction Parameter χ and Solvent Activities

The parameter χ can be obtained by several techniques including osmometry, vapor sorption, gas-liquid chromatography, freezing point depression of solvent, swelling equilibrium, intrinsic viscosity, and critical solution temperatures [15, 16].

Differential solvent vapor sorption procedures to measure χ and the activities of solvent in uncross-linked and cross-linked polymers are described in this Section [42, 43]. A schematic diagram of the apparatus designed by Yen and Eichinger [43] is reproduced in Fig. 3.

A Cahn RG electrobalance is housed in a stainless steel case B. Polymer samples on quartz sample pans are suspended on both sides of the balance with 32-gauge nichrome wires. A strip-chart recorder is used to display portions of the electrobalance read-out. The uncross-linked and cross-linked samples of approximately equal wights are placed on the two sides of the electrobalance. A glass weighing chamber K communicates with the balance case, and a quartz spring is attached to a removable chamber cap D. An uncross-linked polymer sample is suspended on

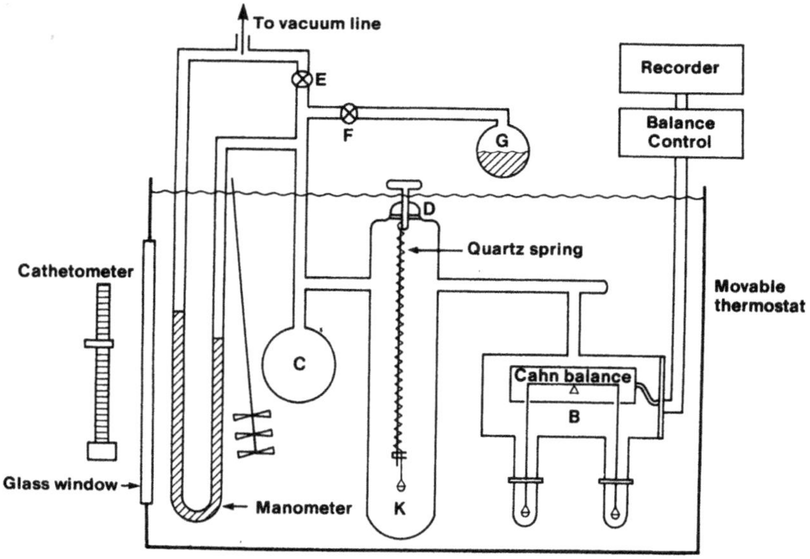

Fig. 3. Schematic diagram of differential solvent vapor sorption measurement apparatus. (Reprinted from Yen, L. Y., Eichinger, B. E.: J. Polym. Sci., Polym. Phys. Ed. *16*, 121, 1978)

the quartz spring, and all dry sample weights are accurately determined. Reservoir C, which serves to cushion sudden pressure changes and acts as a mercury trap, is connected to the chamber K and to one arm of the mercury manometer connected to a vacuum line. A solvent reservoir G can be opened to the system via a mercury float valve F. A similar valve E connects all parts of the apparatus to the vacuum line. A water bath is used as a thermostat for the apparatus. Measurements of the displacement of the quartz spring and manometer are made with a cathetometer, through a glass window.

Valve F is opened to allow solvent vapor into the system. The amount of solvent introduced is controlled by controlling the temperature of the solvent reservoir. Valve F is closed after each addition of vapor. Solvent vapor and polymer samples are then allowed to equilibrate. The integral and differential sorptions are observed on the respective balances and the pressure of solvent vapor is recorded. It is then possible to calculate the volume fraction of polymer in the cross-linked and uncross-linked swollen samples, v_2^c and v_2^u respectively, at a solvent pressure $p_1^c = p_1^u$ and thus to know p_1^c and p_1^u at equal volume fractions $v_2^c = v_2^u$ and to calculate the ratio of activities $a_1^c/a_1^u = p_1^c/p_1^u$. From knowledge of p_1^0, p_1^u, v_2^u and x, $\chi(v_2^u)$ is deduced through equations 3, 4 and 5.

5 Typical Experimental Data

A plot of $\lambda \ln (a_1^c/a_1^u)$ versus $\lambda^2(\lambda = v_2^{-1/3})$ for the system poly(dimethylsiloxane) + benzene is given in Fig. 4 [43]. A similar curve was obtained by Gee and co-workers [42]. Data are compared with theoretical predictions given by Eq. (44), (47), (79) and (80)

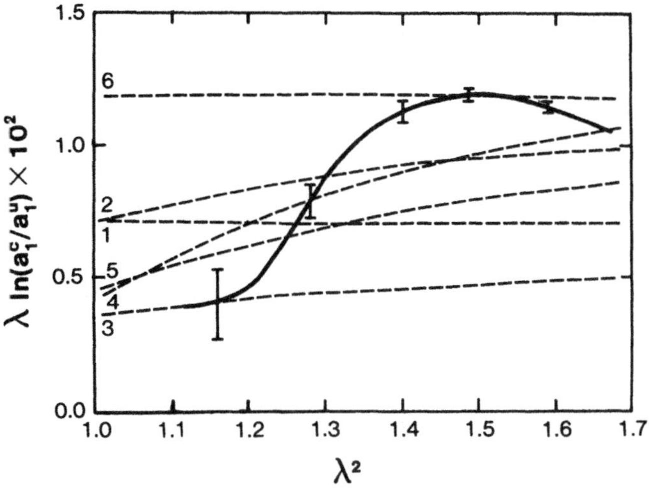

Fig. 4. Plot of $\lambda \ln (a_1^c/a_1^u)$ versus λ^2 as a smooth function for PDMS + benzene at 25 °C (solid curve). Dashed lines are calculated from the reduced elastic component of chemical potential $(\mu_1^c - \mu_1^u)_{el}/RT$ given by various elastic equations. Curve 1: eq. 44, $(\xi/V_0)V_1 = 0.007$; curves 2 and 3: eq. 47, $\mu/\xi = 1$, $(\xi/V_0)V_1 = 0.007, 0.0035$; curve 4: eq. 80, $C_1V_1/RT = 0.007$, $C_2V_1/RT = 0.005$; curves 5 and 6: eq. 79, $m = 1/2, 1$, $C_1V_1/RT = 0.007, 0.0044$, $C_2V_1/RT = 0.005, 0.0072$ (Reprinted from Yen, L. Y. and Eichinger, B. E.: J. Polym. Sci., Polym. Phys. Ed., *16*, 121, 1978)

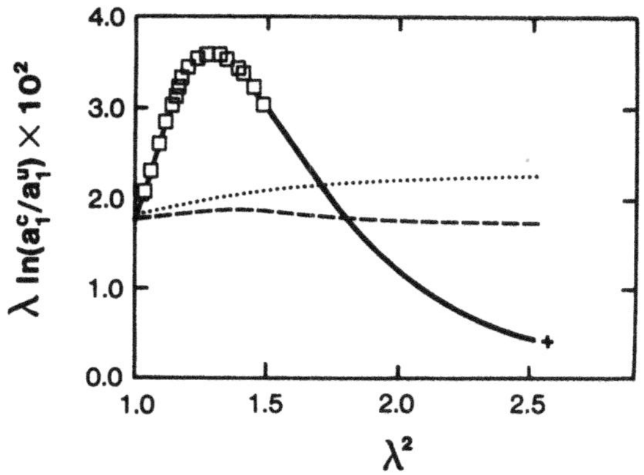

Fig. 5. Plot of $\lambda \ln(a_1^c/a_1^u)$ versus λ^2 as a smooth function for PDMS + benzene at 30 °C (solid curve). The + represents the swelling equilibrium point. Dashed and dotted curves represent Eq. (42) ($\mu/\xi = 1$, $(\xi/V_0)V_1 = 0.017$) with respective parameters $\varkappa = 10$, $\zeta = 0.4$, and $\varkappa = 10$, $\zeta = 0$. (Reprinted from Brotzman, R. W., Eichinger, B. E.: Macromolecules *15*, 531, 1982)

with adjusted parameters. None can account for the maximum exhibited by $\lambda \ln(a_1^c/a_1^u)$. An improvement is obtained if the Flory-Erman elastic contribution is used for comparison (Eq. (43)), since a maximum is predicted by this theory [10]. However, agreement with experiment is only qualitative, as is shown in Fig. 5 [11]. Although it is possible to find parameters $\varkappa$ and ζ such that the theoretical curve exhibits a maximum for the same value of λ^2 as the experimental curve, this maximum is not sharp enough to reproduce the data. Gottlieb and Gaylord [13] have showed that the Flory-Erman, Graessley [44] and Marrucci-Gaylord [45] models may give a better fit of the data, but only with unrealistic values of the parameters.

6 Conclusions

Although swelling is a very useful method for elucidating the molecular structure of networks and theory in this field has now had some qualitative success, it is difficult to use for quantitative purposes. The reported qualitative disagreement between experiments and theoretical predictions has lead some workers to question the separability of elastic and mixing contributions (Eq. (11)), since the Flory-Erman theory of elasticity accounts well for stress-strain data [21], and the Flory-Huggins lattice treatment seems reasonable and reproduces well the features of the mixing process for uncross-linked polymer and solvent [3]. A more nearly complete theory of swelling may also have to include the transition of chain size $(r^2)^{1/2}$ which scales like $N^{1/2}$ (N is the number of subunits or statistical segments or steps in a lattice) in a dense system of chains (melt) and like $N^{3/5}$ for dilute chains swollen in a good solvent [46].

Acknowledgements: It is a pleasure to acknowledge the postdoctoral fellowship generously provided J.P.Q. by the Michelin Tire Company, and the financial support received by J.E.M. from the National Science Foundation through Grant DMR 7918903-03 (Polymers Program, Division of Materials Research).

7 References

1. Huggins, M. L.: J. Chem. Phys. *9*, 440 (1941)
2. Flory, P. J.: J. Chem. Phys. *10*, 51 (1942)
3. Flory, P. J.: Principles of Polymer Chemistry, Chapters 12 and 13, Cornell University Press, Ithaca, New York (1953)
4. Flory, P. J., Rehner, J.: J. Chem. Phys. *11*, 512 (1943)
5. Flory, P. J., Rehner, J.: J. Chem. Phys. *11*, 521 (1943)
6. Flory, P. J.: Chem. Rev. *35*, 51 (1944)
7. Moore, C. G., Watson, W. F.: J. Polym. Sci. *19*, 237 (1956)
8. Flory, P. J.: J. Chem. Phys. *66*, 5720 (1977)
9. Flory, P. J., Erman, B.: Macromolecules *15*, 800 (1982)
10. Flory, P. J.: Macromolecules *12*, 119 (1979)
11. Brotzman, R. W., Eichinger, B. E.: Macromolecules *15*, 531 (1982)
12. Brotzman, R. W., Eichinger, B. E.: Macromolecules *16*, 1131 (1983)
13. Gottlieb, M., Gaylord, R. J.: Macromolecules *17*, 2024 (1984)
14. Queslel, J. P., Mark, J. E.: Polymer Bull. *10*, 119 (1983)
15. Orwoll, R. A.: Rubber Chem. Technol. *50*, 451 (1977)
16. Eichinger, B. E., Flory, P. J.: Trans. Farad. Soc. *64*, 2035, 2053, 2061, 2066 (1968)
17. Flory, P. J., Tatara, Y. I.: J. Polym. Sci., Polym. Phys. Ed. *13*, 683 (1975)
18. Yen, L. Y., Eichinger, B. E.: J. Polym. Sci., Polym. Phys. Ed. *16*, 117 (1978)
19. James, H. M., Guth, E.: J. Chem. Phys. *15*, 669 (1947)
20. Flory, P. J.: Proc. R. Soc. Lond. *A 351*, 351 (1976)
21. Queslel, J. P., Mark, J. E.: Adv. Polym. Sci. *65*, 135 (1984)
22. Erman, B., Flory, P. J.: J. Polym. Sci., Polym. Phys. Ed. *16*, 1115 (1978)
23. Pak, H., Flory, P. J.: J. Polym. Sci., Polym. Phys. Ed. *17*, 1845 (1979)
24. Erman, B.: J. Polym. Sci., Polym. Phys. Ed. *19*, 829 (1981)
25. Erman, B., Flory, P. J.: Macromolecules *15*, 806 (1982)
26. Erman, B.: J. Polym. Sci., Polym. Phys. Ed. *21*, 893 (1983)
27. Scanlan, J.: J. Polym. Sci. *43*, 501 (1960); L. C. Case, J. Polym. Sci. *45*, 397 (1960)

28a. Pearson, D. S., Graessley, W. W.: Macromolecules *11*, 528 (1978); b. P. J. Flory, Macromolecules *15*, 99 (1982)

29. Mark, J. E.: Rubber Chem. Technol. *54*, 809 (1981); Mark, J. E.: Adv. Polym. Sci. *44*, 1 (1982)
30. Andrady, A. L., Llorente, M. A., Sharaf, M. A., Rahalkar, R. R., Mark, J. E., Sullivan, J. L., Yu, C. U., Falender, J. R.: J. Appl. Polym. Sci. *26*, 1829 (1981)
31. Leung, Y. K., Eichinger, B. E.: results to be published
32. Erman, B., Flory, P. J.: Macromolecules *16*, 1607 (1983)
33. Flory, P. J., Rehner, J.: J. Chem. Phys. *12*, 412 (1944)
34. Gee, G.: Trans. Farad. Soc. *42*, 33 (1946)
35. Treloar, L. R. G.: The Physics of Rubber Elasticity, Chapter 7, 3rd Ed., Clarendon Press, Oxford (1975)
36. Mooney, M.: J. Appl. Phys. *11*, 582 (1940)
37. Rivlin, R. S.: Phil. Trans. R. Soc. *A 241*, 379 (1948)
38. van der Hoff, B. M. E.: Polymer *6*, 397 (1965)
39. Mullins, L.: J. Appl. Polym. Sci. *2*, 257 (1959)
40. Booth, C., Gee, G., Holden, G., Williamson, G. R.: Polymer *5*, 343 (1964)
41. Allen, G., Kirkham, M. J., Padget, J., Price, C.: Trans. Farad. Soc. *49*, 1495 (1971)
42. Gee, G., Herbert, J. B. M., Roberts, R. C.: Polymer *6*, 541 (1965)
43. Yen, L. Y., Eichinger, B. E.: J. Polym. Sci., Polym. Phys. Ed. *16*, 121 (1978)
44. Graessley, W. W.: Adv. Polym. Sci. *46*, 67 (1982)
45. Marrucci, G.: Macromolecules *14*, 434 (1981)
46. de Gennes, P. G.: Scaling Concepts in Polymer Physics, Chapter 1, Cornell University Press, Ithaca, London (1979)

J. P. Kennedy (Editor)
Received June 26, 1984

Author Index Volumes 1–71

Subject Index

Polymerizations and Polymer Properties

1982. 94 figures. V, 252 pages
(Advances in Polymer Science/
Fortschritte der Hochpolymeren-
Forschung, Volume 43)
ISBN 3-540-11048-8

Contents:

Springer-Verlag
Berlin
Heidelberg
New York
Tokyo

Poly-diacetylenes

Editor: **H.-J. Cantow**

1984. 87 figures, 11 tables.
XIII, 149 pages
(Advances in Polymer Science/
Fortschritte der Hochpolymeren-
Forschung, Volume 63)
ISBN 3-540-13414-X

Contents:

Springer-Verlag
Berlin
Heidelberg
New York
Tokyo

GPSR Compliance

The European Union's (EU) General Product Safety Regulation (GPSR) is a set of rules that requires consumer products to be safe and our obligations to ensure this.

If you have any concerns about our products, you can contact us on ProductSafety@springernature.com

In case Publisher is established outside the EU, the EU authorized representative is:

Springer Nature Customer Service Center GmbH
Europaplatz 3
69115 Heidelberg, Germany

Batch number: 10761697

Printed by Printforce, the Netherlands